T3-BUV-772

ENVISIONING NATURE, SCIENCE,

AND RELIGION

ENVISIONING NATURE, SCIENCE, AND RELIGION

EDITED BY *James D. Proctor*

TEMPLETON PRESS

Templeton Press
300 Conshohocken State Road, Suite 550
West Conshohocken, PA 19428
www.templetonpress.org

Designed and typet by Kachergis Book Design

Library of Congress Cataloging-in-Publication Data
Envisioning nature, science, and religion / edited by James D. Proctor.
p. cm.
Includes bibliographical references and index.
ISBN 978-1-59947-314-7 (hardcover : alk. paper) 1. Philosophy of nature—Congresses. 2. Religion and science—Congresses. I. Proctor, James D., 1957–
BD581.E58 2009
113—dc22
2009006182

Printed in the United States of America

09 10 11 12 13 14 10 9 8 7 6 5 4 3 2 1

CONTENTS

ACKNOWLEDGMENTS

From the first gathering of contributors to this volume in fall 2004 in Santa Barbara, California, to the last of four North American and European regional workshops held in Wageningen, the Netherlands, in summer 2007, a vast support network made possible the project known as New Visions of Nature, Science, and Religion. I primarily acknowledge the generosity of the John Templeton Foundation and the University of California, Santa Barbara, which jointly provided necessary funding and in-kind support. Paul Wason of the John Templeton Foundation and the staff of the Institute for Social, Behavioral, and Economic Research at UC Santa Barbara deserve a special note of thanks for their able assistance and encouragement. Other key support entities at UC Santa Barbara included the College of Letters and Science, Office of Institutional Advancement, and Office of Instructional Resources. Thanks go to the New Visions faculty steering committee at UC Santa Barbara, an admixture of expertise in fields such as the geological sciences, psychology, biological sciences, religious studies, anthropology, and history, and to external steering committee members Philip Clayton, Nancey Murphy, and Jeffrey Schloss. Steering committee member Michael Osborne ably took over administration of the UC Santa Barbara portion of the program when I left in summer 2005 to serve at Lewis & Clark College in Portland, Oregon. In addition to sponsoring UC Santa Barbara graduate students doing research in nature, science, and religion, the program received key support from a number of graduate student researchers, in particular Jennifer Bernstein, Evan Berry, and Catherine Newell.

I also wish to offer a special word of thanks for the thirteen scholars who dedicated nearly three years of their academic careers to work with each other toward the result that is this volume. It goes without saying that the journey I asked them to take with me was unorthodox: Interdisciplinarity is a gamble, and its fruits taste sweeter to some than others. Yet we fondly remember our times together in Santa Barbara, Boston, Chicago, Oxford, and Wageningen, and sincerely wish that this volume inspires young scholars to stretch beyond their fields of expertise in search of surprising intellectual connections.

As with my previous edited volume, *Science, Religion, and the Human Experience,* I save final thanks for my daughters, Elise and Joy, who have given me support and encouragement in ways far beyond their years. I can see that their heads, too, are filled with crazy visions. If this volume carries a lesson for them, it is that visions can—must—be both lived and questioned, that one needn't nullify the other, that the paradox of blending critical and constructive passion is their delightful challenge as much as it has been mine.

ENVISIONING NATURE, SCIENCE, AND RELIGION

Introduction

VISIONS OF NATURE, SCIENCE, AND RELIGION

James D. Proctor

Nature, Science, and Religion: A Tangled Trilogy

In the popular Japanese hand game, known in the English-speaking world as rock, paper, scissors, a trilogy is defined by the relations between these common items. The relations are generally fixed, as we all know; for instance, if you are a rock, it's best to come up against scissors (which you can crush) versus paper (which can, we have been persuaded, "cover" and thus defeat rock).[1]

In many ways, the rock-paper-scissors trilogy could not be further from the trilogy of nature, science, and religion. Everyone knows the one set of rules governing rock-paper-scissors relations; yet there are innumerable ways in which the relations between nature, science, and religion have been, and could be, envisioned. Even this comparison does not suggest a more fundamental difference: Most of us don't even imagine nature, science, and religion as being related at all! How often do we hear assertions that science has nothing whatsoever to do with religious faith, or that religion has no connection with how nature operates?

Yet nature, science, and religion are not entirely unlike the objects of that simple game: They can indeed be viewed as separate (like your favorite pair of scissors and a big rock in your garden), but when viewed relationally, new insights emerge. Indeed, this volume's contributors find their understandings of nature, science, and religion to be inextricably tangled. Every time we hear someone talking about religion, we hear glimmers of nature and science; every time an assertion is made about nature, science and religion are implicated. It stands to reason, then, that if either nature, science, or religion matters to you, then all three matter to you as well.

We are scholars representing a diverse array of specialties that span the physical and life sciences, the social and behavioral sciences, the humanities, and theology. What we share is a passionate intellectual and personal concern over concepts as politically, culturally, and psychologically significant as these. We want everyone to think a bit more deeply about nature, science, and religion, because each has been invoked to justify some of the most profound as well as pernicious claims advanced by humanity.

The trilogy of nature, science, and religion is a vast and tangled terrain, too much for one book to cover. What focuses our volume is the insight that many of these claims—both profound and pernicious—have primarily been about nature: claims that certain people are naturally of inferior intellect; that science and religion have aligned in decrying the wanton destruction of the natural world; that certain human sexual proclivities are unnatural and thus should be prohibited; that the wonders of nature, long relegated to the realm of religion, are now best understood via the empirical logic of science. In each of these cases, a claim on nature is simultaneously a claim on religion and science, whether made explicitly or (more commonly) not.

Our point of departure, therefore, in working together to shed light on the tangled trilogy of nature, science, and religion was to consider how particular contemporary understandings or "visions" of nature implicate science and religion in important ways. We proceeded with five, expanded below: evolutionary nature, emergent nature, malleable nature, nature as sacred, and nature as culture. These were selected to represent a spectrum of contemporary academic inquiry spanning the physical and life sciences, the social and behavioral sciences, and the humanities. Many classic visions or metaphors for nature, such as nature as machine versus organism,[2] find contemporary expression in the visions we selected for our collaboration, but these classic visions were not themselves included due to the extensive literature already covering them.

Our group of scholars held a series of workshops in Santa Barbara, starting in fall 2004 and culminating in late spring 2006, under the sponsorship of the University of California, Santa Barbara, and with the university's generous support as well as that of the John Templeton Foundation. Our outlook was as diverse as the terrain we explored together; no ready consensus emerged. But we developed a respect for each other's ideas, which led to an extremely productive collaboration. This volume represents the culmination of our joint efforts.

Visions of Nature, Science, and Religion

The term *nature* comes from the Latin *natura*, which is derived from the verb that means "to be born" (*natal* comes from the same root). According to one classic account, there have been three progressive senses of the English use of the word *nature* through time.[3] From the thirteenth century on, *nature* meant the essential quality or character of something, such as the nature of a person or of mortality. Beginning in the fourteenth century, the word was also used to represent the inherent force directing the world and human beings, as in "the way of nature." Not until the seventeenth century—relatively recently in English language usage—did the word *nature* also come to mean the physical world as a whole. Thus it spans a wide variety of meanings in reference to both humans and biophysical reality.

Visions of external (biophysical) and internal (human) nature have been at the heart of theories of science and religion, running from Thomas Aquinas to Isaac Newton, and continuing in the work of notable contemporaries, such as Ian Barbour.[4] In addition to strong scientific interest in external and internal nature, questions of human nature are found in all major religious traditions,[5] and concerns regarding biophysical nature have emerged in many religions as well.[6]

Yet visions of nature have both united and divided science and religion. In its reference to the biophysical world, nature has been invoked by scientists to reject religious or "supernaturalistic" explanation, but it also serves as a common sacred ground for theologians and scientists oriented toward ecospirituality. In its reference to human nature, the concept has been used to explain everything from the theological doctrine of sin to the biological basis of religion. Nature plays a central role in policy concerns of our time, yet still unites and divides science and religion: Consider, for instance, the 1991 joint statement signed by leading scientists and religious leaders declaring their common concern for environmental protection,[7] versus the ongoing dispute—with significant scientific and religious dimensions—over human cloning.

As noted above, the five visions to be considered in this volume include evolutionary nature, emergent nature, malleable nature, nature as sacred, and nature as culture. The first two of these visions have arisen in the physical, life, and behavioral sciences; the final two have arisen in the social sciences, the humanities, and theology, with malleable nature straddling the sciences and the humanities. Taken together, these visions represent a variety of scholarly approaches toward understanding nature, with important assumptions and implications regarding science and religion.

Evolutionary Nature

The evolutionary vision of nature is the predominant contemporary scientific means of addressing questions of the origin and diversity of life, with important parallels to scientific theories of the origin and development of the universe. It links biophysical and human nature in a common naturalistic explanatory framework. Though its supposed challenges to traditional religious belief are well-known, it may offer new theological insights for spirituality. It may also help us reflect on, and reevaluate, some of science's basic metaphysical assumptions.

Evolution is an ancient idea; but the evolutionary vision of nature derives primarily from one of the most far-reaching and influential works in the history of science: Charles Darwin's *On the Origin of Species*.[8] Beginning with the publication of Darwin's work in the mid-nineteenth century, continuing through the twentieth-century modern synthesis with population genetics, and running all the way up to contemporary research, the evolutionary vision of nature has played a powerful, integrative role among life scientists.

Evolutionary theory is far from settled, which is understandable given its considerable power and breadth of explanation. One of its most celebrated recent interpreters, Stephen Jay Gould, released soon before his death a magnum opus on evolutionary theory, reconsidering the basic questions of whether (a) natural selection is the primary mechanism of adaptation, (b) natural selection operates at the genetic, organism, and/or group level, and (c) changes induced by evolutionary mechanisms are incremental or sudden.[9] Yet Gould's take on evolution stands in sharp contrast to that of Richard Dawkins, for whom genetic selection is paramount and the lessons of evolution apply equally to humans and nonhumans.[10] Dawkins' strident position on genetic selection is opposed by more scientists than just Gould, however; the long-celebrated biologist Ernst Mayr has also rejected the implications of genetic reductionism.[11]

The discussion is equally vigorous when evolution is applied to human nature. An example is the field of evolutionary psychology, an approach in which knowledge and principles from evolutionary biology are put to use in research on the structure of the human mind.[12] Researchers in this area have derived results for behaviors as wide-ranging as cooperation, love, incest, and racism. Yet biologist Paul Ehrlich (a staunch defender of evolutionary theory) has argued that it is primarily cultural evolution, rather than biological evolution, environment rather than genes, that is responsible for human behavior.[13] When evolution is applied to morality, questions such as the reality and possibility of empathy emerge.[14]

There are strong philosophical parallels in accounts of the evolution of life and the evolution of the universe. Both are answers to fundamental "origins" questions. Both have traditionally involved recourse to a deity, whether as a Prime Mover or an involved God; yet scientific theories have been advanced by some to suggest that the notion of a deity is unnecessary, perhaps even impossible. It is this thoroughgoing naturalism (or, rather, antisupernaturalism) that has united certain proponents. Thus, for instance, Steven Weinberg has linked evolutionary and cosmological theory as part of a historical process of scientific "demystification," which ultimately suggests "a chilling impersonality in the laws of nature."[15]

It is a popular assumption that the evolutionary vision of nature poses a direct threat to religion; thus, debates over evolution versus creation (or intelligent design) have persisted.[16] Yet evolutionary nature has been seen as a threat by some scholars in the social sciences and humanities as well. As one example, E. O. Wilson's *Consilience* argues for a unity of knowledge based largely on the natural sciences, in particular a model of human nature based on biological evolution;[17] this model predictably finds mixed support in the scholarly community.[18]

In summary, evolutionary nature is a powerful, sweeping vision of biophysical and human nature with significant implications for the relationship between science and religion, and the sciences and the humanities. These implications are far from resolved.

Emergent Nature

A second major scientific understanding of biophysical and human nature hinges on emergence, which has been invoked to explain complex phenomena, ranging from biological diversity to human consciousness; its influence has spread far beyond the sciences as well.[19] Emergent nature is becoming a unifying vision for a vast array of scientific disciplines, and sheds new light on traditional metaphysical questions of order and chaos, parts and wholes. Emergence has also been offered as a way to situate theology in a scientifically valid framework.

Emergent nature champions antireductionist explanation. It has been recognized throughout the ages that nature exists at multiple scales of complexity; yet what is the relationship between these levels? The perennial Great Chain of Being posited a vast hierarchy running from matter to spirit, joining levels of complexity (and, significantly, science and religion) with higher levels ultimately explaining lower levels.[20] But many of the sciences have, especially in the last century, moved in the opposite, reductionist direction, seeking explanation at smaller and smaller levels of reality.

A good example is physics, which arguably encompasses a broader range of scales of complexity than any other science. A well-known advocate of reductionist explanation is Steven Weinberg.[21] Weinberg believes that complex phenomena, such as mind and life, do emerge out of simpler systems, yet "The rules they obey are not independent truths, but follow from scientific principles at a deeper level" (p. 115). Reductionist explanation has generally been the hallmark of physics, but has not gone without criticism.

A key early paper was written by condensed-matter theorist Philip Anderson in 1972, in an essay aptly titled "More Is Different."[22] One of Anderson's main points is that "The ability to reduce everything to simple fundamental laws does not imply the ability to start from those laws and reconstruct the universe" (p. 393). The early work of Anderson and other physicists led to the cross-disciplinary field of complex systems analysis, which is explicitly devoted to establishing nonreductive modes of explanation of complex phenomena. This interest has spawned research centers, such as the Santa Fe Institute and the New England Complex Systems Institute, with significant participation by physicists such as Murray Gell-Mann.[23]

Complex systems research has led to new ways of understanding the age-old question of the relationship between order and disorder in reality, leading to fundamental insights into nature, classically understood as part of an orderly cosmos. Pivotal to this work has been the concept of deterministic chaos, in which apparent disorder emerges from very orderly simple rules, yet this emergent disorder turns out to be quite orderly in other ways. The vision of emergent nature thus challenges the strict separation of cosmos and chaos, order and disorder in the universe. In emergent nature, randomness and pattern are linked; this very different metaphysical way of looking at nature has led to fundamental new insights in natural science fields, such as ecology.[24]

Perhaps the most breathtaking recent publication on emergent nature is *The Emergence of Everything*.[25] In this work, biophysicist Harold Morowitz assembles a continuum of twenty-eight steps of higher levels of emergent complexity, rivaling in scope the classical Great Chain of Being and running from the universe to planets to cells to animals to humans to culture to spirit. Morowitz ascribes much of the recent flurry of scientific discovery around emergence to the advent of high-speed computing, which has presented new opportunities for modeling complexity in nature. Major implications exist for science, as it potentially moves from mathematical to algorithmic modes of explanation (e.g., understanding the emergence of complex behaviors based on simple computational models, such as cellular automata), as championed in Stephen Wolfram's *A New Kind of Science*.[26]

Morowitz's work reaches beyond science to religion in tracing implications of this vision of emergent nature. He advances the radical theological thesis that "Transcendence is an emergent property of God's immanence. . . . We *Homo sapiens* are the mode of action of divine transcendence" (p. 195). Morowitz thus claims that, according to the vision of emergent nature, God is to be understood as the immanent laws of nature; people, who possess emergent consciousness, are the true transcendent agents in the cosmos. Others have also discussed theological implications of emergence with varying degrees of departure from traditional theism: for instance, John Polkinghorne has considered implications of chaos, complexity, and emergence, linking God with the possibility of top-down causation between levels of reality.[27] More recently, Philip Clayton argues that emergence theory in recent science offers an important opening for language about the spiritual dimension of human existence, including the concept of spirit and perhaps even the idea of God.[28] He traces emergentist arguments from the emergence of the classical world out of quantum mechanics, through contemporary debates in evolutionary biology and neurophysiology, and up to the emergence of spirituality and metaphysical concepts.

Emergent nature is thus in many respects an even more far-sweeping vision than evolutionary nature. It is quite recent, may signal major changes in science, and has afforded diverse theological interpretations. Its stronger scientific advocates have, however, not escaped criticism for their ambitious extension of this vision.[29] In its theological extensions, emergence can, if not carefully articulated, become an inspiring but fuzzy "god-of-the-gaps" argument; indeed, its popularity in certain new religious movements bears little resemblance to its scientific origins.[30] Yet these theological extensions suggest ways in which contemporary visions of nature can have significant spiritual dimensions, to be explored later under the cultural and philosophical vision of nature as sacred.

Malleable Nature

The vision of nature as malleable straddles the sciences and the humanities: It arises in the sciences and engineering from pathbreaking research in genetics and development of new genetic technologies over the last several decades,[31] and has arisen in the same time period in conjunction with development in the humanities of poststructural and postmodernist perspectives on the nature of reality and human beings.[32] The vision of malleable nature challenges the boundaries of nature and the natural, as what lies beyond these boundaries—the unnatural, the artificial—is now less easily distinguishable from the realm of nature. As such, it also challenges the bedrock of biophysical and human nature upon which many societal and religious values are based,[33] and has thus engendered serious discussion and debate over its philosophical, theological, and political implications.

Malleable nature encompasses a wide swath of related topics: Examples include human reproduction and enhancement, genetic discrimination, human stem cell research, and food and agriculture in developing countries. But positions taken on these topics by scientists, religious leaders, industry, and the public have generally fallen into one of two camps, reminiscent of the polar "catastrophist" versus "cornucopian" stances Stephen Cotgrove detected in environmental politics some two decades ago.[34] On the catastrophist side, a number of religious denominations, environmental organizations, and sectors of society have denounced biotechnology as an imminent threat to humanity and the natural world; on the cornucopian side, advances in genetic research and biotechnology have been heralded by many scientists and industry as a panacea for problems ranging from birth defects to global food supply.

Much of this academic and popular discussion has focused on developments in science and technology, ranging from the Human Genome Project to current government-sponsored biodefense projects.[35] Proponents address public anxieties regarding risk in contemporary nature-society relations (e.g., pesticide-dependent industrial agriculture) and invoke larger values concerning the proper place of humans in the natural world, in casting biotechnology as a safe human improvement upon nature.[36] Similarly, opponents typically invoke potential environmental risks, coupled with the threat of societal disempowerment as human and biophysical nature becomes corporatized.[37] Others note the religious dimensions on all sides of biotechnology.[38]

In a broader context, these developments have been examined in terms of implied features of science, and its connections with larger political and economic processes. Peter Dickens, for instance, argues that genetic research and technology treat biophysical and human nature as mechanisms comprising subsystems composed of parts that ultimately boil down to bits of information in the genetic code.[39] To Dickens, this fragmented idea of nature serves well its commodification in multiple market niches:

Nature is stuff that can be manipulated to presumably human, and certainly corporate, benefit. Others similarly link genetic research with the increasing emphasis on profitable information in science,[40] as witnessed by the rapid rise of molecular biology.

Yet malleable nature is not wholly restricted to the sciences. In the humanities and popular culture, a related discussion has considered malleable nature from a poststructural and postmodernist perspective. Jean Baudrillard, for instance, has argued that the malleable human genome erases the boundary between natural and the artificial, real and virtual; there is no reality beyond our "Disney World" representations of it.[41] And though some have warned of the dangers of treating human biology as infinitely malleable,[42] others have pointed out the historicity of supposedly biological concepts, such as *woman* in arguing for an embrace of postmodern difference in biotechnology.[43] The upshot of these critiques has been a rejection of appeals to "nature" or "natural" in justifying policy and morality. All of this has taken on new dimensions as the malleability of nature has been reduced to the molecular level in nanotechnology.[44]

In sum, much discussion concerning biotechnology has taken science and religion as givens, rather than provoking a deeper examination of implications of malleable nature for the very science that studies it, and religious bodies that comment on it. Preliminarily, biotechnology paints a mixed picture of contemporary science, and one in which religion has not yet advanced far beyond a simplistic reading of both nature and science. Yet malleable nature is an unsettling notion, in the same way that poststructural and postmodernist notions of malleable reality are unsettling. Malleable nature is hence both sweeping and inconclusive in its implications for science and religion, and must be situated in the context of other visions of nature in order to derive robust indications for progress in religion and science in the future.

Nature as Sacred

In contrast to the notion of biophysical and human nature as thoroughly material entities, distinct from the sacred realm of God or spirit, a more theological vision of external and internal nature has recently arisen in both scholarly and popular circles. This vision of nature, with variants running from theistic ecospirituality to agnostic religious naturalism, may serve as an important metaphysical basis governing ethical behavior, yet raises major challenges for reconciliation with both transcendent religion and scientific rationality.

Scholarly attention has been empirical (involving historical and contemporary studies of concepts of sacredness in nature and sacred space) and philosophical and theological (attempting to systematize this empirical information and understand it in light of religious teachings and sacred texts). As an example of the latter, Ian Barbour has included themes of stewardship, celebration, sacrament, and the Holy Spirit into a theology of nature.[45] An example of the former is the Forum on Religion and Ecology at Harvard University, a cross-cultural project involving a multiyear series of conferences and related publications.[46] Perhaps the fullest scholarly analysis concerns how religious notions of the sacred inform fully secularized transformations of biophysical and human nature in late modernity.[47]

In the American context, Catherine Albanese has identified a perennial "nature religion" in the United States, stretching from early settlement to contemporary spirituality.[48] To Albanese, the Western religious tradition "has placed nature near the top of its short list of major categories by which to make sense of religion. God and humanity [as expressed in organized religion and civil religion] comprise the first two categories. Nature, however culturally diffuse and evanescent, forms the third."[49] Albanese notes four expressions of nature religion in American history: (1) the Transcendentalist legacy inherited by contemporary environmentalism, (2) metaphysical forms of spiritualism (e.g., Theosophy) reaching to contemporary New Age practices, (3) a revitalized emphasis on bodily healing and well-being grounded in nature, and (4) Enlightenment-style natural religion and natural theology, expressed in peculiarly American forms, such as pragmatism.[50] Thus, both biophysical and human nature fall under this broad rubric. Albanese's historical work is validated by contemporary social science research, demonstrating the ubiquity and significance of notions of nature as sacred in contemporary environmental concern.[51]

A much more voluminous literature has been devoted to philosophical and theological dimensions of the vision of nature as sacred.[52] This literature is quite diverse, mixing immanent and transcendent sacredness and exploring related practices in multiple religious traditions. Much of it constitutes a continuing response to Lynn White's famous thesis that the roots of environmental crisis lie in Judeo-Christian attitudes of domination over nature,[53] but some of this literature traces implications for human as well as biophysical nature.

What are the implications of the vision of nature as sacred for science and scientific rationality? Scientific opinion is apparently mixed: Some have strongly supported this vision as a mode of reenchantment of the natural sciences,[54] whereas others have charged that it constitutes a "betrayal of science and reason,"[55] an "assault on reason,"[56] or "nature worship."[57] This discussion suggests different positions on the boundary between science and religion, and many of these contradictions have yet to be resolved. The vision of nature as sacred is thus quite culturally diffuse and important among theologians, humanists, and social scientists; it will surely play an important role in science-religion dialogue in the future. But more attention is needed to systematize and join its empirical and philosophical/theological dimensions, and to rectify potential contradictions with science.

Nature as Culture

A diffuse vision of nature arising in the social sciences and humanities concerns nature as culture. This vision emphasizes nature's inextricable connection with human meaning, in contrast to the prevalent notion of nature as entirely separable from culture. As with the other visions, it poses important challenges and opportunities for rethinking science and religion, in this case as human endeavors as opposed to direct conduits to reality and God.

The separation of nature and culture is one of the most deeply entrenched divides in Western thought.[58] It can be traced back at least to Aristotle, for whom nature (*physis*) is that which is not made by humans, in contrast to *techné*, that which is of human origin. It underscores ideas of objectivity, which arose in the seventeenth-century valorization of scientific rationality, often grounded in nature as an objective referent, as a means of technical ordering of society based on a new, naturalist "religion."[59] The idea of objectivity forced culture into the diminutive category of subjectivity, and forced God into two polar alternatives as either equivalent in status to the objectively verifiable reality explored by science, or merely the subjective projection of a wishful or oppressed people.

The vision of nature as culture has roots in Kantian philosophy and earlier expressions of idealism, but it is best known for its recent flourishing in opposition to naïve notions of objectivism, underscoring the practice and interpretation of natural and behavioral science. It is often called *social constructivism* or the "social construction of nature" thesis,[60] and should be understood primarily as an epistemological assertion concerning our knowledge of nature, rather than an ontological assertion concerning the reality of nature itself. Nonetheless, one of the primary tenets of social constructivism is that biophysical and human nature are incomprehensible outside of culturally based knowledge schemes, so the vision of nature as culture cannot be readily dismissed as merely a vision of ideas of nature versus nature itself.

The vision of nature as culture has been primarily championed among the social science and humanities disciplines—those for which culture is a primary category of analysis. Its most vocal opponents have been scholars working in the natural sciences. This debate, known popularly as the "science wars," has tended to promote philosophical caricatures of naïve realism, asserting the reality and ready knowability of nature, against naïve relativism, questioning the truth-value of all scientific knowledge.[61] Fortunately, an excellent and growing body of scholarly work has refused to accept these polarized terms of the epistemological debate over nature and culture.[62]

The vision of nature as culture, then, resonates with a diffuse epistemological position, characterizing many of the social sciences and humanities. It has been understood by some as standing in fundamental opposition to science, but it need not be, as long as dualistic caricatures are rejected. On the contrary, this vision poses a powerful means of potentially reconciling the "two cultures" problem of the sciences and humanities,[63] and bears important potential for bringing science and religion together.

Comparison

There are some important similarities in these visions. All are strong arguments concerning nature in its entirety, not simply weak arguments concerning certain properties of nature. For example, the evolutionary vision attempts to explain all life through the prism of evolution, not just certain forms or aspects of life; similarly, the vision of nature as culture maintains that all knowledge of nature is filtered through cultural lenses, including scientific as well as popular understandings. This common feature will pose challenges for synthesizing these visions, as none necessarily includes room for the others. Yet what may arise could thus be something entirely new for nature, science, and religion.

As strong arguments, each of the five visions challenges a prevalent metaphysical dichotomy. The evolutionary vision stresses the continuity of all nature, and hence opposes the notion that humans are entirely separate from nature. The emergent vision not only challenges the reductionist notion that nature at all scales of complexity can ultimately be analyzed in terms of its constituent pieces, but more fundamentally revisits the larger opposition between chaos (disorder) and cosmos (the order of nature). The malleable nature vision challenges the dichotomy between natural and artificial, in that genetic manipulations of nature are arguably both. The vision of nature as sacred challenges the distinction between matter (the stuff of which nature is ostensibly composed) and spirit, secular and sacred. The vision of nature as culture challenges the same notion questioned by the evolutionary vision, but takes the opposite tack by means of "culturizing" nature as opposed to "naturalizing" culture.

These five visions of nature are by no means entirely distinct. There has been a good deal of interest, in particular, in bringing together the two scientifically based visions of evolutionary and emergent nature,[64] with important implications for human morality and religion.[65] Similarly, the vision of nature as sacred could be understood as a specific claim made by certain cultural groups, thus falling under the vision of nature as culture. In many ways, the vision of malleable nature is the ontological equivalent of the epistemological argument of nature as culture: In one, nature is literally constructed, whereas in the other it is conceptually constructed. Other linkages are indeed possible: Consider the notion of an embodied mind,[66] which links the seemingly opposing visions of evolutionary nature and nature as culture, or theological work from an emergentist perspective,[67] potentially linking emergent nature and nature as sacred. Yet there are differences. For instance, the vision of nature as culture can have a corrosive effect on the realist epistemological assumptions underlying evolutionary nature and emergent nature.[68] Similarly, evolutionary nature may explain, and hence explain away, the vision of nature as sacred.[69] These differences may suggest important points of departure for a comparative and synthetic effort.

The Essays

The volume proceeds with a broad examination of the very notion of visions of nature, in an essay by Willem B. Drees titled "The Nature of Visions of Nature: Packages to Be Unpacked." Drees notes that visions of nature involve both facts and values, descriptive and prescriptive elements. Though there are many elements to religion, visions of nature relate most closely to what he calls *theologies,* or the cognitive dimension of religion. According to Drees, theologies are creative combinations of cosmologies—how the world is—and axiologies—how the world should be. The natural sciences deal centrally with cosmologies, yet their most significant cosmological insights do not fully determine the nature of reality, so cosmology cannot be reduced to the findings of science. When this schema is applied to the five visions of nature we considered as a point of departure, some (e.g., evolutionary nature) appear to favor cosmological elements, whereas others (e.g., nature as sacred) favor axiological elements. Yet all include important cosmological and axiological assumptions, being both visions "of" and "for" nature. Drees then reviews each of the five visions using this schema, noting their very different implications for science and values. Drees concludes by observing that, though reflection on the nature of visions sounds far removed from the business of living one's life, this practice affords the opportunity to reflect on, integrate, and apply these cosmological and axiological concerns.

In "Visions of Nature through Mathematical Lenses," Douglas Norton takes another broad look at the nature of visions, this time from the perspective of a mathematician. To Norton, mathematics is far deeper and more aesthetically driven than most people would assume, given its typical quantitative caricatures. This impulse to seek elegant explanations of the nature of reality links mathematics centrally with visions of nature, starting in the earliest years of mathematics with a vision not unlike that of nature as sacred. Norton invokes nature as culture to acknowledge the historicity of mathematics, moving from its early times to medieval, Renaissance, and modern instances. What is significant in Norton's review is the extent to which nature, science, and religion are intertwined in the personal lives and professional outlooks of leading mathematicians. Norton then focuses on dynamical systems and chaos theory to suggest their significant and novel perspective on nature, with broader implications for science and religion.

In this contemporary body of mathematics, the relation between order and disorder is central. Norton closes by acknowledging his sense of delight when mathematics provides an elegant vision of nature.

As emphasized above, nature is human as well as nonhuman. In "Between Apes and Angels: At the Borders of Human Nature," Johannes Thijssen considers the Aristotelian heritage of our views of what is special about the human being. These views share many of the qualities of our views of biophysical nature, in that we commonly assume that nature has some special property that sets it apart from nonnature. In the case of biophysical nature, this is generally understood to be its lack of human origin; in the case of human nature, this is generally understood, following Aristotle, to be some peculiar property of reason that sets the human apart from the nonhuman. But Thijssen disputes this notion of nature on two levels: First, this capacity for reason was understood differently at different times in history, and, more broadly, the influence of history on ideas of human nature validates the nature-as-culture vision that nature, whether biophysical or human, is more construction than essence. Thijssen examines the understandings of human nature, and the Aristotelian capacity for reason in particular, as evidenced in the anatomical ruminations of Edward Tyson (1650–1708), the earlier theological speculations of Saint Augustine and Albert the Great (c. 1200–1280), and debates among Spanish conquerors over native Americans. Thijssen closes by invoking the Aristotelian notion that living according to reason was not only a defining feature of human nature but an ethical norm, which at the time meant becoming godlike but since then has taken many different forms, of which we could readily note the practice of science as one envisioned ethical fulfillment of the capacity for reason. Nature, science, and religion thus closely intertwine in discussions over what it means to be human.

In the next essay, "Locating New Visions," the geographer David Livingstone builds on the nature-as-culture notion adopted by the historically minded Thijssen in arguing that any given vision of nature must be located—that is, geographically situated—in order to be fully understood. He does so in the context of the first of the five visions of nature that launched our collaborative effort, that of evolutionary nature. Livingstone is intrigued by the particular ways in which different communities responded to the evolutionary notions of nature and humanity, following Darwin. He builds on the argument of other contributors to this volume that ideas are inescapably open to multiple interpretations, noting the same to be true with respect to what people often mistakenly take as monolithic Darwinian theory. Livingstone provides a summary review of knowledge as what Edward Said called *traveling theory*, observing that ideas are not immutable abstractions but circulate in material form (e.g., as texts) much the same as other objects, and are thus influenced on each occasion where they are written and read. Turning to the place-based reception of Darwin, Livingstone contrasts first the sense of threat felt among Presbyterians in Belfast with the more dialogic reception among those in Londonderry. He then compares two sites in the U.S. South, where in both cases Darwin was understood as a threat to race relations, but the authority invoked to resist this threat was in one case polygenist science, and in the other (monogenist) biblical literalism. The third comparison examines Dunedin and Wellington, New Zealand, where in differing ways the cultural grip of religious institutions was less than in Ireland and the United States. But in the case of Wellington, Darwinism provided ready justification for the inevitable domination of native Maori by European settlers. Livingstone closes by clarifying his position as distinct from claiming that visions of nature, science, and religion are nothing but products of their location, yet maintaining that any call to fashion "new" visions must itself be aware of its locatedness.

Robert Ulanowicz bridges our first two visions of evolutionary and emergent nature in the essay "Enduring Metaphysical Impatience?" Ulanowicz makes the bold claim that fundamentalism of a sort exists among both scientists and religious advocates, to the extent that they manifest what John Haught has called metaphysical impatience in considering all ultimate questions on the nature of reality to be effectively solved. To Ulanowicz, the necessary entanglement of differing visions of nature suggests a necessity for scientists and theologians to be in patient dialogue with each other. He starts by reviewing some primary metaphysical principles of Newtonian naturalism, suggesting that even though many of these tenets have been updated in science they still hold sway among neo-Darwinists. This contemporary scientific form of metaphysical impatience has been challenged by proponents of intelligent design (ID), an equally impatient movement that talks science but is to Ulanowicz primarily theological. Ulanowicz then critically examines how ID uses information theory–derived notions of complexity to attempt to prove the necessity of theological design in the evolution of life. He then proposes an alternate reading of complexity by posing metaphysical principles arising from the study of ecosystems that run counter to each of Newton's principles. One important implication is that disordered complexity is inevitable yet, given its ambiguity, amenable to explanation from both scientific and theological principles. Ulanowicz closes by briefly remarking on how his ecological metaphysics addresses some major conflicts between science and theology, such as free will and the efficacy of prayer, issues that could potentially result in rich dialogue instead of bitter conflict.

Similar to Ulanowicz's consideration of the first two visions of nature is Barbara King's essay, "God from Nature: Evolution or Emergence?" But King develops her argument from quite a different empirical basis: the African great apes. King defines religion by emphasizing practice and emotional connection with the supernatural, and argues that great apes may help us understand the evolution of religion among humans, given their likely similarities to early hominids. Her approach is decidedly dissimilar to that of scholars who use evolutionary theory to base the human religious impulse on particular genes or brain evolution. For King, these reductionist approaches neglect the social and relational dimensions of how primates, and presumably hominids, evolved. She offers important anecdotes that, corroborated by other empirical evidence, suggest the extent to which great apes display the capacity for empathy and compassion; these, King argues, are precursors for the meaningful ritual practice characterizing religion. King also provides a summary history of hominid evolution to modern human beings, noting how these primate traits were then extended into ritual and symbolic practice. But does this antireductionistic approach suggest the relevance of emergence theory—that very bastion of antireductionism—in explaining how religion came to be among humans? Working with a definition of emergence adopted from Philip Clayton, King seeks the "unpredictable, irreducible, and novel" dimensions of human religion vis-à-vis hominid precursors, closing with a nod toward the difficulty in resolving the possible role of emergence, yet noting that the largely evolution-based explanation she has provided is anything but reductionistic, given its vision of nature as "deeply social, emotional, and creative."

Gregory Peterson's essay "Who Needs Emergence?" summarizes over a century of scholarly interest in emergence, then characterizes contemporary forms by invoking Clayton's taxonomy of *façon de parler*, weak, and strong types. *Façon de parler* emergence, essentially a nonemergentist account of complex realities, generally shares with weak emergence a physicalist view of existence, though the latter admits some reality and causal efficacy to these complex entities. By contrast, strong or radical emergence generally moves beyond strict physicalism in positing the existence of wholes that are in some ways of an entirely different sort than their pieces: Classic examples include the mind or consciousness and the Pauli exclusion principle, typically explained via the example of multiple electron energy states in an atom. Peterson then considers emergence as used in the physical and life sciences, arguing that their application is understandable, given the common need for scientists to account for causal connections between different scales of complexity; in this regard, however, the scientific use of emergence entails less a metaphysical commitment than an explanatory necessity. Indeed, scientists are wary of an "emergence of the gaps," in which emergence serves, miraclelike, to rescue causal explanations between scales of complexity. Peterson contrasts its uses in science with how philosophers and theologians consider emergence: In these latter fields, ontology as well as causality are key, leading, for instance, to the philosophical interest in strong versus weak emergence (given differing accounts of reality), and the theological interest in God as an emergent reality (or conversely, the world as emerging from God), and in the causal sufficiency of nature to account for novelty. Given its various forms and the differences of focus between the sciences and philosophy/theology, Peterson concludes with caution that much more work needs to be done in coming to terms with the role of emergence as a vision of nature.

In "Creativity through Emergence: A Vision of Nature and God," theologian Antje Jackelén further explores emergence, proceeding from the observation that several centuries ago it would be not complexity, novelty, and emergence, but conformity to law, which would have been celebrated, suggesting the culture in which emergent nature has arisen. Jackelén then reviews the long discussion over whether complexity has emerged from nature or God, ultimately suggesting that both may be intertwined; yet this discussion entails a level of vulnerability, given the vagueness of the concept. She notes that the emphasis on levels of reality in emergence theory, with its attendant hierarchy of value, is problematic. Jackelén's emergence reminds us of indeterminacy and potentiality: Important related concepts are design and order, with implications for theological doctrines, such as the creation of order out of nothing versus chaos, suggesting a form of tehomophilia (appreciation of chaos), and linking closely with possible emergentist accounts in science. Jackelén does mention a number of limitations of emergence, most especially the value-ladenness implied among some of its proponents; for Jackelén, emergence does not provide a new proof of God or defeat of materialism. Yet Jackelén closes with the bold move of building a new vision of nature, human nature, God, evil, sacramentality, and ultimately theology in light of emergence.

Martha Henderson's essay, "Rereading a Landscape of Atonement on an Aegean Island," further extends the vision of emergent nature in the context of human-nature interactions on landscapes. To Henderson, the emergent notion of a self-organizing system offers a better understanding of these interactions in the Aegean island of Lesvos than the typical declensionist story, especially as told by George Perkins Marsh in the classic mid-nineteenth-century *Man and Nature*,[70] that human practices, such as overgrazing, have resulted in landscape destruction (e.g., erosion). Henderson views Greek Orthodox religion as playing a key cultural role in the coevolution of these landscapes by providing guidance and necessary correctives to landscape practices. She reviews the literature on self-organizing systems and the history of Aegean landscape, emphasizing scholarly interpretations of the latter that dispute a declensionist reading. Henderson suggests that the environmental damages noted in recent times in these landscapes are not so much the product of ancient practices as of modern pressures, including urbanization. Henderson discusses the religious landscape of Orthodox Christianity and features relevant to land use, such as the need for sacrificial lambs, then moves to a more detailed description of Lesvos, located just off the coast of Turkey. She argues that Marsh misread landscapes, such as Lesvos, because he applied a background combining North American farming and ideas of humans as a destructive force in nature. In between this declensionist reading and a more positive one that similarly privileges human agency, Henderson suggests the possibility of coevolution between humans and biophysical processes, as well as the significance of religious practices in supplying meaning. She closes by noting that landscapes provide tangible evidence of ideas of nature, science, and religion, and ties the proclamations of the "green" Greek Patriarch Bartholomew I into this connection between religion and landscape.

Moving to another broad framework for understanding biophysical and human reality is Andrew Lustig's "The Vision of Malleable Nature: A Complex Conversation." Lustig notes that, more so than the other visions of nature, malleable nature problematizes the human role in altering nature, thus raising the deeper question of the relationship between descriptive and normative statements about nature. Yet, to Lustig, this makes malleable nature open to multiple perspectives and interpretations; he reviews the other four visions of nature to demonstrate a similar potential for plural interpretation. Lustig then considers the philosophical distinction between description and prescription, suggesting that strict readings make science and ethics mutually exclusive in spite of recent accounts that effectively base one on the other. In moving deeper on the question of whether nature or the natural can be used to ground morality, Lustig finds particular relevance in the critical perspective of feminism, noting related points feminists make that call for serious caution. He then turns to a summary of major religious traditions, considering how their doctrine and historical practices shed light on what position may be derived as to the propriety of altering biophysical and human nature. The variety of these perspectives calls in Lustig's mind for care to be taken in making generalizations of any sort about religion per se and its relationship with science. Lustig concludes with a number of questions that nonetheless remain, such as possibilities for interfaith cooperation on guidelines for biotechnology, given these differences.

Fred Ledley's "Visions of a Source of Wonder" suggests how one scholar specializing in the scientific and medical practices underlying malleable nature approaches the relations between nature, science, and religion. Ledley's interest lies in the potential to escape the constraints of both science and religion in visions of nature by attending to the sense of wonder that is often understood to typify religious and scientific experience, yet precedes formal characterization via the languages of science and religion. Ledley notes broad evidence of this sense of wonder in experiencing the natural world—a sense sometimes translated as the sublime. Ledley then seeks further evidence and possible explanation of the experience of the sublime in religion, specifically rabbinic commentary on the first words of Genesis, at the very beginning of the Hebrew Bible: A wide range of Jewish scholars have weighed in on their grammatical complexities, often with an interpretive eye toward aspects of nature lying in the wonder of creation and not toward any straightforward interpretation. Ledley then considers the domain of mysticism, one in which experience of wonder is key, alongside religion and science: Mysticism is often appreciated as a bridge between the realm of the sublime and the realm of religion, and even scientific intuition can be understood as a form of mysticism (certainly of wonder). He next suggests how neurological research indicates a common human capacity for mystical experience, via heritable brain pathways. Yet the presence of these neurobiological structures does not, as some argue, deny the reality of the source of these mystical experiences; indeed, these evolved capacities may have conferred benefits, including the motivation for advancement of religion and science! Ledley closes by noting that this sublime vision of nature need not be a backhanded proof of God, nor need it diminish claims based on scientific and religious understanding.

The final vision of nature is exemplified in Nicolaas Rupke's "Nature as Culture: The Example of Animal Behavior and Human Morality." Here the engagement with concepts of nature seems to concern science more than religion, but ultimately, in serving as an expression of political and other sentiments, both scientific and religious claims to authority are brought into context. Rupke begins by reviewing the nature-as-culture argument, which challenges the moral authority of nature as revealed by science. He prefers a view from somewhere, the notion shared with Livingstone and certain others in this volume, that situates scientific accounts of nature in their geographical-historical contexts. Rupke chooses the case of scientific research into animal behavior in the nineteenth century, first offering an overview of the field, then focusing on the German earth and life scientist Carl Vogt (1817–1895). Rupke provides details on both the scientific and philosophical-political inclinations of Vogt, emphasizing his strong materialist and antimonarchist impulse: For instance, he described bee colonies as little more than organized systems of class violence and oppression. Vogt was internationally recognized for his scientific achievements, but his extension of scientific research into political commentary was blatant, though not unrepresentative of other scientific accounts linking animal behavior and human morality. Rupke closes by providing a brief overview of similar scientific studies of animal behavior that address themes of human sexuality, war, and aggression. His ironic conclusion is that, in attempting to explain culture in terms of the forms of nature, scholars such as Vogt draw unwittingly on culture in their explanatory reliance on nature.

My essay, "Environment after Nature: Time for a New Vision," partly affirms the notion of nature as culture, yet problematizes the binary distinction between nature and culture assumed in support or denial of this vision. I focus on the notion of environment accompanying contemporary environmental concern: I argue that the profound limits of our contemporary understanding of environment as nature call for a renewal of the earlier vision of environment as connectedness, noting that connection is emphasized in a wide variety of essays in this volume. I begin by reviewing the usage of *environment* over the last few centuries, as it moved from a sense of connection with surroundings to the surroundings themselves, and ultimately physical surroundings. I point out that this reduction of environment to physical reality has broader roots, revealed as well in our understandings of science and religion, where their relationship is often portrayed as one of either essential similarity or dissimilarity. In particular, recent statements suggest that science and religion come together in their support of environmental concerns. Yet there is a binary implied in these treatments of science and religion that ultimately boils down to an assumed binary of nature and culture, and prevents the harmonization of science and religion, or of humankind and environment, that is desired. What, then, is the solution? I draw on the work of Bruno Latour as one example of how to supplant the nature-culture binary by counting beyond two, rejecting the duality that subtly colors our understanding of environment, and the dual authorities of science and religion we commonly invoke to justify environmental concerns. Latour's emphasis on connection and hybridity results in a very different sense of science, religion, and environment than is commonly understood, one that resonates in recent "death of environmentalism" claims that similarly cite environmentalism's conceptual foundations as fundamentally flawed. I conclude by asking whether environmentalism will indeed move toward this new/renewed vision.

The last essay in this volume is John Hedley Brooke's "Should the Word *Nature* Be Eliminated?" Brooke's contribution brings the vision of nature as culture to its logical conclusion, as there arguably is nothing at all natural to nature, in this view. What is particularly important and problematic in our cultured notions of nature, Brooke emphasizes, is the dualities on which they are inexorably founded: Nature versus human, natural versus supernatural, nature versus art, and so forth, all largely constitutive of how nature is understood and used in religious and scientific discourse—and Brooke ably points out connections with both. One early account that recognized the resultant lack of conceptual coherence to nature was that of Robert Boyle, who made the case to dispense with the word entirely, given this ambiguity. Brooke then turns to a review of the dichotomies on which our notion of nature is founded; for instance, that between the realm of the natural and the supernatural, which was complexified not only by the introduction of a third domain, that of the preternatural, in theological and early scientific speculation, but by critiques of the long-standing notion of the two books of nature and Scripture, one revealing the natural and the other the supernatural. Similarly, the distinction between nature and art, or the question of whether human artifice could ever resemble the majesty of the natural world itself, has been elided in countless examples of human contrivance. Similar cases are provided by Brooke for the distinction between nature and nurture, and ultimately between nature and culture. If, then, the word *nature* ought to be dispensed with, given its foundation on unsupportable binaries—that is, as defined by that which it is not—then what ought to replace this word *nature*? Brooke suggests that, instead of looking for new visions of nature, we simply look for new visions of reality, of which there are many from which to choose. One advantage Brooke notes is that there is no way we can claim reality to be any less expansive or broad than it really is! If, then, there is also no essential human nature, still an exploration of the reality of the human condition is worthwhile. Yet Brooke reminds us, in conclusion, that any new *vision of* nature is more properly understood as another *voice for* nature; hence we should ask who as well as what is implicated in visions of nature.

These fourteen essays reflect a wide range of scholarly views on envisioning nature, science, and religion. How may they be compared? The initial framework specified five visions of nature, and the resultant structure of the volume largely follows these visions of evolutionary nature, emergent nature, malleable nature, nature as sacred, and nature as culture. Yet, clearly, a set of recurrent themes emerges through the volume. Additionally, a comparison of the essays as well as the essayists themselves may offer us some perspective on the possibility of harmonizing these visions and their underlying philosophical themes, given the wide variety of disciplinary backgrounds and predispositions of the scholars who participated in our collaborative research.

The Afterword presents visualizations of visions of nature, based on a set of empirically derived graphics focusing on the essayists' disciplinary specialties, their inclinations regarding the five visions of nature, and the emphasis on nature, science, and/or religion revealed by key terms in their essays. It also follows the inspiration of John Brooke's concluding essay in constructing a set of four key philosophical binaries, representing common ontological, epistemological, and aesthetic distinctions between nature and antinature, and summarizing essayist positions with respect to these binaries. A common visualization technique employed in this Afterword is correspondence analysis, which offers preliminary glimpses into interesting points of resonance and differences among the essayists and essays. The upshot of this comparative analysis is a set of important tensions underlying the relations between nature, science, and religion, which we believe must be engaged and embraced, rather than resolved, if new visions are indeed to emerge out of the creative soup constituting this book. The volume thus closes not so much with an ending as an invitation for a renewed exploration of the essays herein, as the reader considers the possibility and form of new visions of nature, science, and religion.

Acknowledgments

The author is grateful for permission to adopt for this essay portions of a paper previously published in *Zygon: Journal of Religion and Science*. See James D. Proctor, "Resolving Multiple Visions of Nature, Science, and Religion," *Zygon* 39, no. 3 (2004): 367–57.

Notes

1. RPS suggests a set of relations formally understood as nontransitive, since one cannot, for instance, infer that rock is superior to paper just because rock defeats scissors and scissors defeat paper.

2. Carolyn Merchant, *The Death of Nature: Women, Ecology, and the Scientific Revolution* (San Francisco: Harper & Row, 1980).
3. Raymond Williams, "Nature," *Keywords: A Vocabulary of Culture and Society*, revised ed. (New York: Oxford University Press, 1983).
4. Ian G. Barbour, *Religion and Science: Historical and Contemporary Issues* (San Francisco: HarperSanFrancisco, 1997).
5. Keith Ward, *Religion and Human Nature* (Oxford: Oxford University Press, 1998).
6. Mary Evelyn Tucker and John A. Grim, "Introduction: The Emerging Alliance of World Religions and Ecology," *Daedalus: Proceedings of the American Academy of Arts and Sciences* 130, no. 4 (2001): 1–22.
7. See http://environment.harvard.edu/religion/publications/statements/joint_appeal.html. See also National Religious Partnership for the Environment, *Earth's Climate Embraces Us All: A Plea from Religion and Science for Action on Global Climate Change*, 2004, www.nrpe.org (accessed September 12, 2004).
8. Charles Darwin, *On the Origin of Species by Means of Natural Selection, or, the Preservation of Favoured Races in the Struggle for Life* (London: J. Murray, 1859).
9. Stephen Jay Gould, *The Structure of Evolutionary Theory* (Cambridge, Mass.: Belknap Press of Harvard University Press, 2002).
10. Kim Sterelny, *Dawkins Vs. Gould: Survival of the Fittest*, Revolutions in Science (Cambridge: Icon, 2001).
11. Ernst Mayr, *What Evolution Is* (New York: Basic Books, 2001).
12. Jerome H. Barkow et al., eds., *The Adapted Mind: Evolutionary Psychology and the Generation of Culture* (New York: Oxford University Press, 1992).
13. Paul R. Ehrlich, *Human Natures: Genes, Cultures, and the Human Prospect* (New York: Penguin Books, 2002).
14. James D. Proctor et al., "Primates, Monks, and the Mind: The Case of Empathy," *Journal of Consciousness Studies* 12, no. 7 (2005): 38–54.
15. Steven Weinberg, *Dreams of a Final Theory*, 1st ed. (New York: Pantheon Books, 1992), 245.
16. Michael Ruse, *The Evolution Wars: A Guide to the Debates* (Santa Barbara, Calif.: ABC-CLIO, 2000); Robert T. Pennock, *Intelligent Design Creationism and Its Critics: Philosophical, Theological, and Scientific Perspectives* (Cambridge, Mass.: MIT Press, 2001); Richard Dawkins, *The God Delusion* (Boston: Houghton Mifflin Co., 2006); Daniel Clement Dennett, *Breaking the Spell: Religion as a Natural Phenomenon* (New York: Viking, 2006).
17. Edward O. Wilson, *Consilience: The Unity of Knowledge* (New York: Alfred A. Knopf, 1998).
18. Wendell Berry, *Life Is a Miracle: An Essay against Modern Superstition* (Washington, D.C.: Counterpoint, 2000); Antonio R. Damasio, ed., *Unity of Knowledge: The Convergence of Natural and Human Science* (New York: New York Academy of Sciences, 2001).
19. Danette Paul, "Spreading Chaos: The Role of Popularizations in the Diffusion of Scientific Ideas," *Written Communication* 21, no. 1 (2004): 185–222.
20. Arthur O. Lovejoy, *The Great Chain of Being: A Study of the History of an Idea* (Cambridge, Mass.: Harvard University Press, 1936).
21. Steven Weinberg, *Facing Up: Science and Its Cultural Adversaries* (Cambridge, Mass.: Harvard University Press, 2001), 107–22.
22. W. Anderson, "More Is Different," *Science* 177, no. 4047 (1972): 393–96.

23. Murray Gell-Mann, *The Quark and the Jaguar: Adventures in the Simple and the Complex* (New York: W. H. Freeman, 1994). See http://www.santafe.edu; and http://necsi.org.
24. Robert M. May, *Stability and Complexity in Model Ecosystems,* Monographs in Population Biology 6 (Princeton, N.J.: Princeton University Press, 1973); Simon A. Levin, "The Problem of Pattern and Scale in Ecology," *Ecology* (December 1992): 1943–67; Robert E. Ulanowicz, *Ecology, the Ascendent Perspective,* Complexity in Ecological Systems Series (New York: Columbia University Press, 1997); Jane Molofsky and James D. Bever, "A New Kind of Ecology?" *BioScience* 54, no. 5 (2004): 440–46; James D. Proctor and Brendon M. H. Larson, "Ecology, Complexity, and Metaphor (Introduction)," *BioScience* 55, no. 12 (2005): 1065–68.
25. Harold J. Morowitz, *The Emergence of Everything: How the World Became Complex* (New York: Oxford University Press, 2002).
26. Stephen Wolfram, *A New Kind of Science* (Champaign, Ill.: Wolfram Media, 2002).
27. J. C. Polkinghorne, *Reason and Reality: The Relationship between Science and Theology* (London: SPCK, 1991).
28. Philip Clayton, *Mind and Emergence: From Quantum to Consciousness* (Oxford: Oxford University Press, 2004).
29. Leo Kadanoff, "A New Kind of Science," *Physics Today* (July 2002): 55.
30. See, for example, http://anunda.com/enlightenment/spiritual-emergence.htm; and http://www.sedonajournal.com/sje.
31. Evelyn Fox Keller, *The Century of the Gene* (Cambridge, Mass.: Harvard University Press, 2000).
32. George Robertson et al., eds., *FutureNatural: Nature, Science, Culture* (London: Routledge, 1996); Noel Castree and Bruce Braun, eds., *Social Nature: Theory, Practice, and Politics* (Malden, Mass.: Blackwell Publishers, 2001).
33. Celia Deane-Drummond et al., "Genetically Modified Theology: The Religious Dimensions of Public Concerns about Agricultural Biotechnology," in Deane-Drummond et al., eds., *Re-Ordering Nature: Theology, Society and the New Genetics* (London: T&T Clark, 2003), 17–38.
34. Stephen F. Cotgrove, *Catastrophe or Cornucopia: The Environment, Politics, and the Future* (Chichester Sussex, UK: Wiley, 1982).
35. See http://www.ornl.gov/TechResources/Human_Genome/home.html; http://genewatch.org/bubiodefense.
36. Les Levidow, "Sustaining Mother Nature, Industrializing Agriculture," in George Robertson et al., eds., *FutureNatural: Nature, Science, Culture* (London: Routledge, 1996), 55–71.
37. Jeremy Rifkin, *The Biotech Century: Harnessing the Gene and Remaking the World* (New York: Jeremy Tarcher, 1998).
38. Deane-Drummond et al., "Genetically Modified Theology: The Religious Dimensions of Public Concerns about Agricultural Biotechnology," 17–38; Brian Wynne, "Interpreting Public Concerns About Gmos: Questions of Meaning," in Deane-Drummond et al., eds., *Re-Ordering Nature: Theology, Society and the New Genetics* (London: T&T Clark, 2003), 221–48; Leigh Turner, "Biotechnology as Religion," *Nature Biotechnology* 22, no. 6 (2004): 659–60.
39. Peter Dickens, *Reconstructing Nature: Alienation, Emancipation and the Division of Labour* (London: Routledge Publishers Ltd., 1996), 107ff.

40. Donna J. Haraway, *Modest_Witness@Second_Millennium.Femaleman©_Meets_Oncomouse™* (New York: Routledge, 1997).
41. Jean Baudrillard, "Disneyworld Company," *Libération* (4 March 1996): 7.
42. Francis Fukuyama, *Our Posthuman Future: Consequences of the Biotechnology Revolution* (New York: Farrar, Straus and Giroux, 2002).
43. Nelly Oudshoorn, "A Natural Order of Things? Reproductive Sciences and the Politics of Othering," in Robertson et al., eds., *FutureNatural*, 122–32.
44. Robert Frodeman, "Nanotechnology: The Visible and the Invisible," *Science as Culture* 15, no. 4 (2006): 383–89.
45. Barbour, *Religion and Science*, 102–3.
46. See http://environment.harvard.edu/religion.
47. Bronislaw Szerszynski, *Nature, Technology and the Sacred* (Oxford: Blackwell, 2005).
48. Catherine L. Albanese, *Nature Religion in America: From the Algonkian Indians to the New Age* (Chicago: University of Chicago Press, 1990); "Fisher Kings and Public Places: The Old New Age in the 1990s," *Annals of the American Academy of Political and Social Science* (May 1993): 131–43; Catherine L. Albanese, *Reconsidering Nature Religion* (Harrisburg, Pa.: Trinity Press International, 2002).
49. Albanese, *Reconsidering Nature Religion*, 3.
50. Ibid., 11–24.
51. Thomas R. Dunlap, *Faith in Nature: Environmentalism as Religious Quest* (Seattle: University of Washington Press, 2004); James D. Proctor and Evan Berry, "Social Science on Religion and Nature," in Bron Taylor, ed., *Encyclopedia of Religion and Nature* (New York: Continuum International, 2005); James D. Proctor, "Old Growth and a New Nature: The Ambivalence of Science and Religion," in Sally Duncant and Tom Spies, eds., *Old Growth in a New World* (Washington, D.C.: Island Press, 2007).
52. Michael Horace Barnes, ed., *An Ecology of the Spirit: Religious Reflection and Environmental Consciousness* (Lanham, Md.: University Press of America, 1994); Roger S. Gottlieb, ed., *This Sacred Earth: Religion, Nature, Environment* (New York: Routledge, 1996); Tucker and Grim, "Introduction: The Emerging Alliance of World Religions and Ecology"; Donald A. Crosby, *A Religion of Nature* (Albany: State University of New York Press, 2002); Stephen R. Kellert and Timothy J. Farnham, eds., *The Good in Nature and Humanity: Connecting Science, Religion, and Spirituality with the Natural World* (Washington, D.C.: Island Press, 2002).
53. Lynn White Jr., "The Historic Roots of Our Ecologic Crisis," *Science* 155 (March 1967): 1203–7.
54. Connie C. Barlow, *Green Space, Green Time: The Way of Science* (New York: Copernicus, 1997); Ursula Goodenough, *The Sacred Depths of Nature* (New York: Oxford University Press, 1998).
55. Paul R. Ehrlich and Anne H. Ehrlich, *Betrayal of Science and Reason: How Anti-Environmental Rhetoric Threatens Our Future* (Washington, D.C.: Island Press, 1996).
56. Martin W. Lewis, "Radical Environmental Philosophy and the Assault on Reason," in Paul R. Gross et al., eds., *The Flight from Science and Reason* (Baltimore: Johns Hopkins University Press, 1996), 209–30.
57. Stephen Budiansky, *Nature's Keepers: The New Science of Nature Management* (New York: The Free Press, 1995).
58. Clarence J. Glacken, *Traces on the Rhodian Shore: Nature and Culture in Western Thought*

from Ancient Times to the End of the Eighteenth Century (Berkeley: University of California Press, 1967).

59. Stephen Edelston Toulmin, *Cosmopolis: The Hidden Agenda of Modernity* (Chicago: University of Chicago Press, 1992).
60. Ian Hacking, *The Social Construction of What?* (Cambridge, Mass: Harvard University Press, 1999).
61. Paul R. Gross and Norman Levitt, *Higher Superstition: The Academic Left and Its Quarrels with Science* (Baltimore: Johns Hopkins University Press, 1994); Paul R. Gross et al., *The Flight from Science and Reason* (Baltimore: Johns Hopkins University Press, 1996); Andrew Ross, ed., *Science Wars* (Durham, N.C.: Duke University Press, 1996).
62. William Cronon, ed., *Uncommon Ground: Toward Reinventing Nature* (New York: W. W. Norton & Company, 1995); Noel Castree and Bruce Braun, eds., *Social Nature: Theory, Practice, and Politics*; Bruno Latour, *Politics of Nature: How to Bring the Sciences into Democracy* (Cambridge, Mass.: Harvard University Press, 2004).
63. C. Snow,"The Two Cultures," in Gladys Garner Leithauser and Marilynn Powe Bell, eds., *The World of Science: An Anthology for Writers* (New York: Holt, Rinehart and Winston, 1987), 157–63.
64. See, for instance, the special issue of *Complexity International*: http://www.csu.edu.au/ci/vol02/ci2.html.
65. Ursula Goodenough and Terrence W. Deacon,"From Biology to Consciousness to Morality," *Zygon: Journal of Religion and Science* 38, no. 4 (2003): 801–19.
66. Francisco J. Varela et al., *The Embodied Mind: Cognitive Science and Human Experience*, first MIT Press paperback ed. (Cambridge, Mass.: MIT Press, 1993); George Lakoff and Mark Johnson, *Philosophy in the Flesh: The Embodied Mind and Its Challenge to Western Thought* (New York: Basic Books, 1999).
67. Nancey Murphy et al., eds., *Chaos and Complexity: Scientific Perspectives on Divine Action* (Berkeley, Calif.: Center for Theology and the Natural Sciences, 1995); Clayton, *Mind and Emergence: From Quantum to Consciousness.*
68. N. Katherine Hayles, *Chaos Bound: Orderly Disorder in Contemporary Literature and Science* (Ithaca, N.Y.: Cornell University Press, 1990); Michael Ruse, *Mystery of Mysteries: Is Evolution a Social Construction?* (Cambridge, Mass.: Harvard University Press, 1999).
69. Pascal Boyer, *The Naturalness of Religious Ideas: A Cognitive Theory of Religion* (Berkeley: University of California Press, 1994); Pascal Boyer, *Religion Explained: The Evolutionary Origins of Religious Thought* (New York: Basic Books, 2001); Karl Peters, *Dancing with the Sacred: Evolution, Ecology, and God* (Harrisburg, Pa.: Trinity Press International, 2002).
70. George Perkins Marsh, *Man and Nature; or Physical Geography as Modified by Human Action* (New York: Charles Scribner and Co., 1864).

Bibliography

Albanese, Catherine L. *Nature Religion in America: From the Algonkian Indians to the New Age*. Chicago: University of Chicago Press, 1990.

———."Fisher Kings and Public Places: The Old New Age in the 1990s." *Annals of the American Academy of Political and Social Science* (May 1993): 131–43.

———. *Reconsidering Nature Religion*. Harrisburg, Pa.: Trinity Press International, 2002.
Anderson, W. "More Is Different." *Science* 177, no. 4047 (1972): 393–96.
Barbour, Ian G. *Religion and Science: Historical and Contemporary Issues*. San Francisco: HarperSanFrancisco, 1997.
Barkow, Jerome H., Leda Cosmides, and John Tooby, eds. *The Adapted Mind: Evolutionary Psychology and the Generation of Culture*. New York: Oxford University Press, 1992.
Barlow, Connie C. *Green Space, Green Time: The Way of Science*. New York: Copernicus, 1997.
Barnes, Michael Horace, ed. *An Ecology of the Spirit: Religious Reflection and Environmental Consciousness*. Lanham, N.Y.: University Press of America, 1994.
Baudrillard, Jean. "Disneyworld Company." *Libération* (March 4, 1996): 7.
Berry, Wendell. *Life Is a Miracle: An Essay against Modern Superstition*. Washington, D.C.: Counterpoint, 2000.
Boyer, Pascal. *The Naturalness of Religious Ideas: A Cognitive Theory of Religion*. Berkeley: University of California Press, 1994.
———. *Religion Explained: The Evolutionary Origins of Religious Thought*. New York: Basic Books, 2001.
Budiansky, Stephen. *Nature's Keepers: The New Science of Nature Management*. New York: The Free Press, 1995.
Castree, Noel, and Bruce Braun, eds. *Social Nature: Theory, Practice, and Politics*. Malden, Mass.: Blackwell Publishers, 2001.
Clayton, Philip. *Mind and Emergence: From Quantum to Consciousness*. Oxford: Oxford University Press, 2004.
Cotgrove, Stephen F. *Catastrophe or Cornucopia: The Environment, Politics, and the Future*. Chichester, Sussex, UK: New York: Wiley, 1982.
Cronon, William, ed. *Uncommon Ground: Toward Reinventing Nature*. New York: W. W. Norton & Company, 1995.
Crosby, Donald A. *A Religion of Nature*. Albany: State University of New York Press, 2002.
Damasio, Antonio R., ed. *Unity of Knowledge: The Convergence of Natural and Human Science*. New York: New York Academy of Sciences, 2001.
Darwin, Charles. *On the Origin of Species by Means of Natural Selection, or, the Preservation of Favoured Races in the Struggle for Life*. London: J. Murray, 1859.
Dawkins, Richard. *The God Delusion*. Boston: Houghton Mifflin Co., 2006.
Deane-Drummond, Celia, Bronislaw Szerszynski, and Robin Grove-White. "Genetically Modified Theology: The Religious Dimensions of Public Concerns about Agricultural Biotechnology." In Deane-Drummond, Szerszynski, and Grove-White, eds., *Re-Ordering Nature: Theology, Society and the New Genetics*, 17–38. London: T&T Clark, 2003.
Dennett, Daniel Clement. *Breaking the Spell: Religion as a Natural Phenomenon*. New York: Viking, 2006.
Dickens, Peter. *Reconstructing Nature: Alienation, Emancipation and the Division of Labour*. London: Routledge Publishers Ltd., 1996.
Dunlap, Thomas R. *Faith in Nature: Environmentalism as Religious Quest*. Seattle: University of Washington Press, 2004.
Ehrlich, Paul R. *Human Natures: Genes, Cultures, and the Human Prospect*. New York: Penguin Books, 2002.

Ehrlich, Paul R., and Anne H. Ehrlich. *Betrayal of Science and Reason: How Anti-Environmental Rhetoric Threatens Our Future*. Washington, D.C.: Island Press, 1996.
Frodeman, Robert. "Nanotechnology: The Visible and the Invisible." *Science as Culture* 15, no. 4 (2006): 383–89.
Fukuyama, Francis. *Our Posthuman Future: Consequences of the Biotechnology Revolution*. New York: Farrar, Straus and Giroux, 2002.
Gell-Mann, Murray. *The Quark and the Jaguar: Adventures in the Simple and the Complex*. New York: W. H. Freeman, 1994.
Glacken, Clarence J. *Traces on the Rhodian Shore: Nature and Culture in Western Thought from Ancient Times to the End of the Eighteenth Century*. Berkeley: University of California Press, 1967.
Goodenough, Ursula. *The Sacred Depths of Nature*. New York: Oxford University Press, 1998.
Goodenough, Ursula, and Terrence W. Deacon. "From Biology to Consciousness to Morality." *Zygon: Journal of Religion and Science* 38, no. 4 (2003): 801–19.
Gottlieb, Roger S., ed. *This Sacred Earth: Religion, Nature, Environment*. New York: Routledge, 1996.
Gould, Stephen Jay. *The Structure of Evolutionary Theory*. Cambridge, Mass.: Belknap Press of Harvard University Press, 2002.
Gross, Paul R., and Norman Levitt. *Higher Superstition: The Academic Left and Its Quarrels with Science*. Baltimore: Johns Hopkins University Press, 1994.
Gross, Paul R., Norman Levitt, and Martin W. Lewis. *The Flight from Science and Reason*. Baltimore: Johns Hopkins University Press, 1996.
Hacking, Ian. *The Social Construction of What?* Cambridge, Mass: Harvard University Press, 1999.
Haraway, Donna J. *Modest_Witness@Second_Millennium.Femaleman©_Meets_Oncomouse™*. New York: Routledge, 1997.
Hayles, N. Katherine. *Chaos Bound: Orderly Disorder in Contemporary Literature and Science*. Ithaca, N.Y.: Cornell University Press, 1990.
Kadanoff, Leo. "A New Kind of Science." *Physics Today* (July 2002): 55.
Keller, Evelyn Fox. *The Century of the Gene*. Cambridge, Mass.: Harvard University Press, 2000.
Kellert, Stephen R., and Timothy J. Farnham, eds. *The Good in Nature and Humanity: Connecting Science, Religion, and Spirituality with the Natural World*. Washington, D.C.: Island Press, 2002.
Lakoff, George, and Mark Johnson. *Philosophy in the Flesh: The Embodied Mind and Its Challenge to Western Thought*. New York: Basic Books, 1999.
Latour, Bruno. *Politics of Nature: How to Bring the Sciences into Democracy*. Cambridge, Mass.: Harvard University Press, 2004.
Levidow, Les. "Sustaining Mother Nature, Industrializing Agriculture." In George Robertson et al., eds., *FutureNatural: Nature, Science, Culture*, 55–71. London: Routledge, 1996.
Levin, Simon A. "The Problem of Pattern and Scale in Ecology." Ecology (December 1992): 1943–67.
Lewis, Martin W. "Radical Environmental Philosophy and the Assault on Reason." In Paul R. Gross, Norman Levitt, and Martin W. Lewis, eds., *The Flight from Science and Reason*, 209–30. Baltimore: The Johns Hopkins University Press, 1996.

Lovejoy, Arthur O. *The Great Chain of Being: A Study of the History of an Idea*. Cambridge, Mass.: Harvard University Press, 1936.

Marsh, George Perkins. *Man and Nature; or Physical Geography as Modified by Human Action*. New York: Charles Scribner and Co., 1864.

May, Robert M. "Stability and Complexity in Model Ecosystems." *Monographs in Population Biology* 6. Princeton, N.J.: Princeton University Press, 1973.

Mayr, Ernst. *What Evolution Is*. New York: Basic Books, 2001.

Merchant, Carolyn. *The Death of Nature: Women, Ecology, and the Scientific Revolution*. San Francisco: Harper & Row, 1980.

Molofsky, Jane, and James D. Bever. "A New Kind of Ecology?" *BioScience* 54, no. 5 (2004): 440–46.

Morowitz, Harold J. *The Emergence of Everything: How the World Became Complex*. New York: Oxford University Press, 2002.

Murphy, Nancey, Robert John Russell, and Arthur R. Peacocke, eds. *Chaos and Complexity: Scientific Perspectives on Divine Action*. Berkeley, Calif.: Center for Theology and the Natural Sciences, 1995.

National Religious Partnership for the Environment. "Earth's Climate Embraces Us All: A Plea from Religion and Science for Action on Global Climate Change," (2004), www.nrpe.org (accessed September 12, 2004).

Oudshoorn, Nelly. "A Natural Order of Things? Reproductive Sciences and the Politics of Othering." In George Robertson et al., eds., *FutureNatural: Nature, Science, Culture*, 122–32. London: Routledge, 1996.

Paul, Danette. "Spreading Chaos: The Role of Popularizations in the Diffusion of Scientific Ideas." *Written Communication* 21, no. 1 (2004): 32–68.

Pennock, Robert T. *Intelligent Design Creationism and Its Critics: Philosophical, Theological, and Scientific Perspectives*. Cambridge, Mass.: MIT Press, 2001.

Peters, Karl. *Dancing with the Sacred: Evolution, Ecology, and God*. Harrisburg, Pa.: Trinity Press International, 2002.

Polkinghorne, J. C. *Reason and Reality: The Relationship between Science and Theology*. London: SPCK, 1991.

Proctor, James D. "Resolving Multiple Visions of Nature, Science, and Religion." *Zygon: Journal of Religion and Science* 39, no. 3 (2004): 637–57.

———. "Old Growth and a New Nature: The Ambivalence of Science and Religion." In Sally Duncan and Tom Spies, eds., *Old Growth in a New World*, 122–32. Washington, D.C.: Island Press, 2007.

Proctor, James D., and Evan Berry. "Social Science on Religion and Nature." In Bron Taylor, ed., *Encyclopedia of Religion and Nature*, 1571–77. New York: Continuum International, 2005.

Proctor, James D., and Brendon M. H. Larson. "Ecology, Complexity, and Metaphor (Introduction)." *BioScience* 55, no. 12 (2005): 1065–68.

Proctor, James D., Frans de Waal, and Evan Thompson. "Primates, Monks, and the Mind: The Case of Empathy." *Journal of Consciousness Studies* 12, no. 7 (2005): 38–54.

Rifkin, Jeremy. *The Biotech Century: Harnessing the Gene and Remaking the World*. New York: Jeremy Tarcher, 1998.

Robertson, George, et al., eds. *FutureNatural: Nature, Science, Culture*. London: Routledge, 1996.

Ross, Andrew, ed. *Science Wars*. Durham, N.C.: Duke University Press, 1996.
Ruse, Michael. *Mystery of Mysteries: Is Evolution a Social Construction?* Cambridge, Mass.: Harvard University Press, 1999.
———. *The Evolution Wars: A Guide to the Debates*. Santa Barbara, Calif.: ABC-CLIO, 2000.
Snow, C. "The Two Cultures." In Gladys Garner Leithauser and Marilynn Powe Bell, eds., *The World of Science: An Anthology for Writers*, 157–63. New York: Holt, Rinehart and Winston, 1987.
Sterelny, Kim. *Dawkins Vs. Gould: Survival of the Fittest. Revolutions in Science*. Cambridge, Mass.: Icon, 2001.
Szerszynski, Bronislaw. *Nature, Technology and the Sacred*. Oxford: Blackwell, 2005.
Toulmin, Stephen Edelston. *Cosmopolis: The Hidden Agenda of Modernity*. Chicago: University of Chicago Press, 1992.
Tucker, Mary Evelyn, and John A. Grim. "Introduction: The Emerging Alliance of World Religions and Ecology." *Daedalus: Proceedings of the American Academy of Arts and Sciences* 130, no. 4 (2001): 1–22.
Turner, Leigh. "Biotechnology as Religion." *Nature Biotechnology* 22, no. 6 (2004): 659–60.
Ulanowicz, Robert E. *Ecology, the Ascendent Perspective*. Complexity in Ecological Systems series. New York: Columbia University Press, 1997.
Varela, Francisco J., Evan Thompson, and Eleanor Rosch. *The Embodied Mind: Cognitive Science and Human Experience*, 1st MIT Press paperback ed. Cambridge, Mass.: MIT Press, 1993.
Ward, Keith. *Religion and Human Nature*. Oxford: Oxford University Press, 1998.
Weinberg, Steven. *Dreams of a Final Theory*, 1st ed. New York: Pantheon Books, 1992.
———. *Facing Up: Science and Its Cultural Adversaries*. Cambridge, Mass.: Harvard University Press, 2001.
White, Lynn, Jr. "The Historic Roots of Our Ecologic Crisis." *Science* 155 (March 1967): 1203–7.
Williams, Raymond. "Nature." In *Keywords: A Vocabulary of Culture and Society*, revised ed., 157–63. New York: Oxford University Press, 1983.
Wilson, Edward O. *Consilience: The Unity of Knowledge*. New York: Alfred A. Knopf, 1998.
Wolfram, Stephen. *A New Kind of Science*. Champaign, Ill.: Wolfram Media, 2002.
Wynne, Brian. "Interpreting Public Concerns About Gmos: Questions of Meaning." In Deane-Drummond, Szerszynski, and Grove-White, eds., *Re-Ordering Nature: Theology, Society and the New Genetics*, 221–48.

1

THE NATURE OF VISIONS OF NATURE: PACKAGES TO BE UNPACKED

Willem B. Drees

What is the nature of "visions of nature"? When labeling visions of nature with terms such as *evolution* or *emergence,* it seems as if they are indeed visions *of nature,* slightly expanded from the natural sciences. However, in speaking of *visions,* the ambition seems to be more far-reaching, as a vision is to guide our actions, and thus to incorporate goals or values and a sense of who we are. This is most obvious when the vision speaks of nature as sacred. This transfer from a descriptive and analytical sense of a "vision of nature" toward an evaluative or normative interest seems to be the significance of thinking through our visions of nature. Visions *of* nature may turn out to be visions *for* nature as well.

However, philosophers have long challenged such a transfer from descriptive to prescriptive claims. Describing a shampoo in advertising copy as "purely herbal" may promote sales, but as a recommendation it is logically fallacious—many herbal substances may well be harmful to one's hair. Philosophers speak of the naturalistic fallacy, or the is-ought distinction. Thus, visions of nature that serve as visions for nature seem to combine in a single package facts and values. Are such packages intellectually illegitimate? In the second part of this chapter I hope to begin by unpacking the five visions under consideration, to understand to some extent the appeal of these visions and some of their limitations.

Before focusing on the five visions, I will introduce a model for understanding religious convictions in relation to the natural sciences and to our moral intuitions, in a brief formula:

a theology = a cosmology + an axiology

My claim is that this model may clarify why disagreements on visions of nature are not likely to be resolved by additional scientific evidence. Thus, the conclusion will be that visions of nature are of intellectual interest, as possible ways of looking at reality, that they may also be of religious or moral interest by nourishing certain "moods and motivations," but that precisely for that reason, science is not qualified to decide on competing visions of nature. Addressing the nature of our "visions of nature" may clarify our moral disagreements, but will not resolve them.

Theologies as Packages

What Might Religion Be?

The anthropologist Clifford Geertz has studied the cultural function of religious symbols. He concluded that[1]

> sacred symbols function to synthesize a people's ethos—the tone, character, and quality of their life, its moral and aesthetic style and mood—and their world view—the picture they have of the way things in sheer actuality are, their most comprehensive ideas of order.

This insight regarding the role of symbols in synthesizing ethos and worldview brought him to an oft-quoted definition:[2]

> a religion is (1) a system of symbols which acts to (2) establish powerful, pervasive, and long-lasting moods and motivations in men by (3) formu-

> lating conceptions of a general order of existence and (4) clothing these conceptions which such an aura of facticity that (5) the moods and motivations seem uniquely realistic.

This definition is, as a definition of the empirical phenomenon of religiosity, not perfect. Perhaps it neglects the role of rituals, and puts too much emphasis on the cognitive role of symbols as contributing to conceptions of the order of existence, thus bypassing ritual, social, and other noncognitive roles of religion. Perhaps the definition fails to shed light on complexities of representation and truth.[3] Besides, the definition suggests a causal arrow from symbols via conceptions to moods and motivations, whereas the symbols may also express moods and motivations—the passage quoted before, about synthesizing ethos and worldview, has less of this causal suggestion in it.

However, insofar as the definition is about the cognitive side of a religion, what could be called a "theology," the definition highlights the observation that, in religious thought, conceptions of the order of existence are intertwined with the appreciation of reality and norms for our behavior. To speak of the world as God's creation has a descriptive and a prescriptive aspect to it. In the same article, Geertz also speaks of models *of* the world and models *for* the world; that is, of models that seem to be descriptive and models that articulate a normative orientation and transformative ambition. In order to acknowledge the concentration on cognitive and normative dimensions, I will speak here not of a "religion," as Geertz does, but of a "theology."

Theologies

Theologies, as systematic positions, seem to offer a particular view of the way the world is *and* of the way the world should be, of the True and the Good, of the real and the ideal. Each theology expresses a particular relationship between a cosmology—in the metaphysical sense as a view of the way reality is—and an axiology, a view of the values that should be realized. Thus, as a heuristic to clarify and explore a complex area of discussion, I suggest the following "formula" for understanding the nature of theologies (plural):

a theology = a cosmology + an axiology

with the + sign not being a mere addition, but the crucial issue: how the two are brought together. *Axiology* is a grandiose word for a theory of values and of what we value, that is, a philosophically reflected articulation of our moral intuitions.[4]

Theologies can be quite different in the way they relate and prioritize cosmological and the axiological aspects. The definition allows one to concentrate on *existential* issues, which become prominent when our reality is not in accord with what we think ought to be, thus stressing the tension that might be involved in the *and* in the formula. But it may also be about *supernatural* or *magical* elements, as particular claims regarding the cosmological order. Within the Christian tradition, there are—based on my definition—various theologies. When the emphasis is on God's saving activity, the tension between the way the world is and the way it will be is prominent, whereas in creation-oriented views (whether ecologically inspired or as natural theologies) cosmology and axiology stand less starkly in contrast; the prophet emphasizes the tension, whereas the mystic stresses how we fit in to the larger reality. Whiteheadian process thought is one particular articulation of the interplay of axiological and causal elements. This way of integrating regulative ideals into cosmology has required particular, and in my opinion problematic, choices in cosmology: choices regarding panexperientialism and regarding the place of physics in the order of the sciences. However, it is an interesting and relevant attempt to integrate valuational and causal elements in a single categorical scheme.

In my opinion, the attempt to combine "is" and "ought" statements is what makes theology problematical *and* valuable. Again and again, the difficulty of the combination finds expression in the problem of evil, which typically concerns the relationship or tension between the two main components. The significance shows up precisely in the same context, when a prophet speaks up against evil in the name of a higher good: "Thus speaks the Lord."

Underdetermination

The definition of theology as "cosmology and axiology" does not yet refer to the natural sciences. A cosmology is, of course, related to the sciences. However, the relation is not straightforward. One may distinguish between science and any interpretation of science as a view of reality; that is, any cosmology, metaphysics, or philosophy of nature. A cosmology, in this sense, is a view of what the world (with its substances and relations, and conceptions of space, time, matter, forces, causality, etc.) might be like, given what we know (and what we know not to be the case; science may well be stronger in what it excludes than in what it includes). Any such metaphysics is an interpretation of scientific knowledge, constrained by the sciences but also underdetermined by them.

It may be useful to distinguish further between various aspects of the sciences. For our current purposes, three levels may be sufficient: (1) theories, (2) taxonomies and empirical generalizations, and (3) observations and experiments. At the "high" end, there are theories that describe vast domains of realities. That is where the integration provided by the sciences is most clearly visible. Among these are Newton's understanding of forces and motion, the atomic theory of matter with the Periodic System, electromagnetism, neo-Darwinian evolutionary theory, quantum physics, Einstein's improved understanding of forces, motion, and space and time, and so on. At the "low" end, there are the manifold observations and experiments that connect scientific ideas to the world.

Relations between the high end and the low end run a large gamut. Some of it is inductive in kind, generalization toward overarching rules or statistical expectations. Major relations are, however, hypothetical-deductive, where the creative researcher postulates entities, forces, or other causal factors that may explain the observed phenomena. Certainty is always limited. Inductive generalization may be of limited validity beyond the phenomena that formed the basis for the generalization, as conditions not yet taken into account may be essential. Hypothetical-deductive approaches, such as those advocated by Karl Popper, acknowledge openly the creative and provisional nature of the hypothesis—which may be refuted if deduced consequences do not match observations. However, even such refutations are not final, since it may be that one does not consider the theory itself refuted, but rather one of the additional hypotheses involved, such as hypotheses regarding the measuring equipment or the initial conditions—an insight regarding the underdetermination of refutations and theories that has come to be known in philosophy as the Duhem-Quine thesis.

Even if one accepts a particular theory, metaphysical or cosmological inferences drawn from it may be highly disputable. Does quantum mechanics force upon us an indeterministic metaphysics? Not at all, as witnessed by the continuing debates over the interpretation of quantum mechanics. Does the Special Theory of Relativity that describes space-time as a four-dimensional space-time force upon us the idea that time is spatial and the future is just as much already there as the past? Again, interpretations vary. Furthermore, precisely where the cosmologically most interesting issues are, such as the nature of time and of causality, of matter and of interactions, science is at its most speculative. Chemists work with the Periodic Table, and atoms consist of protons, neutrons, and electrons. The first two of these are made up of quarks, but what is the end of the line—superstrings? And the Big Bang theory is a very successful theory describing the evolution of the universe, but precisely when we get to the very beginning, beyond the Planck horizon, the theory becomes unreliable, as we need a quantum theory of space and time—an area of quantum cosmology where approaches are as much guided by preconceived philosophical ideas and preferences in mathematics as by observations. In a review article, Jeremy Butterfield and Christopher Isham wrote about theory construction in this field of quantum gravity and quantum cosmology:

> In this predicament, theory-construction inevitably becomes much more strongly influenced by broad theoretical considerations, than in mainstream areas of physics. More precisely, it tends to be based on various *prima facie* views about what the theory *should* look like—these being grounded partly on the philosophical prejudices of the researcher concerned, and partly on the existence of mathematical techniques that have been successful in what are deemed (perhaps erroneously) to be closely related areas of theoretical physics. . . .
>
> The situation . . . tends to produce schemes based on a wide range of philosophical motivations, which (since they are rarely articulated) might be presumed to be unconscious projections of the chtonic psyche of the individual researcher—and might be dismissed as such! Indeed, practitioners of a given research programme frequently have difficulty in understanding, or ascribing validity to, what members of a rival programme are trying to do. This is one reason why it is important to uncover as many as possible of the assumptions that lie behind each approach: one person's "deep" problem may seem irrelevant to another, simply because the starting positions are so different.[5]

Thus, underdetermination seems a real issue within the sciences, and especially so when one comes to metaphysical or cosmological conclusions regarding the nature of nature. However, underdetermination need not be understood as "Everything goes," as lower levels and requirements of consistency considerably constrain the options.

A Scheme

So far, I have described theological convictions as combining cosmological and axiological ones. Each of these is related to underlying disciplines, such as ethics and the natural sciences, with these, in turn, related to observations, experiments, and moral intuitions. Deep down, of course, both of these series relate to the world in which we live and the experiences we have. Thus, in a sense there are two levels where integration occurs—at the theoretical level of theology (or worldview) and in practice, in life as lived.[6]

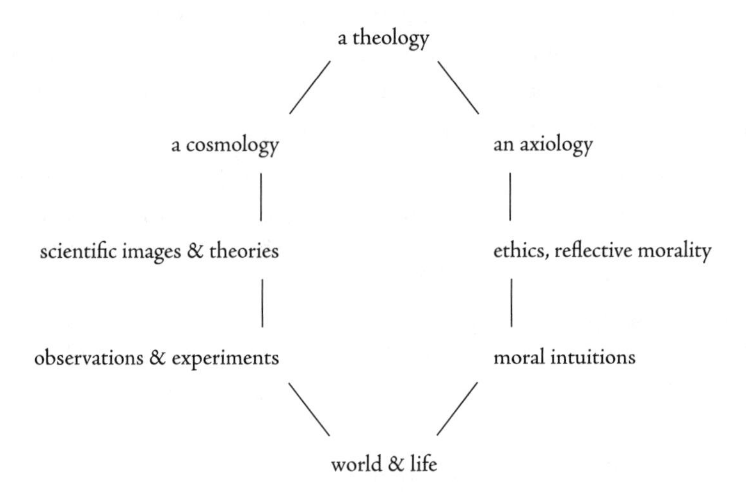

Visions of Nature

Where should we locate our topic "visions of nature"? Some visions fit nicely in the left-hand column, as a scientific image of nature. When *evolution* or *emergence* is the central notion, science seems to have primacy. However, when nature is considered "malleable," there is a cosmological dimension, as reality is in some way flexible, but also an axiological one, or at least this vision of nature that requires a moral orientation—for how would we otherwise have a sense of direction? When nature is considered "cultural," we seem to leave the plane of discussion, as the whole structure is a human scheme. However, it may also be understood within this framework as emphasizing underdetermination in the relationships, and thus reminding us of the distance between empirical evidence and all theoretical constructions, whether still within science (theories) or beyond science in interpretations and visions. When nature is referred to as being "sacred," we clearly have moved toward the axiological side of the scheme, stressing the valuation of nature. It could even be considered a particular theological perspective, excluding other models, such as those that would reserve the concept of sacredness for a transcendent realm or entity beyond nature.

Thus, the emphasis of the five visions discussed here is not uniform—as alternatives addressing issues at the same locus in the scheme. Rather, they move across the whole scheme. However, in seeing them as visions *of* nature, with the connotation that they are also visions *for* nature, visions that should guide our behavior, they have a similar integrative function as theologies as understood here. Thus, in the next section, the five visions will be considered, each in relation to science and to the wider meanings that may be attached to it.

Visions of Nature

Evolutionary Vision: Continuing Disputes

Evolutionary biology does provide an encompassing vision of nature. It offers explanations for similarities between organisms, the functionality of organs and behaviors, the biogeographical distribution of species and extinct species, and much more. Its explanatory scheme is of a remarkable simplicity, with hereditary variation and selection at its core, supplemented with knowledge of underlying chemical processes (DNA), and various variations on the basic scheme (e.g., sexual selection).

Evolution as an understanding of nature is acceptable to many Christians, from the Catholic pope, John Paul II, to the liberal Anglican Arthur Peacocke; David Livingstone could write a book on certain American evangelicals in the late nineteenth century as *Darwin's Forgotten Defend-*

ers. However, some Christians strongly oppose evolution, resulting in controversies surrounding "creationism" and "intelligent design." In the period spanning 1960–1990 or so, this may have been primarily about the inerrancy of Scripture. However, if one takes a longer view, it seems to me that two other issues are at stake; namely, the loss of the argument from design and the association of evolution with modernism.[7]

As for the argument from design: In the seventeenth and eighteenth centuries, some modern Christians did not appeal to Scripture but rather to our understanding of nature, as enriched by science. This was not only an argument for faith but also a legitimization of science; the historians John Brooke and Geoffrey Cantor titled a section of their book *Reconstructing Nature* on this topic: Natural Theology and the Promotion of Science.[8] When in the second half of the nineteenth century Darwin's evolutionary theory and other naturalistic theories came to be, those who used the argument from design faced a choice. Either one could accept the new theories, and thereby one would have to give up the argument from nature to God. Of course, one could relocate the argument as one about the remarkable conditions that allowed for evolution to take place, such as the expanses of time and the laws of nature. The natural explanation would be accepted, though itself understood from a theistic or a deistic perspective. The other option was to stick to the argument from design, and thus to reject Darwin's explanations. In this context, teleological-evolutionary schemes can be understood, but also the evolution of various forms of creationism, including the contemporary "intelligent design" movement. Science, as understood by the mainstream, must be incomplete; an extra organizing principle is needed. There is a logical fallacy here, in that it seems assumed that if the argument from design fails, the religious beliefs would fall away. However, normally when a particular argument is not successful, the idea for which it was an argument is thereby not necessarily abandoned. Thus, a question is why the argument has become so central to the believers involved.

The way I perceive the controversies over evolution and design, the argument from design and against evolutionary explanations is not really what moves believers involved. The concept of evolution is significant because it has an enormous symbolic value; it has become a shibboleth in a struggle. If that hypothesis is correct, the obvious question is what it is a symbol for.

In my perception, it is a shibboleth for a controversy about modernism. In the nineteenth century, the more troubling enemy for conservative believers was historical criticism, the study of the Bible as a collection of historical documents. In more recent times, the central issue is at least as much a concern about society. After the shooting at Columbine High School in Littleton, Colorado, in April 1999, there was a debate in Washington in the U.S. House of Representatives on the treatment of juvenile offenders. The leader of the Republicans in the House, Tom DeLay (R-Tex.), read from a letter that said the issue was not that parents should have kept their children away from guns. The letter offered a variety of alternative explanations:

> It couldn't have been because half of our children are raised in broken homes. . . . It couldn't have been because we place our children in day care centers where they learn their socialization skills among their peers under the law of the jungle. . . . It couldn't have been because we allow our children to watch, on average, 7 hours of television a day filled with the glorification of sex and violence that isn't even fit for adult consumption. . . . It couldn't have been because we have sterilized and contracepted our families down to sizes so small that the children we do have are so spoiled with material things that they come to equate the receiving of the material with love. . . . It couldn't have been because our school systems teach the children that they are nothing but glorified apes who have evolutionized out of some primordial soup of mud.[9]

Teaching evolution is, apparently, part of a package with other ills, such as contraception, sterilization, and abortion; television and computer games; child care and divorce. That the struggle over evolution is not just about biology is also obvious from the strategic document of the ID movement, "the wedge," where the effort to discredit evolution and to present ID as a respectable alternative is the thin end of a wedge that serves to realize a particular cultural transformation.

Thus, the controversy over evolution and intelligent design can be understood as a struggle not over a particular scientific theory or vision *of nature*, but over values that have come to be associated with evolutionary theory. Evolution is seen as part of a package, and it is the package that is contested.

The package is not merely an invention of antievolutionists. Time and again, evolution has been appealed to by advocates of other causes, from Herbert Spencer and John D. Rockefeller up to the present with Richard Dawkins, Daniel Dennett (*Darwin's Dangerous Idea*), and E. O. Wilson's *Consilience* and *Biophilia*. The evolution of evolutionary theory itself has been loaded with cultural values, as Michael Ruse has pointed out on many occasions (*Monad to Man*), though over time the role of nonscientific values has decreased significantly (*Mystery of Mysteries*).

Evolution as a vision of nature is contested. I happen to think that the scientific credentials of evolutionary theory are flawless, and thus that it should be part of our understanding of reality, at least up to the level of scientific theories. However, it cannot be an all-encompassing theory, as if we would offer an evolutionary explanation of the conditions that make evolution possible—the laws of nature, the material upon which evolution works. Thus, it cannot serve as our cosmology, at least not as a complete cosmology.[10] At that level, there may be further explanatory moves within the natural realm, or an appeal to a "primary Cause" who has set up the world so that evolution could bring forth complex life forms.

As a vision of nature, however, evolution is especially contested when it comes in a package with values. Some expect that an evolutionary understanding will deliver such values, whether capitalistic, due to the survival of the fittest (Herbert Spencer), or more sympathetic, due to the values of cooperation and empathy (Frans de Waal, and earlier predecessors; E. O. Wilson on biophilia). Some expect a new religious framework that serves us all, the "evolutionary epic" (e.g., Loyal Rue). In the opposite direction is the position of Thomas H. Huxley, who in his lecture on evolution and ethics in 1893 suggested that we have evolved, but should not make evolution into our moral framework. To the contrary, "brought before the tribunal of ethics, the cosmos might well seem to stand condemned."[11]

As a philosopher it seems to me that we should be especially aware that evolution has brought us so far, but need not justify particular values. Perhaps it might be evolutionarily intelligible that we use skin color as a indication of relatedness, and thus tend to be inclined toward racism. Or it may be evolutionarily intelligible why human parents are more protective of daughters in puberty than of sons of the same age. But once we realize that we tend to do so, we may ask ourselves whether we consider it right. The distinction between explication and justification is important.

That leaves, of course, fully open that evolution serves as a repository of images and epics that allow us to speak of ecological responsibility or any other moral issue. But when serving as a vision of nature with moral consequences, it is no longer just the science that delivers the conclusions—it is a moral vision that draws on science to promote certain "moods and motivations."

Emergence: A New Kid on the Block?

Alongside evolution, "emergence" has been introduced as a vision of nature. The rhetoric is that of antireductionism,[12] even though speaking of emergence suggests that "higher" entities or phenomena come about by the organized interplay of "lower" entities. Thus, even though it may be that the analysis of the behavior of higher entities needs a vocabulary of its own, their existence is understood as material in kind. A well-known example: "Paying someone money" is always a physical process where material objects change places or the state of the computer at the bank is modified, but the economics of paying is not intelligible when described in these physicalist terms.

In my perception, *emergence* is a useful notion within science to correct an unwarranted extrapolation from successful cases of intertheoretical reduction with uniform bridge principles (e.g., the reduction of thermodynamics to statistical mechanics) to all cases where interlevel relationships may seem to invite intertheoretical reductions. *Emergence* is a word that indicates that "higher-level" phenomena are fruits of "material processes" even though our description of the "higher-level" phenomena cannot be reduced in a straightforward way to a description of the underlying processes.

What is *emergence* to deliver as a vision of nature? One expectation seems to be that this provides a way to go beyond the domain of the sciences, just as Samuel Alexander in the early twentieth century suggested that the deity was always "the next level up."[13] However, this usage of *emergence* brings with it the value judgment that *higher*, in the sense of reduction and emergence, would imply being of greater value. It is hard to see, anyhow, why that which emerges is also in an ultimate sense more fundamental. One might even argue the contrary: that what emerges is to some extent derived, a consequence of conditions and components.

In his *Theology for a Scientific Age,* Peacocke wrote that God as the source of the personal has to be at least personal.[14] If one takes evolution and emergence seriously, as Peacocke does, the argument seems misguided, as evolution and emergence imply that something may come to be that has genuinely new properties. Thus, such an argument for the personal characteristics of God as the source of personal being seems to fail. When challenged at this point, and offered an alternative, Peacocke agreed that this argument did not work, and that his consideration was Anselmian rather than causal. That is, God as the supremely worshipable being has to have the highest perfections available, and we consider personal characteristics to be additional perfections not found in, say, stones.[15] That may be a fine argument, but it raises the question of whether the emphasis on emergence does the theological work it is assumed to do.

Emergence stands for a vision that appreciates nature as a creative process, and seems to accept the scientific understanding of this process as well. As a vision, it is opposed to *reductionism* or *mechanicism* or even *materialism,* even though it is hard to see the difference between materialism and emergence, since emergence tells us how complex behavior arises in and through material structures. A major difference between "emergence" and "reductionist" visions seems to be that *emergence* serves as label for an axiological order that values "higher" over "lower" structures, "complexity" over "simplicity," and often "culture" over "nature."

Malleable Nature: Technology

Speaking of nature as malleable fits our life as active beings. With the help of technology we have continuously modified nature, from early farming to modern high tech. Andrew Lustig gives a nice analysis of the manifold aspects of this perspective in his contribution to this volume. Malleability has a cosmological aspect. It assumes that reality is not fully determined. If one distinguishes between humans and nature, nature may be seen as a deterministic system, except for the input humans give. Thus, the vision of nature as malleable seems to rely on an ontological distinction between humans and (the rest of) nature, with humans as being outside the system, so as to modify it. Perhaps a more unified picture may be possible in terms of Neurath's boat. Otto Neurath suggested that one may reconstruct a ship while at sea—but only piecemeal, and not by taking the whole apart, as one could do while in a dock.[16] Similarly, nature as malleable may be understood as reconstruction of some parts of nature, while assuming the given nature of other aspects—which may be changed at a different moment. We use tools to make tools—what is made at some time is means to make at another moment.

In discourses on technology, humans express attitudes toward human creativity, but thereby also may come to speak about divine creativity and the nature of creation. Technology may be seen as part of the human calling to care for those in need, to heal, perhaps even as the calling to counter negative consequences of our fallen world, as in Bacon's notion of science and technology as the "Great Instauration."[17] Or humans with their technological concerns may be considered to be overstepping boundaries, and thus their actions may be rejected as hubris, as "playing God." In response to the understanding of nature as malleable, we find religious vocabularies and notions from mythic material and fiction (*Frankenstein*, the golem) alongside metaphors from the political domain (boundaries) to praise and to condemn, to encourage and to slow down. Within the ambience of Christian thought, one finds reference to humans as *stewards* and as *cocreators*. To explore the difference between these two images, let me offer a simplistic Christian summary of the Bible, in a single sentence. According to the Bible, the world begins on high, with paradise, which is followed by a long and troublesome journey through history, with the expectation of final salvation. The liturgy reflects this U-shaped trajectory in the emphasis on memory and on hope. The Sabbath recalls creation and Exodus and is a foretaste of fulfillment.[18] This U-shaped profile implies that images of the good are present in two varieties, as images of the past (paradise) and as images of a City of God, a new heaven and a new earth, the world to come. If humans are considered stewards, one looks back in time, to a good situation, which has to be kept and preserved. If humans are addressed as cocreators, the eyes are mainly on the future, on that which might come. In biblical language the good is to be found not only in the past but also in the future. Humans, even when considered as stewards (as in Matthew 25:14–30), can be active and even ought to be active, although the initiative is with God, and this activity is normatively determined as care for the weak and needy. If nature is understood as "malleable," human responsibility is emphasized.

Our task becomes to make God present in the world. Theology is not rooted in positive experiences of beauty and goodness, but rather in engagement with justice and love, a vision of this world made better. Rather than *order*, the central theological theme is *transformation*. In natural theology, there is the tendency to appreciate the actual state of affairs as one deserving of wonder. Natural theologies arising out of experiences with the natural world mostly lack interest in transformation; that is why, in the discourses of natural theology, chemistry and technology are not as prominent as biology or physics.[19] However, theology should, in my opinion, attempt to disclose the possibilities for transformation of the natural order. Unavoidably, this also introduces questions about aims, goals, or norms—issues of values separate from facts, of axiology separate from ontology. Not only natural theologies, but also a theology with a strong liberationist tendency and one that acknowledges the depth of the human technological ability to transform reality, requires a metaphysics, which is adequate relative to what we know and to what we find ourselves able to do. I do not have such a metaphysics and axiology, but want to indicate here that technology does raise issues for cosmology, for axiology, and for the way these two are combined in a theology or religious worldview.

With respect to cosmology, technology requires us to envision not only the real but also the possible, not just order and laws but also flexibility (but not only flexibility, as technology is victorious over the laws by obeying them—think of flying and gravity). With respect to axiology, technology requires us to consider the expansion of the domain of choices, an issue to which we will return below. Focusing on technology might make us sensitive to elements neglected when we focus on science mainly as source of understanding, rather than as a source of transformative power.

Even nonbelievers find "playing God" a useful metaphor in criticizing new technologies. The American philosopher Ronald Dworkin suggested that this is because those new technologies do not merely raise ethical issues, but create insecurity by undermining a distinction that is vital to ethics. Underlying our moral experience is a distinction between what has been given and what is a matter of choice and responsibility. What is given is the stable background of our actions. We cannot change those issues. Traditionally, this has been referred to as fate, nature, or creation: domains of the gods or of God. We assume a clear demarcation between who we are, whether the product of Divine Providence or of blind chance, and what we do in the situation we find ourselves in. When new technologies expand the range of our abilities, and thus shift the boundary between what is given and what is open to our actions, when nature becomes more malleable, we become insecure and concerned. It is especially in such circumstances that the phrase *playing God* arises. There is a reference to "God" when something that was experienced as a given becomes part of the domain of human considerations. We accuse others of playing God when they have moved what was beyond our powers to our side of the boundary. The fear of "playing God" is not the fear of doing what is wrong (which is an issue on our side of the boundary), but rather the fear of losing a grip on reality through the dissolution of the boundary.

New technologies imply a different range of human powers, and thus a changing experience of fate, nature, creation, or God. At least, if God is associated with that which has been given, identified with creation. If God is seen thus, our technological activity will be seen as pushing God back to the margin. Antibiotics and contraceptives have contributed more to secularization in Western cultures than Darwin; practices are more important than ideas. This God who is pushed to the margin is a god-of-the-gaps, not so much the gaps in our knowledge as the gaps in our skills. If we do not accept this god-of-the-gaps, how to proceed? Theism, with its root pair of metaphors of power (on the side of the transcendent God) and dependence (on our side), is challenged to rethink itself in the light of our powers over malleable nature.

Cultural Nature: The Risk of Dismissing Science

Treating nature as cultural is of a different order. It emphasizes not so much the nature of nature, but rather the nature of human understanding: Our concepts of nature are cultural, human constructs. Thus, a cultural vision of nature is not a particular vision (such as evolutionary nature might be), since all visions are seen as cultural, including the evolutionary one.

This is what I am arguing here, to some extent. Any package of values and understanding is constructed, is cultural. In terms of the scheme offered earlier, the emphasis on "nature" as a cultural product corresponds to remarks about underdetermination, from scientific theories to worldviews. It is not evolutionary science that determines our worldview. Rather, the worldview is a construct that draws on the available resources, such as scientific theories.

However, the understanding of nature as cultural can easily be overdone, as if there were no nature out there as a "given" constraint on our behavior. Malleable nature acknowledges this constraint, while seeking to overcome it—we have learned to fly with the help of machines, even though our biological nature has not endowed us with wings. *Cultural nature* may suggest too much that all constraints are in the eyes of the beholder, rather than out there.

Nature as culture can also be overdone with respect to scientific understanding.[20] Of course, all science is a human product. But, to a large extent, it is a human product that seems to be accepted by scholars from culturally different backgrounds. The Periodic Table was constructed by a Russian, but the data underlying it weren't, and it is used all over the world. It is a theory, but not "just a theory," as if all theories were equal. Against those who want to relativize science as cultural in order to keep unwelcome messages at bay, I see two clusters of considerations.

The epistemic success of the natural sciences—as they developed in the last century or two, resulting in corroborated theories that have a wide scope, unifying the understanding of phenomena in various contexts, in combination with remarkable precision—is totally without equal to any cognitive understanding offered in previous human history, whether in religious myths, theological systems, or in philosophical speculations. This success makes it urgent to take these theories as our best available guides to the understanding of reality.

The natural sciences, in conjunction with technology, have provided us with unprecedented means to act in reality. Actions can be laudable or despicable. However, it is morally preferable to use one's knowledge, rather than to willfully close one's eyes. A few in "religion and science" seek to play down the epistemic significance of science, arguing that it is as much about dogmatic frameworks and prejudices as religion. Not only is such a *tu quoque* argument not of much help for religion, but it may also have morally problematic consequences. In one example, William Stahl and colleagues treated the theory that HIV was the cause of AIDS as just a theory that suited the interests of the pharmaceutical industry and served to promote discrimination.[21] In their rejection of consolidated science, the authors were explicitly voicing support for the South African leader Thabo Mbeki, who by his denial of the viral background of AIDS deprived tens of thousands of people—if not more than a million people—effective treatment. There is a major moral risk in playing down established science. This is related to the "ethics of belief," though the demands articulated by William K. Clifford in his 1879 essay were unrealistic.[22] Playing down established science, even if it is done for morally lofty purposes, may have immoral consequences. Working with the best knowledge available, rather than playing such knowledge down, may well be morally required.

Nature as Sacred: Neglecting Responsibility?

Envisioning nature as sacred is, again, of a somewhat different order. It is not a particular voice on the scientific/cosmological side, but purely on the axiological side. In practice, however, speaking of nature as sacred seems to privilege emergence over reduction, while it also may coincide to some extent with an evolutionary view, as witnessed, for instance, by the book by biologist Ursula Goodenough, titled *The Sacred Depths of Nature.* Value is expected to reside within the natural, and the natural is what is valuable. This position contrasts most easily with an understanding of nature as malleable, where the driving idea is that nature can be improved. Here, nature is accepted and appreciated. *Nature as sacred* seems to advocate a hands-off policy in ecology and genetics. But is such a position morally compelling? It is the opposite of Thomas Huxley's remark that ethical nature, though emerging out of biological nature, goes beyond or even against it.[23] In that sense, seeing nature as sacred fails to do justice to our abhorrence of nature, our empathy with "victims of nature."[24]

Visions and Life: Integration and Integrity

In this essay, I have discussed visions of nature as theologies or worldviews. I suggest that the nature of such visions is that they offer systematic articulations of our understanding of reality in conjunction with our values. As such, discussing visions of nature is a highly intellectual, speculative project—far removed from the daily business of life. However, both strands integrated in our visions of nature—namely, our knowledge of nature and our values—are rooted in life as it is lived. They rest on moral intuitions and human experiences, filtered via reflection, testing, and the imaginative construction of theories. Any vision of nature is also to be judged by the way it shapes our lives; that is, by the way it relates us to this level of life as lived. As David Livingstone in his essay in this volume, "Locating New Visions," shows with various historical examples, visions of nature in their mix of moral and scientific elements are shaped by the context in which they arise and serve.[25] We have to make do with theoretical integration in a vision of nature, but this integration reflects and serves the integrity of our lives.

Notes

1. Clifford Geertz, "Religion as a Cultural System," in Geertz, *The Interpretation of Cultures* (New York: Basic Books, 1973), 89.
2. Ibid., 90.
3. Nancy K. Frankenberry and Hans H. Penner, "Clifford Geertz's Long-Lasting Moods, Motivations, and Metaphysical Conceptions," *The Journal of Religion* 79 (1999): 617– 40.
4. Though there are some superficial resemblances with a scheme proposed by Nancey Murphy and George Ellis in their book *On the Moral Nature of the Universe* (Minneapolis: Augsburg Fortress, 1996), where they see two branches of learning, the one cosmological and the other social and moral, culminating in theology, there are major differences. My scheme is a heuristic for exploring the diversity of theologies, rather than a substantive thesis about the proper view of the relationship between theology, ethics, and the sciences, as it is for Murphy and Ellis. Besides, I do not want to pronounce in this context on the "moral nature of the universe"; my formula can also be used to describe positions of those who consider the universe to be amoral, whether indifferent or evil. Unlike Murphy and Ellis, for whom each level of understanding requires a higher one until it finally includes a doctrine of God, I do not consider atheists to be necessarily deficient in understanding; rather, they hold a different existential position.
5. Jeremy Butterfield and Christopher J. Isham, "Spacetime and the Philosophical Challenge of Quantum Gravity," in Craig Callender and Nick Huggett, eds., *Physics Meets Philosophy at the Planck Scale: Contemporary Theories in Quantum Gravity* (Cambridge: Cambridge University Press, 2001), 38.
6. A somewhat similar scheme, distinguishing the "context of life" and "metaphysics," was presented by Jürgen Hübner at the Second European Conference on Science and Religion, in Enschede, the Netherlands, March 10–13, 1988. See Jürgen Hübner, "Science and Religion Coming Across," in Jan Fennema and Iain Paul, eds., *Science and Religion: One World—Changing Perspectives on Reality* (Dordrecht: Kluwer, 1990), 173–81, esp. 177.
7. See, e.g., Ronald L. Numbers, *The Creationists: The Evolution of Scientific Creationism* (New York: Alfred A. Knopf, 1992). It might also be claimed that eschatological expectations (premillennialism) may have had something to do with it, as suggested, under the influence of Livingstone and Numbers, in Michael Ruse, *The Evolution-Creation Struggle* (Cambridge, Mass.: Harvard University Press, 2005), 157 and 319.

8. John Brooke and Geoffrey Cantor, *Reconstructing Nature: The Engagement of Science and Religion* (Edinburgh: T&T Clark, 1998).
9. Debate on the "Consequences for Juvenile Offenders Act of 1999," *Congressional Record—House*, June 16, 1999, H4366; accessed at http://thomas.loc.gov on March 1, 2005.
10. There are some scientific speculations that use an evolutionary type of explanation, also at the cosmological level, explaining the natural laws of our universe as the product of selection in a process that produces universes (e.g., Lee Smolin, *The Life of the Cosmos* [New York: Oxford University Press, 1997]). It seems to me that one can put my objection against the completeness of evolutionary explanations again at that level—the rules by which this universe-producing process goes are themselves not part of that process. However, we may run into an infinite regress here—either another layer of evolutionary explanations (and so on, all the way down) or at some point a nonevolutionary "given" character of the basic rules.
11. Thomas H. Huxley, "Evolution and Ethics," in James Paradis and George C. Williams, eds. *T. H. Huxley's* Evolution and Ethics *with New Essays on Its Victorian and Sociobiological Context* (Princeton, N.J.: Princeton University Press, 1989), 117; reprinted from T. H. Huxley, *Evolution and Ethics* (London: Macmillan, 1894), 59.
12. For a presentation that works with the (in my opinion mistaken) opposition of emergence and reduction, one can consult the interesting contribution by Barbara J. King, "God from Nature: Evolution or Emergence?" in this volume, where *reductionism* comes to stand for all-too-simple reductions, say, in speaking of a gene for religion. Of course, given the wider focus of my contribution, I cannot go as much in depth as Gregory R. Peterson in his "Who Needs Emergence?" in this volume, which I read as supporting the suspicion that the rhetoric of *emergence* is often antireductionist, even though emergentist theories affirm that "higher-level realities" are rooted in and brought forth by lower-level realities, and that the difference is thus more one of description than of reality.
13. Samuel Alexander, *Space, Time and Deity*, 2 vols. (London: Macmillan and Co., 1920).
14. Arthur Peacocke, *Theology for a Scientific Age: Being and Becoming—Natural, Divine, and Human*, enlarged edition (London: SCM, 1993), 106–12.
15. The exchange between Peacocke and me occurred at one of the January workshops organized by the Ian Ramsey Centre at the University of Oxford in the late 1990s.
16. Andrew Lustig, "The Vision of Malleable Nature: A Complex Conversation," in this volume; Otto Neurath used the image of reconstructing a ship while at sea in a different context, when he argued against the existence of pure protocol sentences in A. J. Ayer, ed., *Logical Positivism* (Glencoe, Ill.: Free Press, 1959); originally, this paper appeared in *Erkenntniss* III (1932f.).
17. Francis Bacon used the expression the "Great Instauration," referring to the great renewal expected to come through science, as the encompassing title of a (never completed) six-part work, of which *Novum Organon* was published in 1620 as the second volume. Francis Bacon, *The New Organon*, edited by Lisa Jardine and Michael Silverthorne, Cambridge Texts in the History of Philosophy (Cambridge Cambridge University Press, 2000), xiii, 1–24.
18. The U-profile is an image from Northrop Frye, *The Great Code: The Bible and Literature* (San Diego: Hartcourt Brace Jovanovich, 1982), 169; the reflection on creation and fulfillment has been inspired by Samuel Terrien, *The Elusive Presence: Toward a New Biblical Theology* (San Francisco: Harper & Row, 1978).

19. John Brooke and Geoffrey Cantor, *Reconstructing Nature: The Engagement of Science and Religion* (Edinburgh: T&T Clark, 1998), chap. 10; John Brooke, "Improvable Nature?" in Willem B. Drees, ed., *Is Nature Ever Evil? Religion, Science and Value* (London: Routledge, 2003), 149–69.
20. Nicolaas Rupke, "Nature as Culture: The Example of Animal Behavior and Human Morality," in this volume, seems to lean more toward the relativist side than I consider warranted and adequate. Even if any historical account were constructed with presentist interests, it seems to me dubious to apply this to the same extent to the natural sciences, as he does when he suggests that the story of cognitive-methodological progress in understanding animal behavior can be subsumed under a nonprogressive story of appropriations of animal behavior on behalf of contemporaneous socio-political purposes. I would object even more strongly if one went on to suggest that this insight could be generalized as applying to all the natural sciences. It would also be self-referentially incoherent if relativism were presented as a universally valid position. It may perhaps itself be understood well in relativist terms, as serving a particular apologetic purpose, namely the legitimization of the humanities in contexts dominated by the natural sciences.
21. William A. Stahl, Robert A. Campbell, Yvonne Petry, and Gary Driver, *Webs of Reality: Social Perspectives on Science and Religion* (New Brunswick, N.J.: Rutgers University Press, 2002), 110.
22. William K. Clifford, "The Ethics of Belief," in *Lectures and Essays*, Vol. 2 (London: Macmillan and Co., 1879), an essay to which William James responded with a less demanding view of the right to believe, in an essay titled "The Will to Believe" (1897).
23. Thomas H. Huxley, "Evolution and Ethics," in James Paradis and George C. Williams, eds. *T.H. Huxley's* Evolution and Ethics *with New Essays on Its Victorian and Sociobiological Context* (Princeton, N.J.: Princeton University Press, 1989), 117; reprinted from T. H. Huxley, *Evolution and Ethics* (London: Macmillan, 1894), 59.
24. *Victims of nature* was the term used by the clinical geneticist Leo ten Kate in his contribution to a conference on value judgments regarding nature. Leo ten Kate, "Victims of Nature Cry Out," in Willem B. Drees, ed., *Is Nature Ever Evil? Religion, Science and Value* (London: Routledge, 2003), 170–72.
25. David N. Livingstone, "Locating New Visions," in this volume.

Bibliography

Alexander, Samuel. *Space, Time and Deity*, 2 vols. London: Macmillan, 1920).

Brooke, John Hedley. "Improvable Nature." In *Is Nature Ever Evil? Religion, Science and Value*, 149–69. London: Routledge, 2003.

Brooke, John Hedley, and Geoffrey Cantor. *Reconstructing Nature: The Engagement of Science and Religion*. Edinburgh: T&T Clark, 1998.

Butterfield, Jeremy, and Chris J. Isham, "Spacetime and the Philosophical Challenge of Quantum Gravity." In Craig Callender and Nick Huggett, eds., *Physics Meets Philosophy at the Planck Scale: Contemporary Theories in Quantum Gravity*, 33–89. Cambridge: Cambridge University Press, 2001.

Clifford, William K. "The Ethics of Belief." In *Lectures and Essays*, vol. 2 (London: Macmillan and Co., 1879).

Dawkins, Richard. *The Selfish Gene*. London: Granada, 1978.
de Waal, Frans. *Good Natured: The Origins of Right and Wrong in Humans and Other Animals*. Cambridge, Mass.: Harvard University Press, 1996.
Dennett, Daniel. *Darwin's Dangerous Idea: Evolution and the Meanings of Life*. New York: Simon & Schuster, 1995.
Drees, Willem B., ed. *Is Nature Ever Evil? Religion, Science, and Value*. London: Routledge, 2003.
Dworkin, Ronald. "Playing God." *Prospect* 41 (May 1999).
Frankenberry, Nancy K., and Hans H. Penner. "Clifford Geertz's Long-Lasting Moods, Motivations, and Metaphysical Conceptions." *The Journal of Religion* 79 (1999): 617–40.
Geertz, Clifford. "Religion as a Cultural System." In Geertz, ed., *The Interpretation of Cultures*. New York: Basic Books, 1973.
Goodenough, Ursula. *The Sacred Depths of Nature*. New York: Oxford University Press, 1998.
Hübner, Jürgen. "Science and Religion Coming Across." In Jan Fennema and Iain Paul, eds., *Science and Religion: One World—Changing Perspectives on Reality*, 173–81. Dordrecht: Kluwer, 1990.
Huxley, T. H. "Evolution and Ethics." In James Paradis and George C. Williams, eds. *T. H. Huxley's Evolution and Ethics with New Essays on Its Victorian and Sociobiological Context*, 57–174. Princeton, N.J.: Princeton University Press, 1989; reprint from T. H. Huxley, *Evolution and Ethics*. London: Macmillan, 1894.
King, Barbara. "God from Nature: Evolution or Emergence?" in this volume.
Livingstone, David N. *Darwin's Forgotten Defenders: The Encounter between Evangelical Theology and Evolutionary Thought*. Grand Rapids, Mich.: Eerdmans, 1987.
———. "Locating New Visions," in this volume.
Lustig, Andrew. "The Vision of Malleable Nature: A Complex Conversation," in this volume.
Murphy, Nancey, and George F. R. Ellis. *On the Moral Nature of the Universe*. Minneapolis: Augsburg Fortress, 1996.
Numbers, Ronald L. *The Creationists: The Evolution of Scientific Creationism*. New York: Alfred A. Knopf, 1992.
Peacocke, Arthur R. *Theology for a Scientific Age: Being and Becoming—Natural, Divine, and Human*, enlarged edition. London: SCM, 1993.
Peterson, Gregory. "Who Needs Emergence?" in this volume.
Rue, Loyal D. *Everybody's Story: Wising Up to the Epic of Evolution*. Albany: State University of New York Press, 2000.
Rupke, Nicolaas. "Nature as Culture: The Example of Animal Behavior and Human Morality," in this volume.
Ruse, Michael. *Monad to Man: The Concept of Progress in Evolutionary Biology*. Cambridge, Mass.: Harvard University Press, 1996.
———. *Mystery of Mysteries: Is Evolution a Social Construction?* Cambridge, Mass.: Harvard University Press, 1999.
———. *The Evolution-Creation Struggle*. Cambridge, Mass.: Harvard University Press, 2005.
Smolin, Lee. *The Life of the Cosmos*. New York: Oxford University Press, 1997.

Stahl, William A., Robert A. Campbell, Yvonne Petry, and Gary Driver. *Webs of Reality: Social Perspectives on Science and Religion*. New Brunswick, N.J.: Rutgers University Press, 2002.

ten Kate, Leo, "Victims of Nature Cry Out." In *Is Nature Ever Evil? Religion, Science and Value*, 170–72. London: Routledge, 2003.

Wilson, E. O. *Consilience: The Unity of Knowledge*. New York: Alfred A. Knopf, 1998.

———. *Biophilia*. Cambridge, Mass.: Harvard University Press, 1984.

2

VISIONS OF NATURE THROUGH MATHEMATICAL LENSES

Douglas E. Norton

Prelude

The door at the end of the building opens onto the attached two-story parking ramp; I step out after a late evening of paper revisions (re-visions?) in my office and stop in my tracks. Straight ahead, low in the sky, but not so low as to be obscured by a single tree, streetlight, or power line, is the oversized bright disk of a full moon, sharp against the backdrop of a dark, cloudless sky. "Wow," I actually mutter aloud to myself, lips moving as my feet hold still. "What a moon." The first reaction, the knee-jerk reaction before conscious thought can take over, is aesthetic. This thin sliver of a story is a peek at a vision of nature through mathematical lenses.

The aesthetic is an unexpected mathematical perspective for the non-mathematician, and yet it has proven a significant factor in the motivation and implementation of mathematical work throughout the history of mathematics, on at least three levels. First, it has often been the order, the predictability, the interconnectedness of the natural world that appeal aesthetically and compel the mathematician to discover the mathematics to describe the patterns and connections observed. There seem to be laws of nature. This is not inconsistent with theology; as Brooke observes in this volume ("Should the Word *Nature* be Eliminated?"), laws of nature may simply be "how God normally chooses to act in the world." The daily sunrise, the phases of the moon, the temporal patterns of seasons, the spatial patterns of flower petals and rock strata and zebra stripes, the unfailing pull of gravity, from footsteps to falling apples to flowing rivers, all call the mathematician to reason with the *why* and try to quantify the *how.* As Ledley suggests in his contribution to this volume ("Visions of a Source of Wonder"), a sense of wonder at nature can serve as a source of inspiration for philosophy, religion, and science. Whether implemented by a sacred other or simply inherent in the natural system, there seem to be laws; the mathematician wants to both quantify and reason why.

This aesthetic carries over into the most abstract of pure mathematics. The practitioner sees an inherent beauty in the relationships among classes of numbers, geometrical shapes, and objects in abstract spaces. The combination of the simplicity of statement of Fermat's Last Theorem and the deep and challenging complexity of its proof, which required the passage of centuries and the creation of entire new areas of mathematics, brings aesthetic delight to the mathematician's imagination. The unexpected recurrence in disparate natural settings of numbers such as *pi,* the exponential base *e,* and the golden mean *phi,* speak of beauty to the mathematician.

Finally, the ultimate aesthetic for the mathematician is the sheer beauty of a well-constructed logical argument. To prove a theorem using just enough assumptions but no more, utilizing just the right theoretical tools and mathematical machinery needed but no more, saying exactly what needs to be said but no more can earn the provider of the proof that ultimate of mathematical compliments: "Your proof is elegant." It is an appreciation of the reasoning process, the beauty of knowing the right assumptions with which to start, the right conclusions to draw, and a sound logical pathway from one to the other, that describe the ultimate mathematical aesthetic.

It is this aesthetic of reason that causes me to label as *mathematicians* all those throughout history who have struggled to reason some sense out of the functioning of the universe around them. From looking for geometrical or numerical patterns to the simple reasoning of hows and whys, there is at least some part mathematician in each natural philosopher, naturalist, scientist, engineer, and every child who ever looks at some new wonder in the world with those always fresh and questioning eyes and asks, "Why?"

It is true that after the original aesthetic response to the moon described above, my next thought was, "Is it a full moon or just almost, full yesterday or tomorrow?" In other words, is the disk a full, filled circle or shaded a bit on one side or the other? Geometrize. Quantify. If I hadn't been drilled with all the facts from a young age, the next questions would have been, as they were to those who did not yet have the answers so long ago: How does that thing stay up there? How do the nightly travels and the monthly phases work? How are they related? Is that daily traveler substantively different in some way related to or independent of its lack of phases? Why don't they fall, or why doesn't everything else fly above us in distant arcs? The mathematician, or the mathematician in each of those questioners, wants to move beyond the quantifying of the how to the reasoning of the why. For each question, they—we—seek the ultimate resolution of *Q.E.D.: quod erat demonstrandum.*

Mathematics, then, can be defined not only as the study of numbers and patterns but also as the study of logical arguments and deductive reasoning. In his essay "The Unreasonable Effectiveness of Mathematics," R. W. Hamming claims that, even in its early development, mathematics had four faces: (1) the ability to carry on chains of logical reasoning, (2) geometry, (3) number, and (4) a sense of beauty.[1] The logical arguments stretch from the particular properties of numbers and shapes to questions of the functioning and interacting of the basic constituents of nature, from the microscopic to the astronomical. Some of these arguments are local in scope, some global, rendering an understanding of nature as dichotomy: universal in some of its characteristics, a patchwork of particular properties in others. In the section to follow, we present a brief glimpse of the contributions of mathematical thought to philosophical and theological views of nature, from the Pythagoreans through Isaac Newton. Then we consider more closely some contemporary interpretations of chaos theory.

Brief Overture

There is a long tradition of the application of mathematical arguments to questions of philosophy and theology. In a very real sense, as a part of the scientific view of reality, mathematics goes back to Thales of Miletus and the founding of natural philosophy. Thales is credited with the notion that multitudes of phenomena could be explained by a small number of hypotheses: he proposed the marriage of physical observation with deductive logical reasoning for explanations of the ways things work.[2] There is evidence that the concept of *number* goes back tens of thousands of years, and both geometry and number notation go back at least five thousand. While these were instances of observation and description of the world, with Thales came the beginnings of something different: The "main tool for carrying out the long chains of tight logical reasoning required by science is mathematics. Indeed, mathematics might be defined as being the mental tool designed for this purpose."[3]

The Pythagoreans linked order, number, and divine intellect. The claim of Thales that "All is water" gave way to the Pythagorean claim that "All is number." Plato's world of mathematical objects and ideas is real, though invisible; the allegory of the divided line in Book 6 of the *Republic* links mathematical objects and God, the natural world an imperfect reflection of the perfection to be found in the realm of the eternal.[4]

As both Brooke and Jackelén observe in this volume ("Should the Word *Nature* be Eliminated?" and "Creativity through Emergence: A Vision of Nature and God," respectively), the early church fathers saw no disconnect between nature and God, but rather saw God as working "in, with and through" nature. For Augustine, writing of ultimate truth and rational reasoning, that which is eternal and true, can be used to illuminate the faith;[5] he can be said to have "assimilated aspects of the Platonic view of divinity, namely the doctrine of the immutability of number and truth, into Christian theology."[6] His argument for the existence of God presages those of Anselm, Aquinas, Descartes, Pascal, Leibniz, Spinoza, Berkeley, Hume, and Kant centuries later. These all link, in one guise or another, nature, reason, truth, and the divine.

Johannes Kepler is best known for his three laws of planetary motion, which not only linked observational astronomy and mathematics but demonstrated his strong belief that God had created the universe according to a mathematical plan, in the Platonic and Pythagorean tradition. In the *Mysterium Cosmographicum,* nature is machine following a divine plan; his "why" for planetary orbits is a link with the mathematical ideal of the Platonic solids, but his laws of planetary motion are the beginning of the physics to explain "how" planetary motion works.[7]

Pascal's Wager and Descartes' Ontological Argument bring tools of logic and the new field of probability to bear on the question of the existence of God. Of larger import was Descartes' vision of knowledge "as a system of interconnected truths that can ultimately be abstracted into mathematics,"[8] his view of the universe as a complex mechanism designed by a clockmaker God. We could understand all the workings of a deterministic universe if we could discover all the rules of the machinery. This deterministic view has driven scientific inquiry for centuries, with and without inclusion of the divine architect and rule maker.

The analytic geometry of Descartes led directly to the calculus of Newton, at the pinnacle of mathematical theology. For Newton, mathematics and physics were tools to "read the mind of God," to provide a better understanding than ever before of the workings of the well-ordered universe of the clockmaker God. While Newton claimed "that God is needed to adjust the planetary orbits from time to time . . . Laplace, by a refinement of the calculus, famously showed that 'we have no need of that hypothesis.'"[9] Theologian Keith Ward observes that "One of the methodological presuppositions of the rise of the experimental method was the desacralization of nature. That is, physical nature had to be seen as a neutral ground which could be controlled, manipulated, and experimented upon, rather than as a part of the Divine being or the abode of spirits who might resent such human activities. Further, all formal and final causes had to be banished from physics, which would deal henceforth only with efficient and material causes."[10] Newton's greater understanding of the mechanism of nature led others to a mechanistic philosophy that omitted God. Reminiscent of Ludwig Wittgenstein's image of throwing away a ladder once one has used it to scale a height, Newton's universe was sufficiently well-ordered to be understood without invoking God.[11]

Theme: Contemporary Mathematics and Chaos Theory

Bridges between the mathematical and philosophical realms continue to be built in modern times. Some more recent examples include logical positivism, probabilistic causation, and philosophical implications of statistical mechanics. In particular, modal logic has been invoked to revive the ontological argument. One area of contemporary mathematics that has lent itself to an exploration of philosophical implications is dynamical systems. As the name suggests, *dynamical systems* is the study of the properties of mathematical models of systems undergoing change. Its questions are more qualitative than quantitative. Its objects of study are typically either systems of iterated functions or differential equations, dating to Newton's invention of calculus and equations of motion. Its tools range from traditional techniques of classical analysis to various branches of topology born in the twentieth century at least partially in response to some dynamical systems questions. Its applications range from ecology to meteorology, from chemical kinetics to population genetics, from economics to celestial mechanics.

A dynamical system consists of two elements: a rule, in the form of a function to iterate or a system of differential equations, which specifies how the system evolves with time; and the space on which the rule acts. The basic lines of inquiry are suggested by the following questions:

- What happens in the space in the long run, given a particular initial condition, or starting value?
- How does "what happens" depend on the starting value?
- Do certain points in the system remain fixed? Exhibit periodic behavior?
- Do certain points approach a limit, in finite time or asymptotically?
- Do different starting values have qualitatively different long-term behaviors?
- How does "what happens" depend on the function or differential equation?
- How does "what happens" depend on parameters within the function or differential equations?

The questions have a clear potential for translation into philosophical ones: If humans, human social structures, and human physical contexts are points in the universe as a complex dynamical system, what are the ultimate fates of the elements of the space and of the system as a whole? French mathematician Henri Poincaré is generally acknowledged as the founder of the field of dynamical systems with his 1889–1890 work on a question that both is typical of the field and has philosophical overtones: Is the solar system stable?

The basic theme, then, of dynamical systems is change. While seeds of the field were sown in the latter days of the nineteenth century, its true beginnings were in the last quarter of the twentieth. By then, science and culture had evolved, with evolution no longer the new intellectual kid on the block. The mechanistic view of nature as machine expanded to include rules that allow for change in the constituent parts of the machine. Animals, cultures, ecosystems, galaxies, intelligence itself—from Darwin through analogies with other systems, from the nineteenth through the twentieth and into the twenty-first centuries—evolve. The great machine of nature has rules that allow for its various subsystems and pieces to evolve, to become what they were not before: nature as evolving. As Brooke observes in this volume, nature is seen no longer as "a fixed and finished product" but rather as "nature transforming itself."

One area within the field of dynamical systems that has both generated vigorous research activity and crossed over into popular culture is chaos theory. It is an area in which the answers to the questions above differ from what previous approaches, and mathematical intuition, would have suggested. At first glance, *chaos theory* may seem to be an oxymoron: systematically organized knowledge applied to total disorder or confusion. In fact, chaos theory provides a framework for a better understanding of some complex, irregular phenomena, from ecosystems to social systems to the solar system. Examples of applications to disciplines farther afield than the traditional ones include psychology[12] and history.[13]

Despite the centrality of precision in the field of mathematics, the novelty and quick growth of chaos theory have left no universally accepted definition of *chaos*. (It is ironic that the process of defining mathematical chaos has been somewhat chaotic.) The one feature appearing in anyone's definition is "sensitive dependence on initial conditions." Any uncertainty in the initial state of a given system can lead to rapidly growing errors in any effort to predict a future state of the system. That is, in a chaotic system, any point has another point arbitrarily close to it whose future is eventually drastically different from its own. Two neighboring water molecules have different fates when they're the ones going separate ways at a fork in a river. In a chaotic system, *every* point in the system exhibits that sort of eventual divergence from nearby points.

Although it was not called *chaos* for decades, Poincaré described the phenomenon in his essay *Science et méthode* in 1908:

> A very small cause which escapes our notice determines a considerable effect that we cannot fail to see, and then we say that the effect is due to chance. If we knew exactly the laws of nature and the situation of the universe at the initial moment, we could predict exactly the situation of that same universe at a succeeding moment. But even if it were the case that the natural laws had no longer any secret for us, we could still only know the initial situation *approximately*. If that enabled us to predict the succeeding situation *with the same approximation*, that is all we require, and we should say that the phenomenon had been predicted, that it is governed by laws. But it is not always so; it may happen that small differences in the initial conditions produce very great ones in the final phenomena. A small error in the former will produce an enormous error in the latter. Prediction becomes impossible, and we have the fortuitous phenomenon.[14]

The implications of this phenomenon were not embraced by the mathematical and scientific community until much later in the twentieth century, stimulated by results of computer investigations.

Meteorologist Edward Lorenz rediscovered "sensitive dependence on initial conditions" in 1963 in a system of three linked nonlinear differential equations for modeling weather.[15] The term *butterfly effect* has been widely used to describe this phenomenon, based on a paper presented by Lorenz in 1972, titled "Does the Flap of a Butterfly's Wings in Brazil Set Off a Tornado in Texas?"[16] The motion of the air or water molecule is completely determined, from centuries-old Newtonian mechanics; yet, the slightest change in its starting position, from the flap of wings, can result in a very different itinerary. Since no physical situation can be known to infinite accuracy, behavior in chaotic systems eventually diverges from the predicted, not because the rules are wrong or even incomplete, but because our *knowledge* about the starting position is *necessarily* incomplete. It is not the case of probabilistic effects; chaos is *not* the same as randomness. It is a matter of the rules of the natural world magnifying small inherent differences into observable and significant ones.

The term *chaos* in this mathematical sense was first used in the paper "Period Three Implies Chaos" in a 1975 issue of the widely read *American Mathematical Monthly*.[17] The wide readership of the *Monthly*, the catchy term *chaos*, and increased use of computers to investigate long-term behavior of dynamical systems combined to spread the newly rediscovered topic through the mathematical community.

Two other increasingly accepted requirements for a system to be called *chaotic* are topological transitivity and density of periodic points.[18] Briefly, a dynamical system is called topologically transitive if, given any two open subsets of the space, the image of one subset under the deterministic rules of the system will eventually intersect the other. That is, the system cannot be decomposed into dynamically distinct sets because one will always eventually intersect the other, for *any* pair of open subsets of the space. Periodic points are dense in a space if every point in the space has another point arbitrarily close to it that is periodic, which means it eventually comes back to itself exactly and repeats a pattern periodically. That is, no matter how "chaotic" or seemingly random and pattern-free a given point's orbit may be, *arbitrarily* nearby at *any* point along its trajectory lies a periodic point, one that repeats a finite itinerary indefinitely. Mathematical chaos, then, has three defining characteristics: unpredictability, indecomposability, and an element of regularity. As it turns out, in the mathematical systems that stand as exemplars of chaotic phenomena, these properties seem to be inextricably linked. The popular treatment of chaos theory almost always focuses exclusively on the "sensitive dependence on initial conditions" criterion, the poster child for the uncertainties of contemporary life. What the mathematical analyses suggest is order within chaos. While much of the world satisfies the centuries-old model of orderliness, with no sign of chaotic phenomena, the mathematics suggests that chaotic pockets we may observe must have order either within or as their larger context.

There may be theological insights to be gleaned in an extension of this chaotic description to a broader metaphor. It does not bridge the traditional gap between C. Snow's "Two Cultures" of the sciences and humanities[19] or Stephen Jay Gould's "nonoverlapping magesteria" of science and religion.[20] On the other hand, it does allow for the determinism of Descartes' clockmaker God, tempered by a mechanism of surprising complexity and unexpectedly plentiful possibilities. It is somewhere between random and rigid, mechanical and magical. It is somewhere between, a merging of, the determinism and contingency carefully addressed by Ulanowicz in this volume ("Enduring Metaphysical Impatience?").

Some scientists have projected that historians will see the scientific ideas of the twentieth century as best represented by three aspects of uncertainty: Gödel's Incompleteness Theorem, Heisenberg's Uncertainty Principle, and chaos theory.[21] Gödel's Theorem states that for any mathematical system, there will always be propositions that cannot be proven either true or false using the rules and axioms of that system. Briefly, for any axiom system, provability is a weaker notion than truth. (That sounds like a mathematical result that may lend itself to theological metaphor.) Heisenberg's "Principle of Indeterminacy" rules out a complete knowledge of the position and momentum of any subatomic particle: Knowledge of one necessarily limits what can be known of the other. Gödel's Theorem places some theoretical results out of reach (but not most), and Heisenberg's subatomic particles seem far removed from daily existence. Chaos theory remains as uncertainty ubiquitous yet tempered.

Chaos theory does not apply universally; some systems are chaotic, others are not. Many systems have components, some of which are chaotic and others of which are not. There is a fundamental theorem in dynamical systems theory[22] that states that a system can be decomposed into gradient-like parts and chain recurrent parts; the system evolves unambiguously in a particular direction some of the time, and there are often regions in which the behavior can be chaotic. (Some systems are all one way and some the other, while others have both types of components.) For example, while the atmosphere around the earth is relatively stable within parameters that sustain life on the surface, the individual weather patterns are stereotypically chaotic. Millennia of planetary and ecosystem evolution have led to largely stable systems; yet there are subsystems in which a meteor could throw a moon out of orbit or a weather anomaly could throw a local species into extinction. Sensitive dependence on initial conditions may not pervade, but survives in pockets—possibly large ones.

Mathematician Ralph Abraham sees chaos as "inescapable in the rhythm of the planets, the climate, the metapatterns of history." It is "the essential chaos of life, which is necessary for evolution and is the source of all creativity."[23] Nineteenth-century historian Henry Brooks Adams says, "Chaos often breeds life, when order breeds habit."[24] In fact, two contemporary applications of chaos theory suggest just this necessity of chaos: A *lack* of chaos in either brain-wave[25] or heart-rate[26] readings can indicate pathology.

Throughout cosmogony, from the *Theogony* of Hesiod to the Genesis stories, the contrasted forces are chaos and cosmos. The universe replacing chaos is presented not as order from confusion but as harmonious existence from "a trackless waste, a formless void." In Genesis, the breath of God stirs the primordial waters of chaos into giving birth to all that is and turns a dirt clod into a living soul. Creation comes out of chaos, as addressed by Jackelén in this volume ("Creativity through Emergence: A Vision of Nature and God"). Different cultures have found some version of "chaos" as essential to the origins of the universe as we know it. This dichotomy even carries over into early literary references. For example, in *The Birds* of Aristophanes, from chaos and the darkness comes Eros, or love, which then is the force responsible for all creation. All these references to chaos are, of course, with respect to the traditional interpretation of the word. The mathematical appropriation of the term for its own particular flavor of unpredictability carries with it some of that tension between chaos and order, determinism and unpredictability, creativity and stability. The cardiac and brain-wave applications both suggest that elements of chaos breaking up the monotony of simple repetition are "creative sparks" necessary for the proper functioning of both the physiological and metaphorical centers of the human personality: the brain and the heart.

From creation stories through contemporary cardiology, such chaotic necessity drives the idea of nature as emergent. A whole can be greater than the sum of its parts if sensitive dependence on initial conditions drives a system past a bifurcation point into a new qualitative relationship. See the essays by Jackelén, Peterson, and Henderson in this collection for more on nature as emergent. In fact, one such variable in the evolution of a system may be the actions of those human components of the system to extend their influence beyond that originally thought possible. Some religious traditions posit humans as made in the image of God. If God creates and maintains the great clock of nature, some see humans as living out their nature as made in God's image by being cocreators,[27] comaintainers of the clock, cotinkerers with the mechanisms of nature, from genetic manipulation to ecosystem conversion, from eradicating diseases to terraforming planets. See the article by Lustig in this collection ("The Vision of Malleable Nature: A Complex Conversation") for more on nature as malleable. In particular, Lustig sees malleable nature as a "conceptual midpoint" of the various visions of nature addressed in this volume. What is unclear is where nature as culture calls us to draw the line in taking that evolving, emergent nature into our own hands as malleable nature. Perhaps it is exactly that fifth vision of nature, nature as sacred, that calls us to consider how the other four views should interact.

So what does the development of chaos theory say about the nature of mathematics as a reflection of the natural world? Ward observes that, according to Kant, "All changes in the physical universe were governed by the principle of sufficient or determining reason. . . . If you add to this axiom the admittedly complex hypothesis that there is a finite number of completely describable physical causes, then one arrives at the immensely influential Laplacian model of the physical universe, . . . [that] from the first state of the universe and a knowledge of all the laws of physics, everything that ever happened thereafter in the course of the universe's history could be infallibly predicted."[28] But in a chaotic system—even a chaotic subsystem of a larger system—any imprecision in information about the initial conditions of the system can lead to predictions that go beyond fallible to vastly incorrect. The entire process of mathematical modeling is based on the premise that, while any good model captures some important element of the system being represented, no model gathers every detail about the reality being modeled. The working assumption is that a model that is close in some sense to its real-world referent will yield conclusions about the expected behavior of the system that are close to the actual behavior of the system. Any system in which mathematically chaotic behavior is the appropriate deterministic mechanism should by its very nature *not* yield to the usual variety of useful mathematical representation. Referring to the various forms of mathematical uncertainty in current discourse, scientist/theologian John Polkinghorne says, "The clockwork universe is dead; . . . even at those macroscopic levels where classical physics gives an adequate account, there is an openness to the future which relaxes the unrelenting grip of mechanical determinism."[29] This seems to reflect the famous quote of Einstein: "As far as the laws of mathematics refer to reality, they are not certain; and as far as they are certain, they do not refer to reality."[30]

Do we metaphorically excommunicate the classic claim of Galileo that "The laws of Nature are written in the language of mathematics"? Is the "unreasonable effectiveness of mathematics in the natural sciences" reduced to reasonably hit-and-miss ineffectiveness? No, for at least two reasons. First, we have no indication that Nature writ large constitutes a chaotic system or even that chaotic subsystems are anywhere near ubiquitous. It remains true that, as Eugene Wigner observes, "The mathematical formulation of the physicist's often crude experience leads in an uncanny number of cases to an amazingly accurate description of a large class of phenomena. This shows that the mathematical language has more to commend it than being the only language which we can speak; it shows that it is, in a very real sense, the correct language."[31] As Hamming adds, "Science is composed of laws which were originally based on a small, carefully selected set of observations, often not very accurately measured originally; but the laws have later been found to apply over much wider ranges of observations and much more accurately than the original data justified. . . . [M]athematics provides, somehow, a reliable model for much of what happens in the universe."[32] Newton's Law remains what Wigner calls "a monumental example of a law, formulated in terms which appear simple to the mathematician, which has proved accurate beyond all reasonable expectations," along with elementary quantum mechanics and quantum electrodynamics.[33] The key is the following observation by Wigner: "Every empirical law has the disquieting quality that one does not know its limitations. We have seen that there are regularities in the events in the world around us which can be formulated in terms of mathematical concepts with an uncanny accuracy. There are, on the other hand, aspects of the world concerning which we do not believe in the existence of any accurate regularities."[34] The scattering of chaotic phenomena throughout an otherwise largely orderly universe prompts us to acknowledge the limitations of any model. As Ward notes: "All these tendencies in modern physics suggest a view of the universe very far removed from the clockwork model associated with Newtonian mechanics. The universe remains a supremely elegant rational set of systems."[35] The second statement here is exactly the key to refuting the first. The universe is a set of systems, a patchwork of subsystems, each containing its own internal, logical consistency. Deterministic chaos provides a mechanism for accurate descriptions of local subsystems within the larger system. Chaotic systems remain deterministic; the mechanism of the clock is both more complex and more subtle than the earlier purveyors of the timepiece metaphor dreamed possible. Wigner's warning of knowing the limitations of a model suggests than the interesting question is not whether chaotic models are appropriate at all, but rather in which contexts various types of models apply.

When faced with the implications of chaos theory, the second reason to retain confidence in the role of mathematics as a reflection of the real world is that there are myriad contexts in which chaos theory is proving useful as a model for formerly less tractable problems. Along with certain physical systems, the literature is exploding with applications to, for example, management, epilepsy, social sciences, consciousness, biological oscillators, hydrology, compressed video, schizophrenia, heart-rate variability, geomagnetic activity, predator-prey models, El Niño, ventricular fibrillation, atmospheric science, creativity, geophysics, biodiversity, medicine, fishing, tourism, the visual cortex, family systems, evolution, plant symbiosis, ecology models, curriculum and teaching, organizations, social evolution, water surges in the North Sea, hypnosis, antibiotic action, veterinary practice, and aging.[36–68] Mathematics has extended its reach to problems formerly intractable in the face of more traditional modeling attempts. This speaks to the depth, power, and flexibility of mathematics as the language of nature. Much as Hilbert space and even complex numbers are examples of seemingly unnatural tools that prove surprisingly useful and even necessary in understanding quantum mechanics and electromagnetism, respectively, chaos theory provides another example of mathematical results whose utility is plumbed after its abstract mathematical origins are established.

Rather than cause to bemoan the loss of mathematics as a reliable tool, chaos theory provides a reason to savor a broader application of mathematics, another tool in the mathematical toolbox. The natural world is no more subtle or complex than before; we simply continue to learn more and more about how subtle and complex it is. As our understanding and appreciation of that subtlety and complexity grow, increasingly subtle and complex mathematical structures find application and reaffirm that mathematics is indeed the language of nature. Just as a microscope, telescope, and any of various other optical tools each has particular utility in physically viewing appropriate portions of the universe, so mathematical lenses come in many shapes and sizes, each useful in its own way for clarity of vision of the natural world.

Coda

The door to my office opens as a student enters, papers in hand; computer code and graphs neatly cover some, while pencil scribblings in multiple directions cover others. I look up and look forward to discussing the latest revisions of a project modeling HIV infection at the immune system level. Systems of differential equations represent the changes in the numbers of T-cells in the bloodstream in the face of HIV infection: those uninfected, latently infected, and actively infected, as well as the number of virus particles present. The computer-generated graphs from earlier versions of the model simply have not matched the graphs of clinical data. Before explaining the latest revisions, the student shoves a graph forward, and my eyes fix on the shape of the curve. "Wow," I actually mutter aloud, more to myself than to the student. "It worked." The first reaction, the knee-jerk reaction before conscious thought can take over, is delight at a long-elusive match between the mathematics that seemed to make such sense and the real-world data that seemed so resistant to modeling. Before we discuss the details of the infusion of a probabilistic term, representing the mutation rates of the virus, into the otherwise deterministic equations, my first reaction is delight in the combination of tools from the mathematical toolbox reaffirming once again the effectiveness—neither unreasonable nor easily unearthed—of mathematics as the language to describe and the lens through which to view the natural world.

Acknowledgments

The author thanks H. Thijssen for thoughtful suggestions for completing this essay; S. Kellert and G. Ellis for their early contributions to the project; B. King, F. Ledley, and A. Lustig, for interesting individual conversations; M. Henderson, G. Peterson, N. Rupke, and B. Ulanowicz for significant contributions to the group discussions; J. Brooke, W. Drees, A. Jackelén, and D. Livingstone for their consummate professionalism and almost startling intellects; and J. Proctor for the vision and invitation at the outset and helpful insights and patience throughout the project. Some of you belong in all these categories; thank you all for the opportunity to work on this project together.

Notes

1. R. W. Hamming, "The Unreasonable Effectiveness of Mathematics," *American Mathematical Monthly* 87, no. 2 (1980): 81–90.
2. Jonathan Barnes, *Early Greek Philosophy, Second Revised Edition* (London: Penguin Books, 2001), 9–18.
3. Hamming, "The Unreasonable Effectiveness of Mathematics."
4. W. K. C. Guthrie, *Plato: The Man and His Dialogues—Earlier Period. A History of Greek Philosophy*, vol. IV (Cambridge: Cambridge University Press, 1986).
5. Augustine, *On Christian Doctrine*, http://ccat.sas.upenn.edu/jod/augustine/ddc.html (accessed April 14, 2006).
6. David J. Stucki, "Mathematics as Worship," in Robert L. Brabenec, ed., *Proceedings of the Thirteenth Conference of the Association of Christians in the Mathematical Sciences* (Wheaton, Ill.: Association of Christians in the Mathematical Sciences, 2001), 49–56.
7. Bernard I. Cohen, *The Birth of a New Physics*, revised and updated ed. (New York: W. W. Norton, 1991), 127–47.
8. Edward O. Wilson, "Back from Chaos," *The Atlantic Online* (March 1998), http://www.theatlantic.com/issues/98mar/eowilson.htm (accessed April 14, 2006).
9. Keith Ward, *Divine Action: Examining God's Role in an Open and Emergent Universe* (Philadelphia: Templeton Foundation Press, 2007), 80.
10. Ibid., 82.
11. John Cottingham, ed., *The Cambridge Companion to Descartes*, Cambridge Companions to Philosophy series (Cambridge: Cambridge University Press, 1992).
12. Society for Chaos Theory in Psychology & the Life Sciences, http://www.societyforchaostheory.org/ (accessed April 14, 2006).
13. Michael Shermer, "The Chaos of History: On a Chaotic Model That Represents the Role of Contingency and Necessity in Historical Sequences," *Nonlinear Science Today* 2, no. 4 (1993): 1–13.
14. Nicholas B. Tufillaro et al., *An Experimental Approach to Nonlinear Dynamics and Chaos* (Redwood City, Calif.: Addison-Wesley, 1992).
15. Edward N. Lorenz, "Deterministic Nonperiodic Flow," *Journal of the Atmospheric Sciences* 20 (1963): 130–41.
16. Edward N. Lorenz, *The Essence of Chaos* (Seattle: University of Washington Press, 1993).
17. T. Y. Li and J. A. Yorke, "Period Three Implies Chaos," *American Mathematical Monthly* 82 (1975): 985–92.

18. Robert L. Devaney, *An Introduction to Chaotic Dynamical Systems,* 2nd ed. (Redwood City, Calif.: Addison-Wesley, 1989).
19. C. Snow, *The Two Cultures* (Cambridge: Cambridge University Press, 1993).
20. Stephen Jay Gould, *Rocks of Ages: Science and Religion in the Fullness of Life* (New York: Random House, 1998).
21. Richard McGehee, personal communication.
22. Douglas E. Norton, "The Fundamental Theorem of Dynamical Systems," *Commentationes Mathematicae Universitatis Carolinae* 36 (1995): 585–97.
23. Ralph Abraham, *Chaos, Gaia, Eros* (San Francisco: HarperSanFrancisco, 1994).
24. Charles E. A. Finney, ed., "Chaos and Related Quotations," http://www-chaos.engr.utk.edu/chaosquotes.html (accessed July 7, 2006).
25. Walter J. Freeman, "Searching for Signal and Noise in the Chaos of Brain Waves," in S. Krasner, ed., *The Ubiquity of Chaos* (Washington, D.C.: AAAS, 1990), 47–55.
26. Ary L. Goldberger, "Is the Normal Heartbeat Chaotic or Homeostatic?" *News in Physiological Sciences* 6 (April 1991): 87–91.
27. Philip Hefner, "Created Co-Creator as Science and Symbol," http://www.metanexus.net/metanexus_online/show_article2.asp?id=8830 (accessed July 7, 2006).
28. Ward, *Divine Action,* 83.
29. John Polkinghorne, *Science and Providence* (London: SPCK, 1989), 33, 30.
30. Albert Einstein, *Sidelights on Relativity* (London: Methuen and Company, 1922), 28.
31. Eugene Wigner, "The Unreasonable Effectiveness of Mathematics in the Natural Sciences," *Communications in Pure and Applied Mathematics* 13, no. 1 (1960): 1–14.
32. Hamming, "The Unreasonable Effectiveness of Mathematics."
33. Wigner, "The Unreasonable Effectiveness of Mathematics in the Natural Sciences."
34. Ibid.
35. Ward, *Divine Action,* 100.
36. David Levy, "Chaos Theory and Strategy: Theory, Application, and Managerial Implications," *Strategic Management Journal* 15 (1994): 167–78.
37. Priscilla Murphy, "Chaos Theory as a Model for Managing Issues and Crises," *Public Relations Review* 22, no. 2 (1996): 95–113.
38. Leonidas D. Iasemidis and J. Chris Sackellares, "Chaos Theory and Epilepsy," *The Neuroscientist* 2 (1996): 118–26.
39. L. Douglas Kiel and Euel W. Elliott, *Chaos Theory in the Social Sciences: Foundations and Applications* (Ann Arbor: University of Michigan Press, 1996).
40. Larry R. Vandervert, "Chaos Theory and the Evolution of Consciousness and Mind: A Thermodynamic-Holographic Resolution to the Mind-Body Problem," *New Ideas in Psychology* 13, no. 2 (1995): 107–27.
41. Michael R. Guevara and Leon Glass, "Phase Locking, Period Doubling Bifurcations and Chaos in a Mathematical Model of a Periodically Driven Oscillator: A Theory for the Entrainment of Biological Oscillators and the Generation of Cardiac Dysrhythmias," *Journal of Mathematical Biology* 14, no. 1 (1982): 1–23.
42. B. Sivakumar, "Chaos Theory in Hydrology: Important Issues and Interpretations," *Journal of Hydrology* 227, nos. 1–4 (2000): 1–20.
43. A. Alkhatib and M. Krunz, "Application of Chaos Theory to the Modeling of Compressed Video," *Proceedings of the IEEE ICC 2000 Conference,* vol. 2 (2000): 836–40.
44. Heng-Chao Li, Wen Hong, Yi-Rong Wu, and Si-Jie Xu, "Research of Chaos Theory and

Local Support Vector Machine in Effective Prediction of VBR MPEG Video Traffic," Intelligent Computing, Lecture Notes in Computer Science (Berlin: Springer, 2006): 1229–34.

45. G. B. Schmid, "Chaos Theory and Schizophrenia: Elementary Aspects," *Psychopathology* 24, no. 4 (1991): 185–98.
46. Federico Lombardi, "Chaos Theory, Heart Rate Variability, and Arrhythmic Mortality," *Circulation* 101 (2000): 8–10.
47. D. N. Baker, A. J. Klimas, R. L. McPherron, and J. Büchner, "The Evolution from Weak to Strong Geomagnetic Activity: An Interpretation in Terms of Deterministic Chaos," *Geophysical Research Letters* 17, no. 1 (1990): 41–44.
48. A. Klebanoff and A. Hastings, "Chaos in One-Predator, Two-Prey Models: General Results from Bifurcation Theory," *Mathematical Biosciences* 122, no. 2 (1994): 221–33.
49. Eli Tziperman, Lewi Stone, Mark A. Cane, and Hans Jarosh, "El Niño Chaos: Overlapping of Resonances Between the Seasonal Cycle and the Pacific Ocean Atmosphere Oscillator," *Science* 264, no. 5155 (1994): 72–74.
50. James N. Weiss, Alan Garfinkel, Hrayr S. Karagueuzian, Zhilin Qu, and Peng-Sheng Chen, "Chaos and the Transition to Ventricular Fibrillation: A New Approach to Antiarrhythmic Drug Evaluation," *Circulation* 99 (1999): 2819–26.
51. Xubin Zeng, Roger A. Pielke, and R. Eykholt, "Chaos Theory and Its Applications to the Atmosphere," *Bulletin of the American Meteorological Society* 74, no. 4 (1993): 631–44.
52. D. Schuldberg, "Chaos Theory and Creativity," in M. Runco and S. Pritzker, eds., *Encyclopedia of Creativity*, vol. 1 (New York: Wiley, 1999), 259–72.
53. B. Sivakumar, "Chaos Theory in Geophysics: Past, Present and Future," *Chaos, Solitons & Fractals* 19, no. 2 (2004): 441–62.
54. J. Huisman and F. J. Weissing, "Biodiversity of Plankton by Species Oscillations and Chaos," *Nature* 402, no. 6760 (1999): 407–10.
55. James E. Skinner, Mark Molnar, Tomas Vybiral, and Mirna Mitra, "Application of Chaos Theory to Biology and Medicine," *Integrative Psychological and Behavioral Science* 27, no. 1 (1992): 39–53.
56. Schwinghamer, J. Y. Guigné, and W. C. Siu, "Quantifying the Impact of Trawling on Benthic Habitat Structure Using High Resolution Acoustics and Chaos Theory," *Canadian Journal of Fisheries and Aquatic Science* 53 (1996): 288–96.
57. Bob McKercher, "A Chaos Approach to Tourism," *Tourism Management* 20, no. 4 (1999): 425–34.
58. D. Hansel and H. Sompolinsky, "Chaos and Synchrony in a Model of a Hypercolumn in Visual Cortex," *Journal of Computational Neuroscience* 3, no. 1 (1996): 7–34.
59. Margaret Ward, "Butterflies and Bifurcations: Can Chaos Theory Contribute to Our Understanding of Family Systems?" *Journal of Marriage and the Family* 57, no. 3 (1995): 629–38.
60. Janet I. Sprent, "Evolution and Diversity in the Legume-Rhizobium Symbiosis: Chaos Theory?" *Plant and Soil* 161, no. 1 (1994): 1–10.
61. Peter Nijkamp and Aura Reggiani, "Non-linear Evolution of Dynamic Spatial Systems: The Relevance of Chaos and Ecologically-Based Models," *Regional Science and Urban Economics* 25, no. 2 (1995): 183–210.
62. Ron Iannone, "Chaos Theory and Its Implications for Curriculum and Teaching," *Education* 115, no. 4 (1995): 541–47.

63. Stephen J. Guastello, *Chaos, Catastrophe, and Human Affairs: Applications of Nonlinear Dynamics to Work, Organizations, and Social Evolution* (Mahwah, N.J.: Lawrence Erlbaum Associates, 1995).
64. D. Solomatine, C. J. Rojas, S. Velickov, and J. C. Wüst, "Chaos Theory in Predicting Surge Water Levels in the North Sea," *Proceedings of the Fourth International Conference on Hydroinformatics*, Cedar Rapids, Iowa, 2000.
65. P. Bob, "Hypnotic Abreaction Releases Chaotic Patterns of Electrodermal Activity during Dissociation," *International Journal of Clinical and Experimental Hypnosis* 55, no. 4 (2007): 435–56.
66. Paul H. Axelsen, "Chaotic Pore Model of Polypeptide Antibiotic Action," *Biophysical Journal* 94 (2008): 1549–50.
67. Eleanor Craven Brennan, "Chaos in the Clinic: Applications of Chaos Theory to a Qualitative Study of a Veterinary Practice," *The Qualitative Report* 3, no. 2 (1997), http://www.nova.edu/ssss/QR/QR3-2/bren.html (accessed July 5, 2008).
68. M. Kyriazis, "Practical Applications of Chaos Theory to the Modulation of Human Ageing: Nature Prefers Chaos to Regularity," *Biogerontology* 4, no. 2 (2003): 75–90.

Bibliography

Abraham, Ralph. *Chaos, Gaia, Eros*. San Francisco: HarperSanFrancisco, 1994.

Alkhatib, A. and M. Krunz. "Application of Chaos Theory to the Modeling of Compressed Video." *Proceedings of the IEEE ICC 2000 Conference*, vol. 2 (2000): 836–40.

American Mathematical Society. "The Mathematical Heritage of Henri Poincaré." *Proceedings of Symposia in Pure Mathematics*, vol. 39, Parts 1 and 2. Providence: American Mathematical Society, 1983.

Augustine. "On Christian Doctrine." http://ccat.sas.upenn.edu/jod/augustine/ddc.html (accessed April 14, 2006).

Axelsen, Paul H. "Chaotic Pore Model of Polypeptide Antibiotic Action." *Biophysical Journal* 94 (2008): 1549–50.

Baker, D. N., A. J. Klimas, R. L. McPherron, and J. Büchner. "The Evolution from Weak to Strong Geomagnetic Activity: An Interpretation in Terms of Deterministic Chaos." *Geophysical Research Letters* 17, no. 1 (1990): 41–44.

Barnes, Jonathan. *Early Greek Philosophy*, 2nd rev. ed. London: Penguin Books, 2001.

Bass, Gary. "Nonlinear Man: Chaos, Fractal & Homeostatic Interplay in Human Physiology." http://www.dchaos.com/portfolio/dchaos1/new_nonlinear_man_article.html (accessed July 7, 2006).

Bob, P. "Hypnotic Abreaction Releases Chaotic Patterns of Electrodermal Activity during Dissociation." *International Journal of Clinical and Experimental Hypnosis* 55, no. 4 (2007): 435–56.

Brennan, Eleanor Craven. "Chaos in the Clinic: Applications of Chaos Theory to a Qualitative Study of a Veterinary Practice." *The Qualitative Report* 3, no. 2 (1997). http://www.nova.edu/ssss/QR/QR3-2/bren.html (accessed July 5, 2008).

Bugliarello, George. "A New Trivium and Quadrivium." Presented at the 2001 Sigma Xi Forum, Raleigh, N.C. http://www.sigmaxi.org/meetings/archive/forum.2001.shtml (accessed April 14, 2006).

Butz, Michael R. "Chaos Theory: Philosophically Old, Scientifically New." *Counseling and Values* 39, no. 2 (1995): 84–98.
Cambel, A. B. *Applied Chaos Theory: A Paradigm for Complexity*. San Diego: Academic Press, 1993.
Cohen, I. Bernard. *The Birth of a New Physics*, rev. and updated ed. New York: W. W. Norton, 1991.
Cottingham, John, ed. *The Cambridge Companion to Descartes*. Cambridge Companions to Philosophy Series. Cambridge: Cambridge University Press, 1992.
Davis, Philip J., and Reuben Hersh. *The Mathematical Experience*. New York: Springer-Verlag, 1995.
Devaney, Robert L. *An Introduction to Chaotic Dynamical Systems*, 2nd ed. Redwood City, Calif.: Addison-Wesley, 1989.
Dyson, Freeman J. "Is God in the Lab?" Review of *The Meaning of It All: Thoughts of a Citizen Scientist*, by Richard Feynman, and *Belief in God in an Age of Science*, by John Polkinghorne. *New York Review of Books* 45, no. 9 (May 28, 1998). http://www.nybooks.com/articles/article-preview?article_id=835 (accessed July 5, 2008).
Eigenauer, John D. "The Humanities and Chaos Theory: A Response to Steenburg's 'Chaos at the Marriage of Heaven and Hell.'" *Harvard Theological Review* 86 (1993): 455–69.
Einstein, Albert. *Sidelights on Relativity*. London: Methuen and Company, 1922.
Field, Michael, and Martin Golubitsky. *Symmetry in Chaos*. Oxford: Oxford University Press, 1992.
Finney, Charles E. A., ed. "Chaos and Related Quotations." http://www.chaos.engr.utk.edu/chaosquotes.html (accessed July 7, 2006).
Freeman, Walter J. "Searching for Signal and Noise in the Chaos of Brain Waves." In S. Krasner, ed., *The Ubiquity of Chaos*, 47–55. Washington, D.C.: AAAS, 1990.
———. "Strange Attractors That Govern Mammalian Brain Dynamics Shown by Trajectories of Electroencephalographic (EEG) Potential." *IEEE Transactions on Circuits & Systems* 35 (1988): 781–83.
Ginzburga, Lev R., Christopher X. J. Jensen, and Jeffrey V. Yule. "Aiming the 'Unreasonable Effectiveness of Mathematics' at Ecological Theory." *Ecological Modelling* 207, nos. 2–4 (2007): 356–62.
Gleick, James. *Chaos: Making a New Science*. New York: Penguin, 1987.
Goldberger, Ary L. "Chaos and Fractals in Human Physiology." *Scientific American* 262, no. 2 (February 1990): 43–49.
———. "Is the Normal Heartbeat Chaotic or Homeostatic?" *News in Physiological Sciences* 6 (April 1991): 87–91.
———. "Nonlinear Dynamics for Clinicians: Chaos Theory, Fractals and Complexity at the Bedside." *Lancet* 347, no. 9011 (May 11, 1996): 1312–14.
Gould, Stephen Jay. *Rocks of Ages: Science and Religion in the Fullness of Life*. New York: Random House, 1998.
Gray, Doug. "Toward a Theology of Chaos: The New Scientific Paradigm and Some Implications for Ministry." http://www.wiscongregational.net/1997_09_18.pdf (accessed April 14, 2006).
Guastello, Stephen J. *Chaos, Catastrophe, and Human Affairs: Applications of Nonlinear Dynamics to Work, Organizations, and Social Evolution*. Mahwah, N.J.: Lawrence Erlbaum Associates, 1995.

Guevara, Michael R., and Leon Glass. "Phase Locking, Period Doubling Bifurcations and Chaos in a Mathematical Model of a Periodically Driven Oscillator: A Theory for the Entrainment of Biological Oscillators and the Generation of Cardiac Dysrhythmias." *Journal of Mathematical Biology* 14, no. 1 (1982): 1–23.

Guthrie, W. K. C. *Plato: The Man and His Dialogues—Earlier Period. A History of Greek Philosophy*, vol. IV. Cambridge: Cambridge University Press, 1986.

Hamming, R. W. "The Unreasonable Effectiveness of Mathematics." *American Mathematical Monthly* 87, no. 2 (1980): 81–90.

Hansel, D., and H. Sompolinsky. "Chaos and Synchrony in a Model of a Hypercolumn in Visual Cortex." *Journal of Computational Neuroscience* 3, no. 1 (1996): 7–34.

Hefner, Philip. *The Human Factor: Evolution, Culture, and Religion*. Minneapolis: Augsburg Fortress Publishers, 1993.

———. "Created Co-Creator as Science and Symbol." http://www.metanexus.net/metanexus_online/show_article2.asp?id=8830 (accessed July 7, 2006.)

Howell, Russell W., and James Bradley, eds. *Mathematics in a Postmodern Age: A Christian Perspective*. Grand Rapids, Mich.: Wm. B. Eerdmans Publishing, 2001.

Huisman J., and F. J. Weissing. "Biodiversity of Plankton by Species Oscillations and Chaos." *Nature* 402, no. 6760 (1999): 407–10.

Iannone, Ron. "Chaos Theory and Its Implications for Curriculum and Teaching." *Education* 115, no. 4 (1995): 541–47.

Iasemidis, Leonidas D., and J. Chris Sackellares. "Chaos Theory and Epilepsy." *The Neuroscientist* 2 (1996): 118–26.

Kernick, D. "Migraine: New Perspectives from Chaos Theory." *Cephalalgia* 25 (2005): 561–66.

Kiel, L. Douglas, and Euel W. Elliott. *Chaos Theory in the Social Sciences: Foundations and Applications*. Ann Arbor: University of Michigan Press, 1996.

Klebanoff, A., and A. Hastings. "Chaos in One-Predator, Two-Prey Models: General Results from Bifurcation Theory." *Mathematical Biosciences* 122, no. 2 (1994): 221–33.

Kraut, Richard. "Plato." In Edward N. Zalta, ed., *The Stanford Encyclopedia of Philosophy* (Summer 2004 ed.). http://plato.stanford.edu/archives/sum2004/entries/plato/ (accessed July 7, 2006).

Kyriazis, M. "Practical Applications of Chaos Theory to the Modulation of Human Ageing: Nature Prefers Chaos to Regularity." *Biogerontology* 4, no. 2 (2003): 75–90.

Laplace, S. *A Philosophical Essay on Probabilities*. Translated by F. W. Truscott and F. L. Emory. New York: Dover, 1951.

Leibniz, G. W. "On the Ultimate Origin of Things." In *Philosophical Writings*. Translated by M. Morris. London: Dent, 1973.

Levy, David. "Chaos Theory and Strategy: Theory, Application, and Managerial Implications." *Strategic Management Journal* 15 (1994): 167–78.

Li, Heng-Chao, Wen Hong, Yi-Rong Wu, and Si-Jie Xu. "Research of Chaos Theory and Local Support Vector Machine in Effective Prediction of VBR MPEG Video Traffic." *Intelligent Computing: Lecture Notes in Computer Science*. Berlin: Springer, 2006: 1229–34.

Li, T. Y., and J. A. Yorke. "Period Three Implies Chaos." *American Mathematical Monthly* 82 (1975): 985–92.

Lombardi, Federico. "Chaos Theory, Heart Rate Variability, and Arrhythmic Mortality." *Circulation* 101 (2000): 8–10.

Lorenz, Edward N. "Deterministic Nonperiodic Flow." *Journal of the Atmospheric Sciences* 20 (1963): 130–41.

———. *The Essence of Chaos*. Seattle: University of Washington Press, 1993.

McInerny, Ralph, and John O'Callaghan. "Saint Thomas Aquinas." In Edward N. Zalta, ed., *Stanford Encyclopedia of Philosophy* (Spring 2005 ed.). http://plato.stanford.edu/archives/spr2005/entries/aquinas/ (accessed July 7, 2006).

McKercher, Bob. "A Chaos Approach to Tourism." *Tourism Management* 20, no. 4 (1999): 425–34.

McLean, Jeffrey T. "Mathematics and Theology: A Conversation." In *Curriculum in a Catholic University: Summer Seminar, 1995. Catholic Colleges and Universities in the 21st Century*. St. Paul, Minn.: University of St. Thomas, 1996.

Mendell, Henry. "Aristotle and Mathematics." In Edward N. Zalta, ed., *Stanford Encyclopedia of Philosophy* (Summer 2004 ed.). http://plato.stanford.edu/archives/sum2004/entries/aristotle-mathematics/ (accessed July 7, 2006).

Mendelson, Michael. "Saint Augustine." In Edward N. Zalta, ed., *Stanford Encyclopedia of Philosophy* (Winter 2000 ed.). http://plato.stanford.edu/archives/win2000/entries/augustine/ (accessed July 7, 2006).

Murphy, Priscilla. "Chaos Theory as a Model for Managing Issues and Crises." *Public Relations Review* 22, no. 2 (1996): 95–113.

Newton, I. *Principia Mathematica*. Translated by A. Motte. Berkeley: University of California Press, 1962.

Nijkamp, Peter, and Aura Reggiani. "Non-linear Evolution of Dynamic Spatial Systems: The Relevance of Chaos and Ecologically-Based Models." *Regional Science and Urban Economics* 25, no. 2 (1995): 183–210.

Norton, Douglas. "The Fundamental Theorem of Dynamical Systems." *Commentationes Mathematicae Universitatis Carolinae* 36 (1995): 585–97.

———. "Mathematics Among the Arts, Sciences, and Humanities." Presented at the 2001 Sigma Xi Forum, Raleigh, N.C. Online abstract at http://www.sigmaxi.org/meetings/archive/forum.2001.prog.pap.shtml (accessed April 14, 2006).

———. "The Universe, Determinism, and Chaos Theory." The Robert M. Birmingham Colloquia Series, Villanova University, Villanova, Penn., October 24, 2002.

———. "Mathematical Connections in Philosophy and Theology: A Brief Tour." MAA session CP X1 on Mathematical Connections in Art, Music, and Science, Joint Mathematics Meetings, Baltimore, January 18, 2003.

O'Connor, J. J., and E. F. Robertson. "Christianity and the Mathematical Sciences: The Heliocentric Hypothesis." In *Christianity and Mathematics*, http://www-gap.dcs.st-and.ac.uk/~history/HistTopics/Heliocentric.html (accessed April 14, 2006).

———. "Thales of Miletus." http://www-gap.dcs.st-and.ac.uk/~history/Mathematicians/Thales.html (accessed April 14, 2006).

Parry, Richard. "Epistêmê and Technê." In Edward N. Zalta, ed., *Stanford Encyclopedia of Philosophy* (Summer 2003 ed.). http://plato.stanford.edu/archives/sum2003/entries/episteme-techne/ (accessed July 7, 2006).

Polkinghorne, John. *Science and Providence*. London: SPCK, 1989.

Raskin, Jef. "A Reply to Eugene Wigner's paper, 'The Unreasonable Effectiveness of Mathematics in the Natural Sciences' and Hamming's essay 'The Unreasonable

Effectiveness of Mathematics.'" http://jef.raskincenter.org/unpublished/effectiveness_mathematics.html (accessed July 5, 2008).
Russell, Robert J., Nancey Murphy, and Arthur R. Peacocke, eds. *Chaos and Complexity: Scientific Perspectives on Divine Action.* Vatican City: Vatican Observatory Publications, and Berkeley: Center for Theology and the Natural Sciences, 1995.
Sarukkai, Sundar. "Revisiting the 'Unreasonable Effectiveness' of Mathematics." *Current Science* 88, no. 3 (2005): 415–23.
Schmid, G. B. "Chaos Theory and Schizophrenia: Elementary Aspects." *Psychopathology* 24 no. 4 (1991): 185–98.
Schuldberg, D. "Chaos Theory and Creativity." In M. Runco and S. Pritzker, eds., *Encyclopedia of Creativity,* vol. 1, 259–72. New York: John Wiley & Sons, 1999.
Schwinghamer, J. Y. Guigné, and W. C. Siu. "Quantifying the Impact of Trawling on Benthic Habitat Structure Using High Resolution Acoustics and Chaos Theory." *Canadian Journal of Fisheries and Aquatic Science* 53 (1996): 288–96.
Shapin, Steven. *The Scientific Revolution.* Chicago: University of Chicago Press, 1998.
Shermer, Michael. "The Chaos of History: On a Chaotic Model That Represents the Role of Contingency and Necessity in Historical Sequences." *Nonlinear Science Today* 2, no. 4 (1993): 1–13.
Sivakumar, B. "Chaos Theory in Hydrology: Important Issues and Interpretations." *Journal of Hydrology* 227, nos. 1–4 (2000): 1–20.
———. "Chaos Theory in Geophysics: Past, Present and Future." *Chaos, Solitons & Fractals* 19, no. 2 (2004): 441–62.
Skarda, Christine A., and Walter J. Freeman. "How Brains Make Chaos in Order to Make Sense of the World." *Behavioral and Brain Sciences* 10 (1987): 161–95.
———. "Chaos and the New Science of the Brain." *Concepts in Neuroscience* 1, no. 2 (1990): 275–85.
Skinner, James E., Mark Molnar, Tomas Vybiral, and Mirna Mitra. "Application of Chaos Theory to Biology and Medicine." *Integrative Psychological and Behavioral Science* 27, no. 1 (1992): 39–53.
Snow, C. *The Two Cultures.* Cambridge: Cambridge University Press, 1993.
Society for Chaos Theory in Psychology & the Life Sciences. http://www.societyforchaostheory.org/ (accessed April 14, 2006).
Solomatine, D., C. J. Rojas, S. Velickov, and J. C. Wüst. "Chaos Theory in Predicting Surge Water Levels in the North Sea." *Proceedings of the Fourth International Conference on Hydroinformatics,* Cedar Rapids, Iowa, 2000.
Sprent, Janet I. "Evolution and Diversity in the Legume-Rhizobium Symbiosis: Chaos Theory?" *Plant and Soil* 161, no. 1 (1994): 1–10.
Steenburg, David. "Chaos at the Marriage of Heaven and Hell." *Harvard Theological Review* 84 (1991): 447–66.
Stewart, Ian. *Does God Play Dice? The Mathematics of Chaos.* Oxford: Blackwell, 1989.
Stucki, David J. "Mathematics as Worship." In Robert L. Brabenec, ed., *Proceedings of the Thirteenth Conference of the Association of Christians in the Mathematical Sciences.* Wheaton, Ill.: Association of Christians in the Mathematical Sciences, 2001.
Tufillaro, Nicholas B., Tyler Abbott, and Jeremiah Reilly. *An Experimental Approach to Nonlinear Dynamics and Chaos.* Redwood City, Calif.: Addison-Wesley, 1992.
Tziperman, Eli, Lewi Stone, Mark A. Cane, and Hans Jarosh. "El Niño Chaos: Overlapping

of Resonances between the Seasonal Cycle and the Pacific Ocean Atmosphere Oscillator." *Science* 264, no. 5155 (1994): 72–74.
Vandervert, Larry R. "Chaos Theory and the Evolution of Consciousness and Mind: A Thermodynamic-Holographic Resolution to the Mind-Body Problem." *New Ideas in Psychology* 13, no. 2 (1995): 107–27.
Ward, Keith. *Divine Action: Examining God's Role in an Open and Emergent Universe.* Philadelphia: Templeton Foundation Press, 2007.
Ward, Margaret. "Butterflies and Bifurcations: Can Chaos Theory Contribute to Our Understanding of Family Systems?" *Journal of Marriage and the Family* 57, no. 3 (1995): 629–38.
Warm, Hartmut. *Die Signatur der Sphären: Von der Ordnung im Sonnensystem.* Hamburg: Keplerstern Verlag, 2001. See also http://www.keplerstern.com/index.html (accessed April 14, 2006).
Weiss, James N., Alan Garfinkel, Hrayr S. Karagueuzian, Zhilin Qu, and Peng-Sheng Chen. "Chaos and the Transition to Ventricular Fibrillation: A New Approach to Antiarrhythmic Drug Evaluation." *Circulation* 99 (1999): 2819–26.
Wigner, Eugene. "The Unreasonable Effectiveness of Mathematics in the Natural Sciences." *Communications in Pure and Applied Mathematics* 13, no. 1 (1960): 1–14.
Williams, Thomas. "Saint Anselm." In Edward N. Zalta, ed., *Stanford Encyclopedia of Philosophy* (Fall 2005 ed.). http://plato.stanford.edu/archives/fall2005/entries/anselm/ (accessed July 7, 2006).
Wilson, Edward O. "Back from Chaos." *The Atlantic* Online (March 1998). http://www.theatlantic.com/issues/98mar/eowilson.htm (accessed April 14, 2006).
Young, T. R. "Chaos and Social Change: Metaphysics of the Postmodern." *Social Science Journal* 28, no. 3 (1991): 289–306.
Zeng, Xubin, Roger A. Pielke, and R. Eykholt. "Chaos Theory and Its Applications to the Atmosphere." *Bulletin of the American Meteorological Society* 74, no. 4 (1993): 631–44.

3

BETWEEN APES AND ANGELS: AT THE BORDERS OF HUMAN NATURE

Johannes M.M.H. Thijssen

Introduction

As a historian of premodern philosophy and science, I am acutely aware of the tremendous range of meanings that have been attached to *nature* in Western thought, meanings that are often concealed by the ease with which we use the term in various contexts.[1] Disentangling the multiple visions of nature behind the term is the goal of the book precisely.[2] Historians usually take a relativistic approach toward such an endeavor, possibly even unduly relativistic, as Willem Drees indicates in this volume.[3] Yet, there is a generally felt intuition that the very concept of nature is not stable or fixed, but is a product of historical periods and cultural contexts. There is nothing natural about nature. Rather, it is a cultural construct. This approach erodes the divide between the natural and the cultural. Or, to paraphrase Jim Proctor: It is time to count beyond two.[4]

In this chapter, I would like to remind readers that nature is not something independent of us, and the most illuminating way of doing so seems to be by examining the very notion of human nature. It comes so *naturally* to think of human nature as an essence, a collection of exclusive properties that set us off from other species and that are somehow "ingrained" in us. When it is said, for example, that human beings have to follow their nature, or have to become what they are, the implication seems to be that we possess some preordained features that need to be (re)discovered. These features, are, moreover, supposed to be very different from those of other living beings. But what are these features that distinguish us from other species? In other words, what does it mean to be human?[5] In particular, the idea that human beings are "by nature" rational and moral beings is taken for granted in our society, so much so that the ancient roots of this self-definition have disappeared into oblivion.

Before Darwin, the identification of human nature was molded in the following form: What does distinguish humankind *from* the animals? What is the difference between man and beast? What we are looking for is some essential characteristic that differentiates all human beings *from* other animals. Nowadays, we would think that everything is wrong with this question, unless we take a human being to be a machine or an angel.[6] Nowadays we would ask: What distinguishes human beings *among* the animals? After all, in our genes *we* are 98 percent chimpanzee.[7] We are apes, be it naked apes, or moral apes, or apes who aspire to be angels. All humans belong in one single species, that of *Homo sapiens*. So what makes us unique? One of the most familiar answers to that question is that humans have the power of reason. But how does one spot reason?

The problem of detecting the possession of reason posed itself when the first encounters occurred, either in texts or in real life, with the inhabitants of the New World, or with creatures such as apes, "Pygmies," or dogheads. In which category did they belong? Were they humans, subhumans, or nonhumans? What are the outward signs of the ability to reason? Did these creatures resemble us at all? Speech, toolmaking, and social life have all been suggested as sure signs of rational behavior.

Interestingly, primatologists who challenge the distinction between primates and humans do not question this self-definition as humans, either.[8] The upshot of their arguments is that primates basically possess the same rational and ethical capacities as humans. The psychologist Sue Savage-Rumbaugh, for instance, once claimed that, in the case of ape language, "Almost any interpretation of the data leads inevitably to a redefinition of man and the sciences that study man."[9] Frans de Waal, on the other hand, believes that the study of bonobos can reveal something about the ethical qualities of humans, such as being "good natured." The implicit suggestion is that the study of chimpanzees will tell us something about human nature, and not merely about chimpanzees.[10]

The current debate about how much apes are like us reaffirms some of the old criteria for rationality, such as language, skills, and social life. The story of how these criteria came to be significant in the first place is a long one, and has many twists. The presuppositions that we entertain about human nature have their origin in antiquity, in particular in the works of Aristotle.[11] In this chapter, I shall retrace a few episodes of the migration of Aristotle's vision of human nature in space and time and of its application in new contexts. In this way, this chapter complements those by David Livingstone and Nicolaas Rupke in this volume, which also explore reconceptualizations of nature and humanity in different times and places.[12]

Edward Tyson (1650–1708)

Late in 1697, or early in 1698, a ship arrived in the harbor of London, carrying, among other things, a creature that had been captured in Angola. In the spring of 1698, Edward Tyson, England's most famous anatomist, dissected the corpse of this creature, which is still preserved in the British Museum. In 1699, Tyson published his findings in a treatise titled *Orang-Outang, sive Homo Sylvestris: Or the Anatomy of a Pygmie compared with that of a Monkey, an Ape and a Man*.[13]

During the voyage, the "Pygmie," as Tyson had labeled this creature, had stumbled into a cannon, knocking out a tooth and triggering an infection that became fatal. While aboard the ship, the Pygmie had revealed itself as "most gentle and loving." As Tyson later reported:

> those that he knew a ship-board he would come and embrace with the greatest tenderness, opening their bosoms, and clasping his hands about

Edward Tyson's "Pygmie" (juvenile chimpanzee) from *Orang-Outang, sive Homo Sylvestris: Or the Anatomy of a Pygmie compared with that of a Monkey, an Ape and a Man.*

> them; and as I was informed, tho' there were Monkeys aboard, yet 't was observed he would never associate with them, and as if nothing a-kin to them, would always avoid their company. (Tyson, 7)

Tyson noted that, from the head downward, the creature was quite hairy "and the hair so thick, that it covered the skin almost from being seen" (Tyson, 7).

> Nature therefore has cloathed it with hair, as a brute, to defend it from the injuries of the weather; and when it goes on all four, as a quadruped, it seems all hairy; When it goes erect, as a biped, it appears before less hairy, and more like a Man. (Tyson, 8)

After the Pygmie had been captured, it was forced to get used to wearing clothes, but

> it was fond enough of them; and what it could not put on himself, it would bring in his hands to some of the company to help him to put on. It would lie in a bed, place his head on a pillow, and pull the cloaths over him, as a Man would do; but was so careless and so very a Brute, as to do all Nature's occasions there. It was very full of lice when it came under my hands which it may be it got on ship-board, for they were exactly like those on Humane bodies. (Tyson, 8)

Tyson was convinced that the creature that had been presented to him was not a human but a beast. However, one of the purposes of the book was precisely to demonstrate that "this animal" resembled "a Man in many of its parts, more than any of the Apekind, or any other animal in the world" that he knew of. His book did much to establish the method of comparative anatomy and the study of primates, for the creature that Tyson had dissected actually had been a juvenile chimpanzee.

At the time Tyson wrote, very little was known in Europe of primates, and what was known was marred by terminological confusion. The title of Tyson's book strings together terms both human and nonhuman: *Orang-Outang, Homo Sylvestris, Pygmie, Monkey, Ape,* and *Man.*

The terms *ape* and *monkey* were considered synonymous. The term *ape* was not yet reserved for the so-called great apes—chimpanzees, gorillas, gibbons, and orangutans. Note, moreover, that the gorilla was only identified in the nineteenth century. Tyson employed the term *Man* to designate any creature that was not a beast.

Orang-outang is a Malaysian word literally meaning "Man of the forest," or *Homo sylvestris* in Latin. In Tyson's day, it served as a general term for all great apes, both African and Asian, and not just for the true orangutan of Southeast Asia, as today.[14] The term *orang-outang* was first employed by Jacob Bontius (1592–1631) in a work that appeared posthumously in 1658 (*Historiae naturalis et medicae Indiae orientalis*). He provided perhaps the first European description of a true orangutan. At the same time, however, he reported that the natives of Indonesia believed this creature to be the offspring of lustful women with monkeys, and added a confusing illustration. It seems to be a picture of a woman covered with hair.[15]

Another illustration of an "orang-outang" was produced by the Dutch physician Nicholas Tulp (1593–1674), well-known as the central figure in Rembrandt's painting *The Anatomy Lesson*.

Tulp provided the first pictorial representation of a great ape, but it was not that of a true orangutan, but of a chimpanzee (*Observationum medicarum libri tres*, 1641). The creature had been presented to the Prince of Orange in 1630. According to a story reported by Diderot, a cardinal once passed the cage and confided to the animal: "Speak and I will baptise you."[16]

As is clear from references in his own work, Tyson was familiar with the contemporary literature about "orang-outangs." Yet, he choose another designation for the creature that had been presented to him.[17]

Jacob Bontius' *Ourang-Outang* as depicted in *Historia naturalis et medicae Indiae orientalis* (Amsterdam: Elzevirum, 1658).

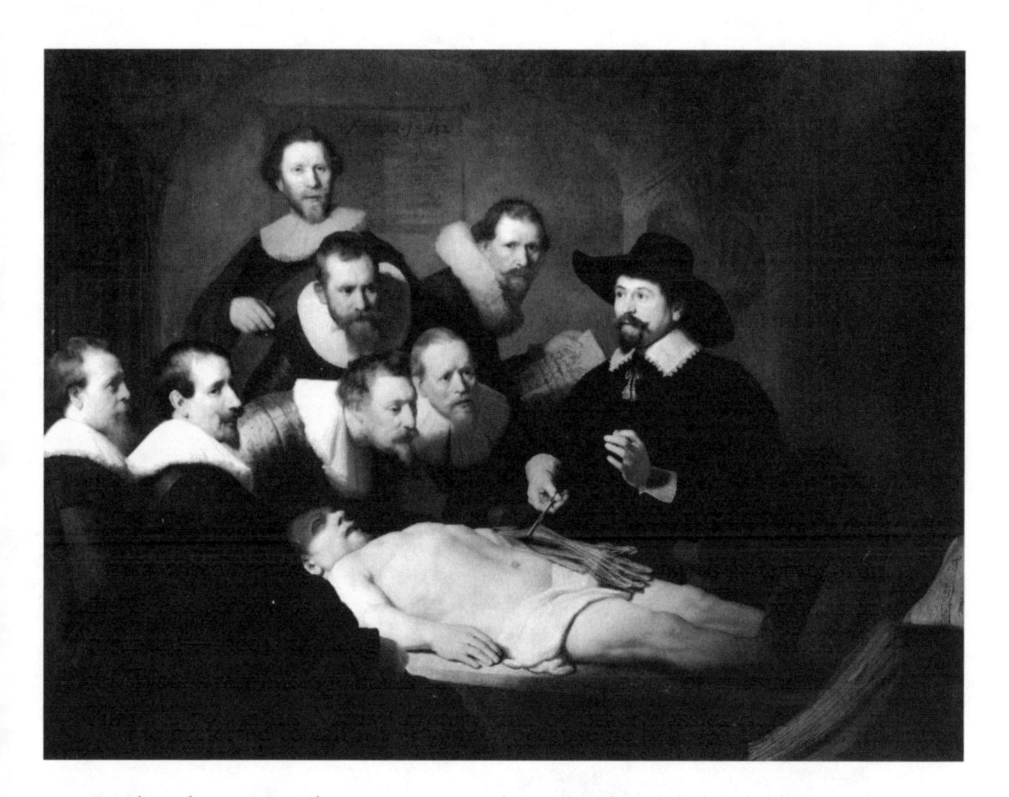

Rembrandt van Rijn, *The Anatomy Lesson by Dr. Nicolaes Tulp* (1632), kept at the Mauritshuis, The Hague. The body is that of Aris Kindt, who had been hanged for theft.

He preferred to call it a "Pygmie," because he believed that this creature was described in the ancient literature about pygmies. But what is a pygmie? Obviously, Tyson was not thinking of pygmie tribes of equatorial Africa, when writing his treatise. The Latin term *pygmeus* is derived from the Greek *pèchmaios,* which is a measure. It refers to a person who is one cubit tall—that is the length from the elbow to the tip of the middle finger. Tyson's terminology harks back to the description of unusual races by Pliny (d. 79), in his *Natural History,* and by other classical authors, such as Homer and Aristotle. He chose the term because he found the stature of the creature that he had been dissecting "Just the same with the stature of the Pygmies of the Ancients."

Nicholas Tulp's *Homo sylvestris* or man of the woods, from his *Observationum medicarum, libri tres* (Amsterdam, 1641).

Emmanuel Hoppius' *Anthromorpha* taken from *Amoenitates Academicae, VI* (Stockholm, 1763). The first manlike creature represents Bontius' "Ourang-Outan" The third figure is Tulp's *Homo sylvestris*.

The Conceptual Framework

The conceptual framework of Tyson's discussion was furnished by two authorities: Saint Augustine and Aristotle. In a particular passage in the *City of God,* Augustine raised the question as to whether the monstrous races described in pagan history had descended from Adam, the first human being.

> Now we are not bound to believe in the existence of all the types of men which are described. But no faithful Christian should doubt that anyone who is born anywhere as a man—that is, as a rational and moral being—derives from that one first-created being [i.e., Adam]. And this is true, however extraordinary such a creature may appear to our senses in bodily shape, in colour, or motion, or utterance, or in any natural endowment, or part, or quality. (Saint Augustine, *De civitate dei,* XVI, c.8)[18]

In other words, even races such as pygmies or dogheads (*Cynocephali*), or other unusual or monstrous races that had been described by the Roman author Pliny, could be human. By the same token, races that did resemble human beings in physical appearance need not be human at all. This is the conclusion that Tyson draws from *his* anatomical comparisons between human and pygmie. The body he was dissecting was more like a human in its anatomical structure than any other species of animal. According to Tyson's counting, there were forty-eight anatomical features in which it resembled human beings, against thirty-four that approximated those of the "ape and monkeykind."[19] Yet, Tyson did *not* conclude that his pygmie was therefore a human being and *not* a beast. On the contrary, Tyson's discovery of the pygmie's humanlike anatomy made him reject the ancient stories that Pygmies were *human beings.* The ancient authors had been misled. The physical shape and form just is not conclusive in determining the nature of a pygmie. According to Tyson, the similarity in organs, and even in the brain, does not correspond to a similarity in actions. For if only the anatomy were relevant, not only would pygmies be elevated to the level of humans, but humans would be turned into

> meer brutes and matter. Whereas in truth Man is part a Brute, part an Angel; and is that Link in the Creation that joyns them both together. (Tyson, 56)

Elsewhere, Tyson states that

> The Animal of which I have given the Anatomy, coming nearest to Mankind; seems the Nexus of the Animal and Rational. (Tyson, iii)

So what was the decisive factor, if not the anatomy? Why was the pygmie not human, but some kind of intermediate link in the chain of creation? The answer that Tyson gives to that question rests in the works of Aristotle.

Aristotle's Legacy

From Aristotle, Tyson and many others borrowed and expanded the concept of what came to be known as the chain of being, or the scale of nature (Leibniz: *scala naturae*).[20] According to this concept, the universe is filled with distinct species that are ordered according to a continuous gradation. Throughout his work, Tyson proves himself to be an ardent proponent of the chain of being. In the preface, he states:

> That from Minerals, to Plants; from Plants, to Animals; and from Animals, to Men; the Transition is so gradual, that there appears a very great Similitude, as well between the meanest Plants, and some Minerals; as between the lowest Rank of Men, and the highest kind of Animals. The Animal of which I have given the Anatomy, coming nearest to Mankind.

This passage clearly echoes Aristotle:

> Nature proceeds from the inanimate to the animals by such small steps that, because of the continuity, we fail to see to which side the boundary and the middle between them belongs. For first after the inanimate kind of things is the plant kind, and among these one differs from another in seeming to have more share of life; but the whole kind in comparison with the other bodies appears more or less animate, while in comparison with the animal kind it appears inanimate. The change from them to the animals is continuous, as we said before. For some of those in the sea might raise for one the question whether they are animal or plant. (Aristotle, *Historia Animalium*, 588b4–13)[21]

Which factor determines the order in the continuum of nature? According to Aristotle, the hierarchical order of nature is determined by the powers of the soul. This view deserves some further explanation. According to Aristotle, whatever is alive has a soul (*psychè; anima*). How do you notice that something is alive? Because it displays certain capacities. But not all living organisms display the same capacities.[22] Some, such as plants, have the capacity to use food, and to reproduce. Others, such as animals, also have the capacity for self-movement, sensation, and desire. But there is only one kind of living thing that also possesses the power of thinking and thought (*De anima*, 414b16–19). In sum, humans share certain features with other living organisms, such as growth or reproduction. But they are unique in that they possess reason. The capacity to think and to reason is linked to their *psychè*, to the specific principle of life that is inherent in humans. Reason as the distinctive feature of humans also plays a crucial role in Aristotle's ethical theory. According to Aristotle, humans have an understanding of what matters in life. They can plan their own lives. There are many wrong plans you can adopt to conduct your life, but there is only one right plan to achieve a good life. The right plan is the plan that aims at the right goal, a goal that is shared by all human beings. According to Aristotle, that goal is happiness or success or flourishing (all common translations of the Greek *eudaimonia*). Happiness is what everyone seeks. Yet, no one can really give a reason for wanting it. In this sense, happiness, success, or flourishing is an ultimate end: We seek it for its own sake. Interestingly, Aristotle believes that happiness is the same for us all. This view rests on the assumption that the ultimate purpose of humans is linked to their "function"; that is, their characteristic activity (*ergon*).[23] The function of a knife, for instance, is to cut. But what can the function of humans be? It is not life.

> Life seems to be common even to plants, but we are seeking what is peculiar to man. Let us exclude, therefore, the life of nutrition and growth. Next there would be a life of perception, but *it* also seems to be common even to the horse, the ox, and every animal. There remains, then, an active life of the element that has a rational principle. (Aristotle, *Ethica Nicomachea*, 1097b32–1098a4).[24]

In other words, only functions that are distinctively human determine human excellence, and these are functions associated with reason. For Aristotle, living well means living one's life under the guidance of reason.

So, both in his psychology and in his ethics, Aristotle presents reason as the essential ingredient of human nature. Animals lack reason. Aristotle explicitly states that animals do not have the capacity for reason, reasoning, thought, intellect, and belief. By this he means that animals are unable to entertain true or false thought and belief. He thus draws a sharp intellectual distinction between animals and humans. Even though the distinction was debated in ancient times, it became the received view.[25] In the context of animal reason, other capacities were discussed, such as speech and skills. In Greek thought, speech and reason are connected. Both are indicated by the same Greek word, *logos*. Moreover, ancient thinkers believed that speech and inner reasoning were two manifestations of the same thing. Thought was considered a form of silent speech. Not surprisingly, Aristotle contended that animals lack the capacity for speech. Some animals can make meaningful sounds, but they do not use words. Ancient adversaries of Aristotle pointed out that animals can understand speech, for they obey our calls. But do they themselves speak? Probably not, because we do not understand them.[26] According to another view, entertained in the eighteenth century, animals remained silent on purpose. A voyager claimed to have overheard conversations between apes in the forests: They pretended to be mute at the prospect of salvation, enslavement, or culture.[27] Skills, too, were considered beyond the ken of animals. They were either merely natural, that is, based on instinct, or they were learned. In either case, however, reason had nothing to do with it.

Albert the Great (c. 1200–1280)

Augustine's view that physical appearance is not very relevant in determining human nature, and Aristotle's idea that the possession of the faculty of reason is crucial, were taken up by Tyson. His pygmie stood at a higher level than either monkeys or apes, and, in fact, resembled a human:

> But at the same time I take him to be wholly a Brute, tho' in the formation of the body, and in the sensitive or brutal soul, it may be, more resembling a man, than any other animal; so that in this chain of the creation, as an intermediate link between an ape and a man, I would place our Pygmie.

In this way, Tyson revived a claim that had first been made in the Middle Ages by the theologian Albert the Great, whom he mentions in passing. In one of his works, *De animalibus*, Albert discusses whether pygmies should be included in the human family.[28] In his view, all soul-endowed beings (*animalia*) are ordered according to their degree of perfection. The powers of the soul provide the criteria for the hierarchical arrangement. Some animals, such as apes and pygmies, have developed their powers of the soul to such a degree that they can imitate human skills; but they will never master any skills (*artes*). Hence, Albert claims that pygmies lack real reason, and only possess a shadow of reason (*umbra rationis*). They are not human, but similar to humans (*similitudines hominis*), because their mental faculties somehow resemble the human intellect. Their position in the chain of being is determined by their degree of perfection. The degree of perfection can be measured and compared with the help of the notion of disciplinability (*disciplinabilitas*), the ability to be instructed. What Albert means by this notion is simply that it is easier to teach tricks to, say, a dog than to a fly. Thus, a dog would be more perfect than a fly, and would be higher up in Albert's scale of being. Albert's conclusion is that pygmies are at a midpoint between humans and animals, but they are still located in the realm of animals. They are just below humans, but within the category of beasts, not in some intermediate category. They can speak, but only about very concrete things.[29] They cannot carry on a discussion. Albert points out that the pygmies' position in the scale of nature just after human beings is also reflected in their behavior. Pygmies do not care for citizenship and laws, but rather follow their natural instincts. Moreover, they cannot distinguish between what is shameful and what is honorable. These latter aspects echo yet another dimension of Aristotle's discussion of reason, which also played a crucial role in later debates about human nature. As we have seen, Aristotle believes that humans are unique in possessing reason and speech. Moreover, he believes that it is precisely these assets that make life in communities possible. The best way for humans to express their human nature, that is, the pursuit of happiness, is in the *polis*, the Greek city-state. Humans on their own can also achieve a certain degree of happiness, but it is limited. A human being is, by nature, a political animal (*politikon zooion*). Animals perceive what is painful or pleasant, and can communicate this by their voice. Human beings, however, can also perceive what is good and bad, and what is just and unjust. In order to communicate these qualities, they have the capacity for speech. In other words, human beings have an innate capacity to engage in complex cooperation in communities that have as their common purpose the flourishing of man. And to achieve this purpose, they use reason and speech.[30] In the context of this social theory, Aristotle develops the idea that animals and natural slaves are unfitted by nature to form a community, since they do not have the power of reason.

> For he who can belong to another (and that is why he does belong to another) and he who participates in reason so far as to apprehend it but not so far as to possess it (for the other animals obey not reason but feelings), is a slave by nature. The use made of them differs little; for from both—slaves and tame animals—comes bodily help in the supply of essentials. (Aristotle, *Politica*, 1254b20–24).[31]

So, in sum, natural slaves can listen to reason, they can understand reason, but they cannot exercise it themselves. Natural slaves need to be governed by others. The contrast with animals is that animals can never apprehend reason, but only follow their instincts. So it is reason that makes the development of communities and life in communities possible.

Bartholomé de las Casas and Juan Ginés de Sepúlveda: A Debate over Inhabitants of the New World

This argument has been extremely influential, as becomes clear, for instance, from the well-known debate between the Spanish theologians Bartholomé de las Casas (1474–1566) and Juan Ginés de Sepúlveda (1490–1573) in the 1550s. At issue was the true nature of the inhabitants of the New World. Sepúlveda argued that the Indians were not really human, but were natural slaves. For this reason, the Spaniards could wage a just war against them, and enslave them. They had no capacity for the Christian religion.

> In prudence, talent, virtue, and humanity, they [the American natives] are as inferior to the Spaniards as children to adults, women to men, as the wild and cruel to the most meek, as the prodigiously intemperate to the continent and temperate, that I have almost said, as monkeys to men.[32]

De las Casas, on the other hand, refuted this idea. They are "not irrational or natural slaves or unfit for government."[33]

He prepared a long treatise *Defense Against the Persecutors and Slanderers of the Peoples of the New World Discovered Across the Seas*. In that treatise, he argues, among other things, that the Indians are not irrational, but clever, that they are adept in grammar and logic, that they are skilled in every mechanical art, and that they have kingdoms, jurisdiction, and lawful government. In brief, they are capable of religion.[34] "I do not know whether there is any people readier to receive the gospel."[35]

He concluded that the Indians were not inferior to the Spaniards at all and that waging war against them was morally wrong. What is interesting here is that the capacity to form communities, based on lawful government, is highlighted as an important feature of human nature.[36]

Conclusion

What now is the upshot of these historical views? In this chapter, I have offered a broad sketch of some of the important issues relating to the debate about human nature. Albert the Great's textual encounter with "Pygmies," las Casas' and Sepúlveda's debate about native Americans, and Tyson's first systematic dissection of a primate were historically situated applications of a conceptual framework that harks back to Aristotle. Understanding what it means to be human was approached from one angle; namely, the ability to reason. This excursion into the past also makes us aware that attempts to define human nature were cast in cultural terms, which basically derive from this same source—Aristotle. Human is as human does, and what human does is supposedly ingrained into our nature: Communication through language, social behavior, and skills are all capacities that indicate the presence of reason. In this sense, these past episodes show the direct impact of an Aristotelian tradition, but can also, in a broader perspective, be seen as its continuation up until the present. Even though the discussion is now framed in evolutionary terms, many of its assumptions can be traced back to Aristotle.

In Aristotle, and in many other ancient philosophers, the idea that the power of reason is the best human quality is an idea that was associated with views about our purpose to lead a good life. The particular nature of humans determines what course of life we have to take, and thus serves as a beacon to be followed, as a compass to the art of living. This ethical dimension to Aristotle's theory of human nature was pushed to the background in the episodes that were studied here. Yet, it is tempting to close this chapter by bringing to mind precisely this interesting side to the definition of humankind. In ancient ethical thought, the fullfillment of human nature was equated with living according to your reason, or even of living according to nature, as the Stoics would maintain. There were many diverse views as to what it involved to live according to your reason, and thus achieve a flourishing life. For Aristotle, as for Plato and the neo-Platonists, living up to your nature involved the spiritual ideal of becoming like God; that is, cherishing the share of the divine in yourself.[37] Other philosophers held other views as to what it means that reason is the key to true happiness.

The ethical implications of the conception of ourselves as rational agents are still being debated to this day, and so is the underlying presupposition that there is such a thing as human nature, responsible for behavior appropriate for all human beings.[38] But even if one were to agree that human nature exists, one is faced with conflicting accounts of what this nature amounts to, or, in other words, in what ways we are similar to or different from nonhumans, such as apes. The historical past of some of these accounts has been traced in this essay.

Notes

1. See, for instance, *La filosofia della natura nel Medioevo* (Milano: Società Editrice Vita e Pensiero, 1966); Kate Soper, *What Is Nature? Culture, Politics and the Non-Human* (Oxford: Blackwell, 1995); *Changing Concepts of Nature at the Turn of the Millennium. Proceedings. Plenary Session of the Pontifical Academy of Sciences*, 26–29 October 1998 (Vatican City: Pontifical Academy of Sciences, 2000); Lorraine Daston and Fernando Vidal, eds., *The Moral Authority of Nature* (Chicago: University of Chicago Press, 2004).
2. See also the mission statement behind our common endeavor and behind this book in James D. Proctor, "Resolving Multiple Visions of Nature, Science, and Religion," *Zygon: Journal of Religion and Science* 39, no. 3 (2004): 637–57.
3. See Willem B. Drees' comments with respect to Nicolaas Rupke's chapter, on 52, n. 20, in this volume.
4. James D. Proctor, "Environment After Nature," in this volume; and also in this volume Willem B. Drees, "The Nature of Visions of Nature: Packages to Be Unpacked," in the section titled "Cultural Nature: The Risk of Dismissing Science," and John Brooke, "Should the Word *Nature* be Eliminated?"
5. A particularly helpful discussion is provided by Felipe Fernández-Armest in *So You Think You're Human?* (Oxford: Oxford University Press, 2004), and also by Raymond Corbey in *The Metaphysics of Apes: Negotiating the Animal-Human Boundary* (Cambridge: Cambridge University Press, 2005).
6. Mary Midgley, *Beast and Man: The Roots of Human Nature* (London: Methuen & Co. Ltd., 1980), 203–4.
7. Jonathan Marks, *What It Means to Be 98% Chimpanzee: Apes, People, and Their Genes* (Berkeley, Los Angeles, and London: University of California Press, 2002).
8. A brief survey of some trends in primatology research can be found in Corbey, *The Metaphysics of Apes*, 145–77.

9. See Sue Savage-Rumbaugh, Stuart G. Shanker, and Talbo J. Taylor, eds., *Apes, Language and the Human Mind* (New York and Oxford: Oxford University Press, 1998), esp. 181–227. See, further, John Dupré, "Conversations with Apes: Reflections on the Scientific Study of Language," in Dupré, ed., *Humans and Other Animals* (Oxford: Clarendon Press, 2002), 236–57, which discusses some of the methodological problems in determining what primates have learned, when it is claimed that they have learned a (humanlike) language, and what one may conclude with respect to the capacities that humans and apes share.
10. Frans de Waal, *Good Natured: The Origins of Right and Wrong in Humans and Other Animals* (Cambridge, Mass.: Harvard University Press, 1996). The approach to base discussions of human nature on the properties of primates is challenged in Marks, *What It Means to Be 98% Chimpanzee*, esp. 179–80.
11. Only occasionally have I referred to specialized literature about the interpretation of Aristotle's texts.
12. See chapters 4 and 12, respectively, in this volume.
13. Edward Tyson, *Orang-Outang, sive Homo Sylvestris: Or the Anatomy of a Pygmie compared with that of a Monkey, an Ape and a Man* (London: Benett, 1699); a facsimile with an introduction by Ashley Montagu (London: Dawsons of Pall Mall, 1966). For what follows, see also Johannes M.M.H. Thijssen, "Reforging the Great Chain of Being: The Medieval Discussion of the Human Status of 'Pygmies' and Its Influence on Edward Tyson," in Raymond Corbey and Bert Theunissen, eds., *Ape, Man, Apeman: Changing Views since 1600* (Leiden, The Netherlands: Leiden University, 1995), 43–51.
14. For a brief history of the discovery of primates, see now also Corbey, *The Metaphysics of Apes*, esp. 36–59.
15. Frank Spencer, "Pithekos to Pithecantropus: An Abbreviated Review of Changing Scientific Views on the Relationship of the Antropoid Apes to Homo," in Corbey and Theunissen, eds., *Ape, Man, Apeman: Changing Views since 1600*, 15 and 23 for illustrations, and Frank Dougherty, "Missing Link, Chain of Being, Ape and Man in the Enlightenment: The Argument of the Naturalists," in the same work, 64–65.
16. This story of Diderot is retold in Robert Wokler, "Enlightening Apes: Eighteenth-Century Speculation and Current Experiments on Linguistic Competence," in Corbey and Theunissen, eds., *Ape, Man, Apeman: Changing Views since 1600*, 96.
17. Tyson also rejected another name that was used by Tulp and Bontius; namely, "Indian Satyr."
18. The translation is taken from Augustine, *Concerning the City of God against the Pagans*, a new translation by Henry Bettenson with an Introduction by David Knowles (Harmondsworth, UK: Penguin Books, 1972), 662.
19. Spencer, "Pithekos to Pithecantropus," 15.
20. At this point, it should be noted that Tyson relies on Aristotle, rather than on Descartes, as one might expect. Descartes, too, singles out the ability to reason as a uniquely human attribute, which also implies the exercise of moral choice, the ability to speak a language, and to live in a society. He posits, however, a much more radical break between animals and humans than Aristotle does, by asserting that every kind of mental or perceptual operation is absent in animals.
21. The translation is taken from Aristotle, *History of Animals*, Books VII–X; edited and translated by David M. Balme; prepared for publication by Allan Gotthelf (Cambridge: Heinemann, 1991).

22. Aristotle, *De anima*, II, 1–3.
23. The intricacies of the *ergon* argument are discussed in C.D.C. Reeve, *Practices of Reason: Aristotle's Nicomachean Ethics* (Oxford: Clarendon Press, 1992), esp. 123–38.
24. The translation is taken from Jonathan Barnes, ed., *The Complete Works of Aristotle: The Revised Oxford Translation*, 2 vols. (Princeton, N.J.: Princeton University Press, 1984).
25. Richard Sorabji, *Animal Minds and Western Morals: The Origins of the Western Debate* (London: Duckworth, 1993), esp. 12–16 and 78–80.
26. As Sextus Empiricus pointed out, this argument was not very convincing, since we do not understand foreigners either, and yet, we do not conclude that they cannot speak. See Sorabji, *Animal Minds and Western Morals*, 83.
27. Wokler, "Enlightening Apes," 96.
28. See Thijssen, "Reforging the Great Chain of Being," 45–46, with quotations and translations of the relevant texts.
29. Tyson observes that Albert correctly guessed that pygmies were a sort of ape, but that he spoiled everything by making them speak.
30. Aristotle, *Politica*, 1253a1–19.
31. The translation is taken from Aristotle, *Politics*, Books I and II, translated with a commentary by Trevor J. Saunders (Oxford: Clarendon Press, 1995).
32. Lewis Hanke, *All Mankind Is One: A Study of the Disputation between Bartolomé de las Casas and Juan Ginés de Sepúlveda in 1550 on the Intellectual and Religious Capacity of the American Indians* (DeKalb: Northern Illinois University Press, 1974), 84.
33. Ibid.
34. Ibid., 74–81.
35. Ibid., 76.
36. More recently, this argument has shown up in Mary E. Clark, *In Search of Human Nature* (London and New York: Routledge, 2002), 58. Among the three basic needs or propensities that specifically "constitute human nature, as distinct from the nature of animals in general," she includes bonding within a social group. According to Clark, that is the most important when it comes to ensuring human survival. Moreoever, just like Aristotle, she claims that "the existence of the functional group precedes the possibility of individual survival. See Aristotle, *Politica*, esp. 1253a20–29, which argues that the community (*polis*) is prior to the individual, in being more complete, and that when humans are separated from the community, they can only exist as beasts. Also, the typically human "propensity for meaning," that is, "valuing choices about what to do, to providing reasons for those choices, and to justifying them with causal explanations" (Clark, 58), can be read according to Aristotelian lines that in a community of humans, speech is the necessary vehicle with which to communicate values and thus connects individuals into a group.
37. See, for instance, Dirk Baltzly, "The Virtues and 'Becoming Like God': Alcinous to Proclus," *Oxford Studies in Ancient Philosophy* 26 (2004): 297–321.
38. See, for instance, Julia Annas, "Virtue Ethics: What Kind of Naturalism?" in Stephen M. Gardiner, ed., *Virtue Ethics, Old and New* (Ithaca, N.Y., and London: Cornell University Press, 2005), 11–30, and the literature cited there.

Bibliography

Annas, Julia. "Virtue Ethics: What Kind of Naturalism?" In Stephen M. Gardiner, ed., *Virtue Ethics, Old and New*, 11–30. Ithaca, N.Y., and London: Cornell University Press, 2005.

Aristotle. *The Complete Works of Aristotle: The Revised Oxford Translation*, 2 vols. Edited by Jonathan Barnes. Princeton, N.J.: Princeton University Press, 1984.

———. *History of Animals*, Books VII–X. Edited and translated by David M. Balme, prepared for publication by Allan Gotthelf. Cambridge: Heinemann, 1991.

———. *Politics*, Books I and II. Translated with a commentary by Trevor J. Saunders. Oxford: Clarendon Press, 1995.

Augustine. *Concerning the City of God against the Pagans*. A New Translation by Henry Bettenson with an Introduction by David Knowles. Harmondsworth, UK: Penguin Books, 1972.

Baltzly, Dirk. "The Virtues and 'Becoming Like God': Alcinous to Proclus." *Oxford Studies in Ancient Philosophy* 26 (2004): 297–321.

Changing Concepts of Nature at the Turn of the Millenium. Proceedings. Plenary Session of the Pontifical Academy of Sciences, 26–29 October 1998. Vatican City: Pontifical Academy of Sciences, 2000.

Clark, Mary E. *In Search of Human Nature*. London and New York: Routledge, 2002.

Corbey, Raymond. *The Metaphysics of Apes. Negotiating the Animal-Human Boundary*. Cambridge: Cambridge University Press, 2005.

Daston, Lorraine, and Fernando Vidal, eds. *The Moral Authority of Nature*. Chicago: University of Chicago Press, 2004.

de Waal, Frans. *Good Natured. The Origins of Right and Wrong in Humans and Other Animals*. Cambridge, Mass.: Harvard University Press, 1996.

Dougherty, Frank. "Missing Link, Chain of Being, Ape and Man in the Enlightenment: The Argument of the Naturalists." In Raymond Corbey and Bert Theunissen, eds., *Ape, Man, Apeman: Changing Views since 1600*, 63–70. Leiden, The Netherlands: Leiden University, 1995.

Dupré, John. "Conversations with Apes: Reflections on the Scientific Study of Language." In John Dupré, ed., *Humans and Other Animals*, 236–57. Oxford: Clarendon Press, 2002.

Fernández-Armest, Felipe. *So You Think You're Human?* Oxford: Oxford University Press, 2004.

La filosofia della natura nel Medioevo. Milano: Società Editrice Vita e Pensiero, 1966.

Hanke, Lewis. *All Mankind Is One. A Study of the Disputation between Bartolomé de las Casas and Juan Ginés de Sepúlveda in 1550 on the Intellectual and Religious Capacity of the American Indians*. DeKalb: Northern Illinois University Press, 1974.

Marks, Jonathan. *What It Means to Be 98% Chimpanzee: Apes, People, and Their Genes*. Berkeley, Los Angeles, and London: University of California Press, 2002.

Midgley, Mary. *Beast and Man. The Roots of Human Nature*. London: Methuen & Co. Ltd., 1980.

Proctor, James D. "Resolving Multiple Visions of Nature, Science, and Religion." *Zygon: Journal of Religion and Science* 39, no. 3 (2004): 637–57.

Reeve, C. David C. *Practices of Reason: Aristotle's Nicomachean Ethics*. Oxford: Clarendon Press, 1992.

Savage-Rumbaugh, Sue, Stuart G. Shanker, and Talbo J. Taylor, eds. *Apes, Language and the Human Mind*. New York and Oxford: Oxford University Press, 1998.
Soper, Kate. *What Is Nature? Culture, Politics and the Non-Human*. Oxford: Blackwell, 1995.
Sorabji, Richard. *Animal Minds and Western Morals: The Origins of the Western Debate*. London: Duckworth, 1993.
Spencer, Frank. "Pithekos to Pithecantropus: An Abbreviated Review of Changing Scientific Views on the Relationship of the Antropoid Apes to Homo." In Corbey and Theunissen, eds., *Ape, Man, Apeman: Changing Views since 1600*, 13–28.
Thijssen, Johannes M.M.H. "Reforging the Great Chain of Being: The Medieval Discussion of the Human Status of 'Pygmies' and Its Influence on Edward Tyson." In Corbey and Theunissen, eds., *Ape, Man, Apeman: Changing Views since 1600*, 43–51.
Tyson, Edward. *Orang-Outang, sive Homo Sylvestris: Or the Anatomy of a Pygmie compared with that of a Monkey, an Ape and a Man*. London: Benett, 1699. A facsimile with an introduction by Ashley Montagu. London: Dawsons of Pall Mall, 1966.
Wokler, Robert. "Enlightening Apes: Eighteenth-Century Speculation and Current Experiments on Linguistic Competence." In Corbey and Theunissen, eds., *Ape, Man, Apeman: Changing Views since 1600*, 87–100.

4

LOCATING NEW VISIONS

David N. Livingstone

The very title of the project that originally drew together the contributors to this venture speaks to a historical problematic that necessarily underlies all our deliberations. The idea of "new" visions is an inescapably historical one—it seeks to transcend earlier conceptualizations of the relations among nature, science, and religion, and to move to some rapprochement between the "multiple visions of biophysical and human nature" that are taken to be the "key unresolved issue central to both science and religion." Of course, there have already been many "new visions" of the relations among science, nature, and religion. The scientific revolution of the seventeenth century put in place a mechanical conception of the natural order that swept into its orbit ideas about the nature of matter, about divine action in the world, and about the appropriate ordering of civil society. The Darwinian revolution of the nineteenth century delivered a new way of thinking about the web of nature, the conduct of science without recourse to teleology, and the character of religious sentiment.

To speak this way, however, is to engage in such historical shorthand as to deliver mere stereotype. What is missing—and what is abundantly present in the work of those historians who have labored long and hard to deconstruct the very idea of either a scientific "revolution" or a Darwinian "revolution"[1]—are the hugely different ways these "new visions" were articulated, appropriated, and mobilized in different settings. What we have come to realize is the need to locate grand theory in the midst of the particulars of its making, to attend to the different ways conceptual revolutions are encountered in different spaces, and to root transcendental visions in mundane soil.

In order to disentangle some of the threads involved in locating new visions, I want to consider how Charles Darwin's theory of evolution by natural selection was read in a number of different sites during the latter decades of the nineteenth century. My argument suggests that there is an inherent instability in at least certain scientific theories, such that they mean—and are made to mean—different things in different venues. Scientific visions, to put it another way, exhibit a good deal of hermeneutic flexibility. In part, this is because conceptions of nature are, as Willem Drees points out, packages of ideas that are inescapably multivalent;[2] in part, it has to do with the volatility of the notion of nature itself, which, as both John Brooke and Andrew Lustig eloquently illustrate in their unveiling of nature's "polyvalency," resists specification in terms of necessary and sufficient conditions.[3] The situation with the attendant notion of "human nature," which Hans Thijssen opens up his chapter, is likely to be yet more tangled.[4] All of this reveals, to echo Antje Jackelén, that it is impossible to map a crisp boundary between the natural and the cultural.[5] The self-referential bite in the following narrative, of course, is clear: Whatever "new vision of nature, science, and religion" toward which we are working in our own time is certain to be every bit as located as those I present below. Whether we are quite able to discern the shape of this influence from our own vantage point, I leave for others to judge; we may need greater historical distance to catch a glimpse of the horizon that casts its formative shadow over us.

In what follows I intend to examine how several communities in different places responded to the new conception of nature and humanity that was promulgated by Charles Darwin. In each case, I believe we can discern the significance of local culture in shaping how evolution was read. Nor is Darwinism an isolated case. As Nicolaas Rupke reveals in his chapter, the study of animal behavior has been conditioned persistently by the sociopolitical locations of its practitioners.[6]

Locating New Visions

Before embarking on this journey, it will be profitable to bear in mind some relevant investigations into the circulation of knowledge. Until relatively recently, it was routinely assumed that ideas, theories, and visions migrate across space and time in some transcendental fashion as nonmaterial entities. Now we have come to acknowledge that ideas, in fact, frequently travel in material form as text. This seemingly simple realization has opened up a vast literature on the history, sociology, and geography of books and journals, and raised a range of intriguing questions rotating around what might best be characterized as textual encounter.[7] At a fundamental level, it is print (and its modern-day surrogates), not simply thought or theory, that is let loose upon the world. And this brings what I call textual space to the forefront of our considerations; namely, those arenas of engagement where moments of hermeneutic encounter are affected.

Let me briefly illustrate something of the significance of these interventions. Take Edward Said's essay on what he calls *traveling theory*. Here he reveals the different ways in which György Lukács' political writing was taken up in different times and places. His point is clear: as theory travels, it is transformed. And this, as he notes, "complicates any account of the transplantation, transference, circulation, and commerce of theories and ideas."[8] Complexities of this stripe face us at a number of different scales. Most obviously, *individuals* respond to new visions very differently from each other. Owen Gingerich's remarkable personal narrative of his "great Copernican chase," in which he set out to track down every extant copy of the early editions of *De Revolutionibus,* is illustrative.[9] Focusing, in part, on the marginal annotations that successive owners inscribed on their copies, he has been able to put together a kind of temporally extended set of reading clusters by which successive meanings of the text were shaped in dialogue with previous readers.

But individual readings are shaped by *cultural* conditions, and these also come in various forms. Nicolaas Rupke's analyses of the way in which cultures of book reviewing shaped how Alexander von Humboldt's works were interpreted operates on the more scholarly end of the spectrum. As Rupke shows, English reviewers of Humboldt's political economy read his work on Mexico very differently from their French counterparts. This project, in my view, convincingly reveals "the extent to which 'local' circumstances were involved in the production of the diverse meanings that [the Humboldt corpus] acquired."[10] In a comparable vein, we have been made aware of the different ways in which the glacial theory was interpreted in a range of different civic and cultural spaces during the Victorian period.[11]

New scientific visions, however, also find their way out into the marketplace of more general intellectual commerce and secure, at least from time to time, more popular audiences. Jim Secord's provocative analysis of the different ways the anonymously published *Vestiges of the Natural History of Creation* was encountered has unearthed an extraordinarily rich tapestry of readings of this pre-Darwinian evolutionary vision. Embraced by some, vilified by others, it at once bemused, infuriated, consoled, and revolted readers in its bold portrayal of the drama of evolution. One thought it a "priceless treasure," another dismissed it as materialist "pigology." Reviewers outdid one another in the metaphors they devised to stage manage the text for readers. Indeed, the book's striking red binding prompted one to "attribute to it all the graces of an accomplished harlot."[12] All in all, very different messages were read in, and read into, *The Vestiges*, depending on local circumstances.

Further reflection on the "geographies of reading" is not necessary to signpost the direction in which I intend to move. What follows is an attempt to locate Darwin's new vision in a range of sites and to discern how it was read, judged, and put to work in the context of circumstantial particulars. Pursuing this line of inquiry connects with the emphasis on reading that snakes its way in various registers through several contributions to this volume. Martha Henderson's project of reading cultural landscapes, Barbara King's reflections on reading primate behavior and indeed her discussion of the ways in which primates themselves engage in social reading, the idea of reading nature through mathematical lenses that comes through in Douglas Norton's essay, Jim Proctor's Latourian-inspired effort to resist reading environment through a sequence of binaries, along with other contributors who engage in one way or another with reading texts, together conspire to underscore the inescapably hermeneutic character of this collective undertaking.[13]

Placing Darwinian Visions

It has long been acknowledged that there is a geography to the reception of Darwin's theory. Thanks to the collection of essays edited by Thomas Glick, we have something of a sense of the regional response to the new biology.[14] There is much of value in this national inspection; but the scale of the nation-state is assuredly *not* the only level of analysis.[15] Inquiring into how Darwin was read in far more local settings is perhaps even more fruitful. In Scotland, for example, the sites associated with popular phrenology marked out a different suite of conceptual spaces from the floor of the General Assembly of the Free Church of Scotland. Again, Edinburgh's Queen Street Hall, where workers heard Huxley taunt the establishment with his talk of human descent from apes, represented a different conceptual venue and social space from the genteel reading rooms of the Scottish literati. In these locations, different conditions prevailed, and different meanings were attached to the new vision of nature, science, and religion that Darwin and his disciples promulgated.

In more specialized scientific venues, location was also crucial. Take the situation in the St. Petersburg Natural History society during the second half of the nineteenth century. Here, a distinctly unorthodox Darwinism surfaced, which minimized the role of selection and struggle in the evolution of species and championed the significance of mutual sociability and cooperation as mechanisms of survival.[16] Developed initially by the St. Petersburg zoologist Karl Kessler, the "mutual aid" tradition was advanced by several naturalists who had carried out field research in Siberia where they believed they had witnessed the survival value of sociability—a markedly different natural world from the teeming tropics, where both Darwin and Alfred Russel Wallace pursued their field investigations.

Here, I want to turn to the way Darwin's vision was read in a number of specific sites on three different continents. In each case, I select two venues that, I believe, highlight the dominating influence of local cultural politics in shaping the rhetorical horizon within which these different communities encountered the new vision of evolutionary nature.

Public Spectacle and Rhetorical Space: Belfast and Londonderry

John Tyndall's presidential address to the British Association for the Advancement of Science when it met in Belfast in 1874 was nothing short of a public spectacle. The coming of the "Parliament of Science" was welcomed as a pleasing respite from "spinning and weaving, and Orange riots, and ecclesiastical squabbles."[17] But it turned out very differently. To a crowded audience, Tyndall delivered an aggressive call to science to "wrest from theology the entire domain of cosmological theory." The implications were plain. All "religious theories, schemes and systems which embrace notions of cosmogony . . . must . . . submit to the control of science, and relinquish all thought of controlling it."[18] The battle cry had been sounded.[19]

The reaction of the local community was immediate. A city that had witnessed little disquiet over the Darwinian vision in the previous fifteen years suddenly became a hive of activity. The Presbyterian hierarchy was spooked into immediate action, and sponsored a series of winter lectures, hoping to stem any tide toward materialism that Tyndall's attack might trigger.[20] Their collective message was clear: The new evolutionary vision was irrevocably intertwined with a dark Epicurean materialism, which had, as one writer put it, "wrought the ruin of the communities and individuals who have acted out its principles in the past; and if the people of Belfast . . . practise its degrading dogmas, the moral destiny of the metropolis of Ulster may easily be forecast."[21] What had lent further fury to the reaction of local Protestant clergy was the fact that its leading conservative champion, the theologian Professor Robert Watts, had been snubbed by the British Association itself when it bluntly refused the offer of a paper from him on reconciling science and religion.[22] A distinctive rhetorical space was in the making.

Measures had to be taken, and that winter a number of clergy, together with the Presbyterian botanist David Moore—keeper of the Glasnevin Botanical Garden in Dublin—worked long and hard to control how Tyndall's vision might be disseminated. The strategy of Josiah L. Porter, theologian and later vice chancellor of Queen's College, was to allocate science and religion to different spheres, and to castigate the wild speculations of

Tyndall, Darwin, and Herbert Spencer. William Todd Martin focused on the moral wing of the question, and judged that the new vision of evolutionary nature ominously presaged the reconstruction of "the whole fabric of personal and social life" by reducing morality to a mere survival strategy. Anticipating eugenic implications, he warned of ominous possibilities connected with "scientific oversight of the question of population."[23]

These interventions disclose the tenor of Belfast's Presbyterian response to Tyndall-style evolution. They simply continued the stream of protest that Tyndall had received from the very day he gave his speech—a circumstance on which he reflected with evident pleasure: "Every pulpit in Belfast thundered of me."[24] The rhetorical space that Tyndall's offensive had opened up very largely defined the cognitive zone in which the Belfast debate about the new biology had to be conducted. It set the terms of what could be *said* about evolution, and what could be *heard* by interlocutors. Something of this adversarial atmosphere was captured in the following representation of the Reverend Professor Robert Watts for the local *Almanack* for 1875:

> This year I faithfully define,
> A learned and orthodox divine
> Of wide and well deserved fame,
> Worthy the Presbyterian name.
> He can uphold our Banner Blue,
> And break a lance with Tyndall too,
> Exposing his fallacious rules
> Respecting atoms—molecules;
> O'erthrows all Huxley's speculations,
> And Darwin's vain imaginations;
> While Spencer's school received a share
> Of our Professor's watchful care.[25]

Forty years later, when the BA returned to Belfast, Todd Martin recollected how the "placid waters were troubled for many days" when "Dr Tyndall . . . took advantage of his position as President to demand in the name of science as its inalienable right complete freedom in speculation and teaching, and . . . illustrated his claim by expounding a materialist theory of the universe."[26]

The rhetorical fervor that typified the Belfast contact zone sharply contrasts with the tone adopted by John Robinson Leebody, professor of mathematics and natural philosophy since 1865 at the Presbyterian Magee College outside Londonderry. During the 1870s, Leebody issued several commentaries on the new visions of nature and religion emanating from the pens of Darwin, Ernst Haeckel, Henry Maudsley, Lionel Beale, John William Draper, Arthur Balfour, and St. George Mivart. Indeed, he had already expressed himself on Tyndall's outlook prior to the Belfast meeting in a review of Tyndall's 1870 discourse to the Liverpool meeting of the BA on "The Scientific Uses of the Imagination." Taking the opportunity to rehearse the views of Pierre-Simon Laplace, August Comte, Edward B. Tylor, John Lubbock, Herbert Spencer, and Carl Vogt, as well as John Tyndall, Charles Darwin, and Ernst Haeckel, Leebody carefully discriminated among various modes in which the theory of evolution might be held. Of course, he objected to its most materialistic rendition, which asserted a radical continuity between matter and mind, and set his face against attempts to reduce "ideas of virtue, truth, and God" to "fictions of the mind, evolved by the ceaseless activity of human thought."[27] Less aggressive versions, however, were welcomed:

> Stated with these restrictions, we do not see that the doctrine of evolution comes in contact with the teachings of Scripture at all, however it may conflict with traditional preconceptions which have become bound up with our religious beliefs. . . . "It is," says Dr McCosh, speaking of its application to account for the origin of species amongst lower animals, "a question to be decided by naturalists and not by theologians, who . . . have no authority from the Word of God to say that every species of tiny moth has been created independently of all species of moths which have gone before."[28]

Thus, even while objecting to Darwinian morality, Leebody welcomed "in the interests of both Science and Religion, . . . the appearance of the 'Descent of Man.' It enriches science by a vast number of valuable facts, and it will stimulate inquiry with regard to the theory of Evolution which may be expected to yield important results."[29]

With these perspectives already in place prior to the theatrics of the Belfast meeting of the British Association, Leebody seems to have been much less shaken by the Tyndall onslaught. Writing in 1876, in a critical commentary on efforts to prosecute what he called "the principle of continuity" and the "correlation of forces," he paused to issue a far more positive commentary on the Darwinian vision than those expressed just two years or so earlier in the 1874 Winter Series:

> The more fully . . . the animal and vegetable worlds are examined, the more fully is the value of Mr Darwin's investigations and speculations seen. . . . there is no one at all familiar with the history of science and the present tendencies of scientific discovery who does not see that Mr Darwin's name is one destined to stand in the first rank among the leaders of intellectual progress. The Newton of biological science has yet, we believe, to arise, but scientific men are pretty generally agreed that Mr Darwin must at least be regarded as the Kepler.[30]

No doubt, there are parallels to be detected between Leebody's anxieties over materialist reductionism and those of his Belfast colleagues. But his discrimination between scientific findings and philosophical speculation was much more carefully prosecuted; his language was much more restrained; his assessments of Darwinism were progressively more tolerant; his reverence for science more thoroughgoing.[31] Indeed, he later came to find value in the application of evolution to the history of religion itself, conceding that "Professor Huxley does show that, to a large extent, various forms of mythology and false religion must be regarded as a natural outcome of the intellectual development of the race."[32] Leebody plainly occupied a different rhetorical space from his Belfast colleagues. Several things seem to have contributed. As a trained scientist, he felt the need to keep scientific inquiry free from unwarranted theological policing. He was well aware, for example, "that some of our ideas with regard to creation and the past history of our planet have recently undergone a change, and that we cannot claim to have been infallible in our interpretations of the opening chapters of Genesis."[33] His concern about maintaining an ongoing dialogue between science and religion thus frequently obtruded:

> The history of the past tells us that we need not dread that the religious beliefs of the community will be enfeebled or destroyed by the advance of science. It may tend to displace some of the traditional beliefs which are the excrescences on Scripture truth rightly formulated, but that will be no loss. A century or two ago it destroyed the belief in witchcraft, which up till that time was considered a crucial test in discriminating between a thoroughly orthodox theologian and one with dangerously rationalistic tendencies. In recent times it has taught us to be slow in interpreting

literally portions of the Old Testament which unquestionably were not left on record as an exposition of cosmological science.[34]

No less significant was the educational space that Leebody occupied. The Magee College, which combined "secular" and "religious" education, enjoyed no state support. Leebody had no desire to change this circumstance, but was passionate in his belief that his students should be placed on an equal footing with other universities in Ireland by being allowed to sit for the University of Ireland examinations for the conferment of a degree. In such a context, the need to promote nonsectarian scientific training was paramount, and Leebody resolutely distanced himself from a Catholic ideology that insisted that *all* subjects must be taught, as Cardinal Paul Cullen put it, "on purely Catholic principles." Leebody vigorously protested: "There is no Protestant Mathematics or Chemistry as distinguished from that taught in a Catholic college."[35] In the educational culture wars of late-nineteenth-century Ireland, Leebody had to occupy a rhetorical space that was seen to preserve the independence of scientific inquiry from too much theological supervision.[36]

Taken together, Leebody's location constituted a social space different from that produced among Belfast Presbyterians in the aftermath of Tyndall's pugnacious address. The new vision of nature, science, and religion that Darwin and his champions put forward was thus talked about in conversational arenas that conditioned what it was taken to mean and what were believed to be appropriate rejoinders.

Sectionalism and Cultural Politics: Charleston and Columbia, South Carolina

A rather different space of encounter was to be found in the U.S. Southern states. We turn first to the Charleston Museum of Natural History and to the naturalists who congregated there during the middle decades of the nineteenth century. Critical here was John McCrady (1831–1881)—mathematician and marine invertebrate naturalist, and a staunch adversary of the Darwinian vision.[37] On the surface, this might seem surprising, since McCrady had long been promoting what he referred to as the "law of development by specialization," which he believed encompassed all life. Indeed, in the early 1860s he even observed that Darwin had "furnish[ed] a most beautiful explanation of the *Modus operandi* which probably characterizes the law of development in the production of specific forms and varieties."[38] Yet the metaphorical trope that most inspired McCrady's notion of development was embryological growth, a sequence that preserved the integrity and identity of the individual rather than, in his view, inviting analogies with species transmutation. Scientifically, Darwin's vision was all wrong for him.

But there were other reasons, too, why McCrady had cause to be anxious over Darwin, and these were grounded in the social and political culture to which he gave lifelong allegiance. McCrady was an energetic apologist for the South, issuing sporadic commentary on the deterioration of civilization in the Northern states. In these circumstances, he was convinced that the only hope for the South was secession, for, as he put it on the eve of the American Civil War, "a slave State never can be a centre of that form of civilization which now flourishes in Europe and at the North."[39] Given these commitments, McCrady took to mobilizing his science in the interests of an independent South and calling on geological metaphor to naturalize, and thereby legimitize, geopolitical disruption. "The great Confederate Republic, founded by our forefathers, is about to break up into two or more confederacies," he remarked in 1861. To be sure, the "separation of this Union will be a convulsion . . . but, like those vast convulsions of geological times, it will be a convulsion of development—a pang and throe of the birth-time of great nations which are yet to be—a grand and majestic step in advance."[40] Political economy and natural history were at one. As McCrady insisted, "If this be the course of our development, then is it in perfect harmony with all other great developments in nature, proceeding as they do by a progress in specialization."[41]

In these circumstances, Darwin's vision of nature was politically abhorrent. Darwin's monogenetic account of human origins just did not work well for McCrady as a Southerner dedicated to the idea of racial superiority. Far better was the natural history of his teacher, Louis Agassiz, who insisted that different races were different *species*.[42] Each race had a separate point of origin and any blurring of its transcendental individuality was both biologically and socially repugnant. McCrady thus repeatedly insisted that it was simply impossible to entertain the notion that the white and black races could have descended from the same origin. To him, Darwin's vision of evolutionary nature was nothing less than a subversive threat to Southern racial and religious culture.

In McCrady's case, the reading of Darwin was shaped by the cultural politics of the interpretive community that congregated at the Charleston Museum of Natural History. There, as Lester Stephens has compellingly revealed, naturalists like Edmund Ravenel, John Holbrook, Lewis Gibbes, and Francis Holmes constructed a distinctively Southern style of science, which was hammered out on the anvil of political ideology. Ravenel declared that the laws of nature could not be obliterated by abolitionists. Holbrook and Gibbes actively connived with Samuel George Morton, the Philadelphia medical practitioner, to marginalize monogenist opposition to their thoroughgoing scientific racism. These men all threw their weight behind Morton and Josiah Nott—and the scientific luminary from whom they drew inspiration, Louis Agassiz—to confirm that the black and white races were different biological species, and thus to legitimize anti-abolitionist sentiment. This was the textual space into which Darwin's work was cast and its manifest unsuitability as a cultural resource conditioned how it was read.

But the Charleston Museum of Natural History was not the only arena in which Southerners encountered Darwin's theory. In a rather different intellectual and social space—the Southern Presbyterian Theological Seminary in Columbia, South Carolina—evolution was also the subject of scrutiny.[43] In 1886, James Woodrow was dismissed from his post for advocating the theological and scientific propriety of Darwin's theory of species change. Woodrow had occupied the Perkins professorship of natural science in connection with revealed religion, a chair that had been established in 1859 for the explicit purpose of rebutting infidel assaults on the integrity of Scripture.[44] Prior to that appointment, he had been a student of Louis Agassiz, held the post of professor of chemical geology and natural history at Oglethorpe University, and studied a range of scientific subjects at the University of Heidelberg in the early 1850s, before his ordination in 1859. His move to Columbia in January 1861 occurred within weeks of South Carolina's secession from the Union. But in 1886 the General Assembly of the Southern Presbyterian Church defined a theological space in which there was no room for Woodrow's teaching.[45]

Woodrow's undoing was on account of his growing conviction that Darwin's theory provided a coherent and persuasive perspective on the history of life. While he eschewed any antiteleological rendering of Darwinism, he laid emphasis on the explanatory power of natural law, the uniformity of nature, and gradualism in earth history—what he considered the Creator's regular *modus operandi*. With such convictions, he found the writings of such harmonizers of Genesis and geology as Hugh Miller, Thomas Chalmers, William Buckland, and Pye Smith congenial.

On the face of it, Woodrow's removal centered on matters of biblical interpretation, which helps to explain why he declared again and again that he accepted the plenary inspiration of the Bible. The Southern Presbyterians' orthodox keeper of its sacred flame, Robert L. Dabney, who had suspicions about the Perkins chair from the start, was not persuaded. He had long harbored doubts about geology, urging that its claims about the age of the Earth fell far short of demonstrated truth.[46] Because to him the Genesis narrative was the record of the one eyewitness who really mattered—God himself—"the circumstantial argument of the evolutionist is superseded. However ingenious, however probable and seemingly sufficient in the light of the known physical facts and laws, this hypothesis yields before the word of this competent witness."[47] To Dabney is was just impossible to believe *both* in Scripture *and* in evolution. He thus resented the implication that "the six days' work of God was not done in six days, but in six vast tracts of time," that "the deluge [covered] only a portion of central Asia," that "man has been living upon the globe . . . for more than twenty thousand years," and that "man is a development from the lowest type of life."[48] John Lafayette Girardeau, then–professor of didactic and polemic theology at the Columbia Seminary and leader of the opposition to Woodrow in the 1880s hearing, in similar tones described it as a contest "between Dr. Woodrow's hypothesis and the Bible as our church interprets it: between this scientific view and our Bible—the Bible as it is to us."[49] Again, George Armstrong, clergyman, former professor of chemistry and geology at Washington and Lee University, and architect of the resolution that outlawed evolutionary theory in Southern Presbyterianism, voiced his opposition to Woodrow with the observation: "We say these teachings of evolution are dangerous errors, because they endanger the plenary inspiration of the scriptures, and leave the Bible no longer worthy to be called the word of God."[50] The fact that evolution kept bad company with materialists didn't help.[51]

Such sentiments were not surprising, of course. Dabney had already derided Hugh Miller's accommodationist *Testimony of the Rocks* as "thoroughly impregnated with the secret virus of rationalistic infidelity."[52] Indeed, Dabney and Woodrow had locked horns during the early 1870s over Dabney's inclination to cry "unbelieving rationalism" at any rapprochement with modern scientific inquiry. For his part, Woodrow was persistently irritated at the ease with which talk of infidel science, atheistic geology, sensualistic philosophy, and the like was on the lips of Dabney and his clan. Thus, in an extended review of a number of Dabney's publications for the *Southern Presbyterian Review* in 1873, Woodrow expressed disquiet over the "alarm" that Dabney had been stirring up through his call "to raise arms against physical science as the mortal enemy of all the Christian holds dear, and to take no rest until this infidel and atheistic foe has been utterly destroyed."[53] Such rhetoric he found frankly "dangerous."[54] While Woodrow had been laboring long and hard to defend geology from its religious adversaries,[55] Dabney had fired broadsides at anti-Christian science, and Woodrow's defense of it.[56] Woodrow snapped back, picturing Dabney as "furiously brandishing his mop against each succeeding wave [of science], pushing it back with all his might." It was, to Woodrow, a pathetic gesture. He went on:

> But the ocean rolls on, and never minds him; science is utterly unconscious of his opposition. If this were all, the contest would be simply amusing. But it is not all. . . . There are numbers, even among our most learned and devoted ministers, who share these views which we regard as so inconsistent with the truth and as so fatal in their consequences. We would fain do something to prevent these terrible consequences by persuading all whom we can influence to review the ground on which they base their present opinions; confident that a fair reëxamination will without fail lead to a change of mind.[57]

Dabney, in turn, continued to insist that the "tendencies of geologists are atheistic" and that supporters of the "science are arrayed in all their phases on the side of scepticism."[58] And in rebuttal of Woodrow's claim that "natural science is silent" on such matters as the existence of a "personal, spiritual God," Dabney quipped that if Woodrow's science was silent on that subject, the same could not be said for the likes of Darwin and Huxley.[59]

Theological and philosophical though these debates seemed to be, dwelling as they did on the character of sacred Scripture, the politics of Southern race relations were significant here, too, albeit in a rather different key from the Charleston polygenists. And in order to map out something of this Southern Presbyterian ideological terrain, we need to situate the entire fracas in its deeper historical context.[60]

Antebellum Southern Presbyterians had long regarded the Bible as the foundation of Southern social order. During the 1850s, this happy coalition came increasingly under threat with the rise of antislavery sentiments and later postbellum Reconstruction. A crucial component in the ideological apologetics of Southern Presbyterians was their conviction that a plain reading of the Bible provided ample justification for the institution of slavery and racial segregation. Abolitionism was seen as a rationalistic assault on the integrity of Scripture and the Christian character of Southern culture. The Bible was thus appealed to as a means of resisting a host of perceived Yankee evils, including radical democracy, emancipation, higher criticism, and modern science.

Woodrow's chief critic—Dabney—was already well known for his opposition to public education on account of its trend toward social leveling, and for his belief that abolitionism was "of infidel tendency" and the product of atheistic theories of human rights.[61] Thus, in an examination of "The Negro and the Common School" in 1876, he told his readers that "God has made a social sub-soil to the top-soil, a social foundation in the dust, for the superstructure."[62] To him, slavery was plainly taught in Scripture and abolitionist attacks on that institution were assaults on the unadulterated Word of God, literally understood. With such convictions, Dabney was constitutionally allergic to theories of evolution that played metaphorical with plain Scripture. Armstrong, the "veteran proslavery polemicist," opponent of Woodrow, and author of *The Christian Doctrine of Slavery* (1857), shared the same convictions and told his readers that abolitionism had sprouted from the infidel philosophy that had inspired the French Revolution. As for Girardeau, who made it clear in his assessment of Woodrow's teaching of evolution that the immediate creation of Adam from literal dust was non-negotiable,[63] he eulogized the work of George Howe, professor of biblical literature at the Presbyterian Seminary in Columbia, who in an unrestrained attack on the polygenist lectures of Josiah Nott repudiated efforts to justify slavery in the language of biology, preferring instead the language of the Bible under which "the slaveholding patriarch, the slaveholding disciple of Moses, and the slaveholding Christian lived, protected and unrebuked."[64]

In the context of an ideological commitment to biblical literalism, it is not surprising that polygenist anthropology, despite its potential to serve the interests of a slavocracy, was excoriated. The kinds of maneuvers that people like McCrady and others engaged in at the Charleston Museum were repudiated in spaces like the Columbia Seminary and the floor of the General Assembly of the Georgia Presbytery. Similarly, attempts to read the Genesis narrative in poetic voice so as to accommodate Darwinian evolution were widely denounced. Southern Presbyterian space, symbolized in the Columbia Seminary, was an anti-Darwinian space like that at the Charleston Museum. But while matters of race relations were critically implicated in the rejection of evolution theory, their justification of Southern ideology was grounded in a radically different way. Whereas anthropological science was used by the Charleston naturalists to legitimize Southern culture, it was a literalist biblical hermeneutic that undergirded it among Southern Presbyterians.[65]

Materialism and Maori: Dunedin and Wellington

Half a world away, in Dunedin, New Zealand, efforts to find ecclesiastical means of banning evolution largely went nowhere. The country's foremost anti-Darwinian, Dr. James Copland, tried to secure synodical approval to outlaw the lectures of the militantly anticlerical social evolutionist Duncan MacGregor, the University of Otago's Scottish-born professor of mental and moral philosophy. Copland's failure is indicative of an antipodean cultural climate altogether different from that in either the American South or the north of Ireland.[66] Here, there was no long-standing church establishment, and "religious institutions were in a delicate state of transplantation from the Old World."[67] Yet this does not imply that the Dunedin community was indifferent to scientific concerns. To the contrary. Dunedin had begun as a Presbyterian church settlement and the newly inaugurated Otago Institute—one of a number of provincial scientific societies affiliated with the New Zealand Institute—hosted a series of lectures on evolution for the general public in 1876. Both pro- and anti-Darwinian sentiments were fully expressed. But the outcome of the altercations owed much to the stature of Frederick Wollaston Hutton, a British geologist, a committed Anglican, and a Darwinian, who had arrived in New Zealand in the mid-1860s. Over the following decades, he established himself as a leading scientific spokesman and the author of major works on New Zealand's wildlife, and was appointed to the university's chair of natural science in 1876. For four decades, John Stenhouse notes, "Hutton disseminated Darwinism to the reading public and was more responsible than any other single figure for making evolution paradigmatic within New Zealand science."[68]

An enthusiastic evolutionist, who accepted the emergence of humans from animal forebears, Hutton worked hard to differentiate between Darwinism as a theory of species change and Darwinism as an out-and-out materialist philosophy, and took on the Anglican Bishop Samuel Nevill, who actually later moved strongly toward theistic evolution, in the Otago Institute's debate.[69] To Hutton, scientific naturalism was the appropriate stance for empirical inquiry, and he believed that supernatural explanations should be excluded from scientific explanation. But he was no *philosophical* naturalist, believing in divine design, rather than randomness and chance. In Dunedin, in fact, his viewpoint won widespread support. Indeed, Hutton's analysis proved so convincing that the Presbyterian professor of divinity, William Salmond, declared to the press that he was persuaded and determined thereafter to reserve aggressive language only for the most extreme evolutionists of the German school. Support for a nonmaterialist rendition of Darwin was also forthcoming from the institute's president, Robert Gilles, an evangelical Presbyterian and a fellow of the Linnaean Society. All of this indicated that moderatism prevailed in Dunedin, and even the Otago debate was conducted in the main with good humor and without factionalism. In Dunedin, key figures in the scientific and religious community read Darwin with sympathetic eyes, and made it a space of encounter where different opinions were tolerated.

The Otago Institute, however, was not the only rhetorical space in New Zealand where Darwin's vision was welcomed. At the Wellington branch of the New Zealand Institute, the kinds of debate that characterized Dunedin's Presbyterian community were conspicuous only by their absence.[70] Here, as at the Charleston Museum of Natural History, matters of racial politics were formative, though with results markedly in contrast to those in South Carolina.

A set of public lectures presented at the Colonial Museum introduced Wellington's scientific community to Darwin's theory of evolution in 1868–1869. The speaker was the New Zealand politician William Travers—an

Irishman from Limerick, a correspondent of Darwin's, a botanist, and a lawyer. As he read *On the Origin of Species,* Travers discerned a theory with immediate political implications for race history. Just as the European rat, honeybee, goat, and other invader species had displaced their New Zealand counterparts, so the "vigorous races of Europe" were wiping out the Maori.[71] Human social history was, evidently, controlled by iron laws of nature: in the struggle for existence, Travers insisted, whenever a "white race comes into contact with an indigenous dark race on ground suitable to the former, the latter must disappear in a few generations."[72] This naturalistic inevitability, moreover, was not to be lamented; it was to be embraced. Whatever the temporary moral disquiet that might seem to attend the prospect of a culture's annihilation, he was sure that the historic successes of European civilization meant that "even the most sensitive philanthropist may learn to look with resignation, if not with complacency, on the extinction of a people which, in the past had accomplished so imperfectly every object of man's being."[73] By thus reading natural selection through the lens of race politics, Travers was simply bringing to Darwin's text the long-standing colonial conviction that the Maori were fated for extinction by Nature. His encounter with Darwin's theory and the meanings he read in it were thus shaped by the contingencies of settler-Maori politics and the desire to enlist enlightened science in the service of colonial policy.

His was not a lone voice, moreover. In a site much less shaped by the Presbyterianism that molded the early Otago community, other members of the Wellington scientific fraternity, no less schooled in the rhetoric of naturalized imperialism, were happy to confirm Travers' construal of Darwin's vision. Thus, the Wellington medical practitioner Alfred K. Newman,[74] in a statistical analysis of the "causes leading to the extinction of the Maori," called on the breeding researches of Darwin, Wallace, and Francis Galton to underwrite his declaration that the "feeble" Maori were "dying out in a quick, easy way, and being supplanted by a superior race." It wasn't, he noted, cause for "much regret."[75] Again, Walter Buller—an ornithologist, a Fellow of the Royal Society, and a Methodist—used the occasion of his 1884 presidential address to the Wellington Philosophical Society to declare that aboriginal peoples must recede in the face of civilization.[76] Pronouncements of this stripe have led Stenhouse, noting that the "scientific establishment . . . favored Darwin from beginning to end," to observe that "New

Zealanders embraced Darwinism for racist purposes."[77] In this context, the idea of struggle as an irresistible primal force became the hermeneutic key to presenting a Darwinian apologia for *pakeha* (white-settler) politics.

And, indeed, cultural politics were deterministic for the scientific fraternity elsewhere in New Zealand as well. The editor of Auckland's *Southern Monthly Magazine* declared, "Whatever may be thought of Mr Darwin's views concerning natural selection and the origin of species, no one will be disposed to deny the existence of that struggle for life which he describes, or that a weak and ill-furnished race will necessarily have to give way before one which is strong and high endowed."[78]

Duncan MacGregor told the readers of his 1876 essay on "The Problem of Poverty in New Zealand" that natural selection had delivered to New Zealand's hardy colonists the possibility of wiping out those social misfits—the thriftless and the stupid—who should be prevented from passing on their "unspeakable curse" to offspring who could only be described as "the common sewer of society."[79]

Putting New Visions in Their Place

If the narrative I have presented here approaches accuracy, it is clear that Darwin's new vision of evolutionary nature was interpreted in very different ways in the different sites where it was encountered. Even interlocutors with remarkably comparable theological convictions read different meanings into—and out of—the new biology. In every case, local circumstances had a crucial role to play in the manufacturing of Darwinian meaning. Cultural politics, ethnic relations, educational policy, public theater, and religious heritage shaped, in one way or another, the debates over Darwin in different venues. All this points to a fundamental hermeneutic instability in the circulation, if not the production, of new scientific visions. None of this, of course, is intended to reduce visions of nature, science, and religion to *nothing but* location, nor is it to imply that locating some particular vision in its space-time setting is inexorably to disable or undermine it. But it is to encourage proponents of new visions to pause, to reflect self-critically on their own location, its cultural preoccupations and political predilections, and to wonder whether contingency is being mistaken for necessity. If nothing else, then, the historical-geographical narrative I have sketched above invites us to ponder its implications for the enterprise of promoting a twenty-first-century reconceptualization of the relations among nature, science, and religion.

Acknowledgments

I am most grateful to Andrew Holmes for much assistance with the Magee material, to John Stenhouse for allowing me access to some of his unpublished work on Dunedin, to Monte Hampton for providing me with a copy of his doctoral thesis on the debates over science and religion among the Southern Presbyterians, and to Ron Numbers for generously providing material on the latter controversy.

Notes

1. See for example, Steven Shapin, *The Scientific Revolution* (Chicago: University of Chicago Press, 1996); Peter J. Bowler, *The Non-Darwinian Revolution: Reinterpreting a Historical Myth* (Baltimore: Johns Hopkins University Press, 1988).
2. Willem Drees, "The Nature of Visions of Nature—Packages to Be Unpacked," in this volume.
3. John Hedley Brooke, "Should the Word *Nature* Be Eliminated?" in this volume.
4. Johannes M.M.H. Thijssen, "Between Apes and Angels: At the Borders of Human Nature," in this volume.
5. Antje Jackelén, "Creativity through Emergence: A Vision of Nature and God," in this volume.
6. Nicolaas Rupke, "Nature as Culture: The Example of Animal Behavior and Human Morality," in this volume.
7. Among many relevant works, see Adrian Johns, *The Nature of the Book: Print and Knowledge in the Making* (Chicago: University of Chicago Press, 1998), and Jonathan R. Topham, "Scientific Publishing and the Reading of Science in Nineteenth-Century Britain: An Historiographical Survey and Guide to Sources," *Studies in History and Philosophy of Science* 31A (2000): 559–612.
8. Edward Said, "Traveling Theory," chapter 10, *The World, the Text and the Critic* (London: Vintage, 1991), 226.
9. Owen Gingerich, *The Book Nobody Read: Chasing the Revolutions of Nicolaus Copernicus* (London: Heinemann, 2004).
10. Nicolaas Rupke, "A Geography of Enlightenment: The Critical Reception of Alexander von Humboldt's Mexico Work," in David N. Livingstone and Charles W. J. Withers, eds., *Geography and Enlightenment* (Chicago: University of Chicago Press, 1999), 321. See also his comparable analysis of the importance of translation in the making of meaning: Rupke, "Translation Studies in the History of Science: The Example of *Vestiges*," *British Journal for the History of Science* 33 (2000): 209–22.

11. Diarmid A. Finnegan, "The Work of Ice: Glacial Theory and Scientific Culture in Early Victorian Edinburgh," *British Journal for the History of Science* 37 (2004): 29–52.
12. James A. Secord, *Victorian Sensation: The Extraordinary Publication, Reception, and Secret Authorship of* Vestiges of the Natural History of Creation (Chicago: University of Chicago Press, 2001), 14.
13. Martha Henderson, "Rereading a Landscape of Atonement on an Aegean Island"; Barbara King, "God from Nature: Evolution or Emergence?"; Douglas Norton, "Visions of Nature through Mathematical Lenses"; and James Proctor, "Environment after Nature: Time for a New Vision," all in this volume.
14. Thomas F. Glick, ed., *The Comparative Reception of Darwinism* (Austin: University of Texas Press, 1974).
15. Pyenson expresses considerable reservation about the value of national scales of analysis for understanding scientific inquiry. See Lewis Pyenson, "An End to National Science: The Meaning and the Extension of Local Knowledge," *History of Science* 40 (2002): 251–90.
16. Daniel Todes, *Darwin without Malthus: The Struggle for Existence in Russian Evolutionary Thought* (Oxford: Oxford University Press, 1989).
17. *Northern Whig* (19 August 1874): 4.
18. John Tyndall, *Address Delivered Before the British Association Assembled at Belfast, with Addition* (London: Longmans, Green, and Co., 1874), 59, 61. See the discussion in Ruth Barton, "John Tyndall, Pantheist. A Rereading of the Belfast Address," *Osiris*, 2nd series, 3 (1987): 111–34.
19. I have discussed this whole episode in detail in David N. Livingstone, "Darwin in Belfast: The Evolution Debate," in John W. Foster, ed., *Nature in Ireland: A Scientific and Cultural History* (Dublin: Lilliput Press, 1997), 387–408. See also Bernard Lightman, "Scientists as Materialists in the Periodical Press: Tyndall's Belfast Address," in Geoffrey Cantor and Sally Shuttleworth, eds., *Science Serialized: Representation of the Sciences* (Cambridge, Mass.: MIT Press, 2004), 199–237.
20. The lectures were collected together and published as *Science and Revelation: A Series of Lectures in Reply to the Theories of Tyndall, Huxley, Darwin, Spencer, Etc.* (Belfast: William Mullan; New York: Scribner, Welford and Armstrong, 1875).
21. Robert Watts, "Atomism—An Examination of Professor Tyndall's Opening Address before the British Association, 1874," in *The Reign of Causality: A Vindication of the Scientific Principle of Telic Causal Efficiency* (Edinburgh: T. & T. Clark, 1888), 28.
22. Watts, "An Irenicum: Or, a Plea of Peace and Co-operation between Science and Theology," in *The Reign of Causality*, 3. See the discussion in David N. Livingstone, "Darwinism and Calvinism: The Belfast-Princeton Connection," *Isis* 83 (1992): 408–28.
23. William Todd Martin, *The Doctrine of an Impersonal God in its Effects on Morality and Religion* (Belfast: William Mullan, 1875), 7, 17.
24. Cited in Ruth Barton, "John Tyndall, Pantheist. A Rereading of the Belfast Address," *Osiris*, 2nd series, 3 (1987): 116.
25. *McComb's Presbyterian Almanack, and Christian Remembrancer for 1875* (Belfast: James Cleeland, 1875), 84.
26. William Todd Martin, "The Recent Meeting of the British Association," manuscript held in the Gamble Library of the Union Theological College, Belfast.
27. J. R. Leebody, "The Theory of Evolution, and Its Relations," *British and Foreign Evangelical Review* 21 (1872): 1–35, on 3.

28. Ibid., 7–8.
29. Ibid., 34–35.
30. J. R. Leebody, "The Scientific Doctrine of Continuity," *British and Foreign Evangelical Review* 25 (1876): 742–74, on 769.
31. Leebody's later support for evolution is clear in "Evolution," *The Witness* (10 October 1890); "Results and Influence of Nineteenth-Century Science," *The Witness* (3 November 1899); "The Influence of Nineteenth-Century Science on Religious Thought," *The Witness* (6 April 1900).
32. J. R. Leebody, *Religious Teaching and Modern Thought: Two Lectures* (London: Henry Frowde, 1889), 19. He went on to argue that "the very success of Professor Huxley's efforts to account for the origin of various false religions on evolution principles brings into greater prominence the utter failure of his attempts to account in like manner for the origin of Christianity."
33. Leebody, "Theory of Evolution," 2.
34. Leebody, "Scientific Doctrine of Continuity," 773.
35. An Irish Graduate, "The Irish University Question," *Fraser's Magazine* (1872): 60, 63. The author was Leebody, who also published a history of his own college. J. R. Leebody, *A Short History of McCrea Magee College, Derry, During its First Fifty Years* (Londonderry: Derry Standard Office, 1915).
36. For background, see T. W. Moody, "The Irish University Question of the Nineteenth Century," *History* 43 (1958): 90–109.
37. See the discussion in Lester D. Stephens, *Science, Race, and Religion in the American South: John Bachman and the Charleston Circle of Naturalists, 1815–1895* (Chapel Hill: University of North Carolina Press, 2000).
38. John McCrady, "The Law of Development by Specialization: A Sketch of its Probable Universality," *Journal of the Elliott Society* 1 (1860): 101–14, on 102.
39. John McCrady, "The Study of Nature and the Arts of Civilized Life," *De Bow's Review*, new series 5 (1861): 579–606, on 604.
40. Ibid., 595.
41. Ibid., 597.
42. See Edward Lurie, "Louis Agassiz and the Races of Man," *Isis* 45 (1954): 227–42, and Mary Winsor, "Louis Agassiz and the Species Question," in William Coleman and Camille Limoges, eds., *Studies in History of Biology* (Baltimore: Johns Hopkins University Press, 1979), 89–117.
43. Monte Harrell Hampton, "'Handmaid' or 'Assailant': Debating Science and Scripture in the Culture of the Lost Cause," PhD thesis, University of North Carolina at Chapel Hill, 2004. From different perspectives, see also T. Watson Street, "The Evolution Controversy in the Southern Presbyterian Church with Attention to the Theological and Ecclesiastical Issues Raised," *Journal of the Presbyterian Historical Society* 37 (1959): 232–50; Frank Joseph Smith, "Calvinism, Science, and Dixie: The Philosophy of Science in Late Nineteenth Century Southern Presbyterianism," PhD diss., City University of New York, 1992; and W. Duncan Rankin and Stephen R. Berry, "The Woodrow Evolutionary Controversy," in Joseph A. Pipa Jr. and David W. Hall, eds., *Did God Create in Six Days?* (Taylors, S.C.: Southern Presbyterian Press, 1999), 53–99.
44. See [Richard S. Gladney], "Natural Science and Revealed Religion," *Southern Presbyterian Review* 12 (1859): 443–67; [James A. Lyon], "The New Theological Professorship—

Natural Science in Connexion with Revealed Religion," *Southern Presbyterian Review* 12 (1859): 181–95.

45. This does not mean that there were *no* voices raised in support of Woodrow. James L. Martin, for example, writing from Memphis, Tennessee, observed that "It is *not* necessary that I shall disprove or disbelieve the doctrine of Evolution, or else disbelieve the Bible; for they are *not* mutually contradictory." Jas L. Martin, *Anti-Evolution: Girardeau vs Woodrow* (n.d., n.p., 1888), 16.
46. See, for example, Robert L. Dabney, "Geology and the Bible," *Southern Presbyterian Review* (July 1861), reprinted in C. R. Vaughan, *Discussions of Robert L. Dabney: vol. III: Philosophical* (Vallecito, Calif.: Ross House Books, 1980), 91–115.
47. Robert L. Dabney, *The Sensualistic Philosophy of the Nineteenth Century Considered* (New York: Anson D. F. Randolph, 1875), 189–90.
48. Robert L. Dabney, *A Caution Against Anti-Christian Science: A Sermon* (Richmond, Va.: James E. Goode, 1871), 3–4.
49. Cited in Hampton, "Debating Science and Scripture," 244.
50. Ibid., 263.
51. Even the most circumscribed interpretation of evolution "in its most limited," he wrote, was unfavorably received by Christians owing "not so much to what it is in itself, as to the company in which they found it." George D. Armstrong, *The Two Books of Nature and Revelation Collated* (New York: Funk & Wagnalls, 1886), 96.
52. Robert L. Dabney, "Memorial on Theological Education," reprinted in C. R. Vaughan, ed., *Discussions by Robert L. Dabney, vol. 1: Theological and Evangelical* (Richmond, Va.: Presbyterian Committee of Publication, 1890), 73.
53. James Woodrow, "An Examination of Certain Recent Assaults on Physical Science," *Southern Presbyterian Review* 24 (July 1873): 327–76. Reprinted in Marion W. Woodrow, ed., *Dr James Woodrow as Seen by his Friends: Character Sketches by His Former Pupils, Colleagues, and Associates* (Columbia, S.C.: R. L. Bryan, 1909), 407–59, on 408.
54. Woodrow, "Examination of Certain Recent Assaults," in Woodrow, *Dr James Woodrow as Seen by his Friends*, 409.
55. Anonymous, "Geology and Its Assailants," *Southern Presbyterian Review* 15 (1863): 549–68.
56. Dabney, "The Caution Against Anti-Christian Science Criticised by Dr. Woodrow," *Southern Presbyterian Review* 24 (1873): 539–85.
57. James Woodrow, "A Further Examination of Certain Recent Assaults on Physical Science," *Southern Presbyterian Review* 25 (1874): 246–91. Reprinted in Woodrow, *Dr James Woodrow as Seen by his Friends*, 460–507, on 506–7.
58. Dabney, "The Caution Against Anti-Christian Science Criticised by Dr. Woodrow," *Southern Presbyterian Review* 24 (1873), reprinted in Vaughan, ed., *Dabney. Vol. 111*, 116–80, on 145.
59. Dabney, "The Caution Against Anti-Christian Science," 147.
60. My understanding of this context owes much to the doctoral thesis of Monte Harrell Hampton, cited above.
61. Robert L. Dabney, *A Defence of Virginia: (and through her, of the South) in Recent and Pending Contests against the Sectional Party* (New York: E. J. Hale, 1867).
62. Robert L. Dabney, "The Negro and the Common School," *Southern Planter and Farmer* (February 1876). Cited in Hampton, "Debating Science and Scripture," 51.

63. John Layfayette Girardeau, *The Substance of Two Speeches on the Teaching of Evolution in Columbia Theological Seminary, Delivered in the Synod of South Carolina, at Greeneville, S.C. Oct 1884* (Columbia, S.C.: William Sloane, 1885).
64. [George Howe], "Nott's Lectures," *Southern Presbyterian Review* 3 (1850): 426–90, on 487.
65. A different story, it seems, is to be told about the situation at another Presbyterian institution in Tennessee, Cumberland University, where there was reportedly much greater tolerance for evolution. See John Noel, "Darwin, Religion, and Reconciliation: Cumberland University and the Teaching of Evolution in the Nineteenth Century," *Journal of Presbyterian History* (Fall 2004): 169–79.
66. Copland is discussed in John Stenhouse, "The Rev. Dr James Copland and the Mind of New Zealand 'Fundamentalism,'" *Journal of Religious History* 30 (1996): 124–40.
67. John Stenhouse, "Darwin's Captain: F. W. Hutton and Nineteenth Century Darwinian Debates," *Journal of the History of Biology* 23 (1990): 411–42.
68. Ibid.
69. This paragraph draws substantially on Stenhouse, "Darwin's Captain."
70. The more general New Zealand response to Darwin is charted in John Stenhouse, "Darwinism in New Zealand, 1859–1900," in Ronald L. Numbers and John Stenhouse, eds., *Disseminating Darwinism: The Role of Place, Race, Religion, and Gender* (Cambridge: Cambridge University Press, 1999), 61–89.
71. William T. L. Travers, "On the Changes Effected in the Natural Features of a New Country by the Introduction of Civilized Races," *Transactions and Proceedings of the New Zealand Institute* 2 (1969): 299–313, on 313.
72. Ibid., 308.
73. Ibid., 313.
74. John Stenhouse, "'A Disappearing Race before We Came Here': Doctor Alfred Kingcome Newman, the Dying Maori, and Victorian Scientific Racism," *New Zealand Journal of History* 30 (1996): 124–40.
75. Alfred K. Newman, "A Study of the Causes Leading to the Extinction of the Maori," *Transactions and Proceedings of the New Zealand Institute* 14 (1881): 459–77, on 475.
76. See Walter L. Buller, "Presidential Address to the Wellington Philosophical Society," *Transactions and Proceedings of the New Zealand Institute* 17 (1884): 443–45.
77. Stenhouse, "Darwinism in New Zealand," 76, 81.
78. J. Giles, "Waitara and the Native Question," *Southern Monthly Magazine* 1 (1863): 209–16, on 215.
79. Quoted in John Stenhouse, "The Evolution Debates in Nineteenth-Century Dunedin," unpublished ms. I am most grateful to Dr. Stenhouse for allowing me to see this analysis.

Bibliography

Armstrong, George D. *The Two Books of Nature and Revelation Collated.* New York: Funk & Wagnalls, 1886.

Barton, Ruth. "John Tyndall, Pantheist: A Rereading of the Belfast Address." *Osiris*, 2nd series, no. 3 (1987): 111–34.

Bowler, Peter J. *The Non-Darwinian Revolution: Reinterpreting a Historical Myth*. Baltimore: Johns Hopkins University Press, 1988.

Brooke, John Hedley. "Should the Word *Nature* Be Eliminated?" in this volume.

Buller, Walter L. "Presidential Address to the Wellington Philosophical Society." *Transactions and Proceedings of the New Zealand Institute* 17 (1884): 443–45.

Dabney, Robert L. *A Defence of Virginia: (and Through Her, of the South) in Recent and Pending Contests Against the Sectional Party*. New York: E. J. Hale, 1867.

———. *A Caution Against Anti-Christian Science. A Sermon*. Richmond, Va.: James E. Goode, 1871.

———. "The Caution Against Anti-Christian Science Criticised by Dr. Woodrow." *Southern Presbyterian Review*, 24 (1873): 539–85.

———. *The Sensualistic Philosophy of the Nineteenth Century Considered*. New York: Anson D. F. Randolph, 1875.

———. "The Negro and the Common School." *Southern Planter and Farmer* (February 1876).

Dabney, Robert L. "Geology and the Bible." *Southern Presbyterian Review* (July 1861). Reprinted in C. R. Vaughan, ed., *Discussions of Robert L. Dabney, Vol. III: Philosophical*, 91–115. Vallecito, Calif.: Ross House Books, 1980.

———. "Memorial on Theological Education." Reprinted in C. R. Vaughan, ed., *Discussions by Robert L. Dabney, Vol. II: Theological and Evangelical*, 63–73. Richmond, Va.: Presbyterian Committee of Publication, 1980.

Drees, Willem. "The Nature of Visions of Nature—Packages to Be Unpacked," in this volume.

Finnegan, Diarmid A. "The Work of Ice: Glacial Theory and Scientific Culture in Early Victorian Edinburgh." *British Journal for the History of Science* 37 (2004): 29–52.

"Geology and Its Assailants." *Southern Presbyterian Review* 15 (1863): 549–68.

Giles, J. "Waitara and the Native Question." *Southern Monthly Magazine* 1 (1863): 209–16.

Gingerich, Owen. *The Book Nobody Read: Chasing the Revolutions of Nicolaus Copernicus*. London: Heinemann, 2004.

Girardeau, John Layfayette. *The Substance of Two Speeches on the Teaching of Evolution in Columbia Theological Seminary, Delivered in the Synod of South Carolina, at Greeneville, S.C. Oct 1884*. Columbia, S.C.: William Sloane, 1885.

[Gladney, Richard S.] "Natural Science and Revealed Religion." *Southern Presbyterian Review* 12 (1859): 443–67.

Glick, Thomas F., ed. *The Comparative Reception of Darwinism*. Austin: University of Texas Press, 1974.

Hampton, Monte Harrell. "'Handmaid' or 'Assailant': Debating Science and Scripture in the Culture of the Lost Cause." PhD thesis, University of North Carolina at Chapel Hill, 2004.

Henderson, Martha. "Rereading a Landscape of Atonement on an Aegean Island," in this volume.

[Howe, George.] "Nott's Lectures." *Southern Presbyterian Review* 3 (1850): 426–90.

Jackelén, Antje. "Creativity through Emergence: A Vision of Nature and God," in this volume.

Johns, Adrian. *The Nature of the Book: Print and Knowledge in the Making*. Chicago: University of Chicago Press, 1998.

King, Barbara. "God from Nature: Evolution or Emergence?" in this volume.
Leebody, J. R. "The Theory of Evolution, and Its Relations." *British and Foreign Evangelical Review* 21 (1872): 1–35.
[Leebody, J. R.] "The Irish University Question." *Fraser's Magazine* (1872): 55–64.
———. "The Scientific Doctrine of Continuity." *British and Foreign Evangelical Review* 25 (1876): 742–74.
———. *Religious Teaching and Modern Thought: Two Lectures*. London: Henry Frowde, 1889.
———. "Evolution." *The Witness* (10 October 1890).
———. "Results and Influence of Nineteenth-Century Science." *The Witness* (3 November 1899).
———. "The Influence of Nineteenth-Century Science on Religious Thought." *The Witness* (6 April 1900).
———. *A Short History of McCrea Magee College, Derry, During its First Fifty Years*. Londonderry: Derry Standard Office, 1915.
Lightman, Bernard. "Scientists as Materialists in the Periodical Press: Tyndall's Belfast Address." In Geoffrey Cantor and Sally Shuttleworth, eds., *Science Serialized: Representation of the Sciences*, 199–237. Cambridge, Mass.: MIT Press, 2004.
Livingstone, David N. "Darwinism and Calvinism: The Belfast-Princeton Connection." *Isis* 83 (1992): 408–28.
———. "Darwin in Belfast: The Evolution Debate." In John W. Foster, ed., *Nature in Ireland: A Scientific and Cultural History*, 387–408. Dublin: Lilliput Press, 1997.
Lurie, Edward. "Louis Agassiz and the Races of Man." *Isis* 45 (1954): 227–42.
[Lyon, James A.] "The New Theological Professorship—Natural Science in Connexion with Revealed Religion." *Southern Presbyterian Review* 12 (1859): 181–95.
Martin, Jas L. *Anti-Evolution: Girardeau vs. Woodrow*. n.p., 1888.
Martin, William Todd. *The Doctrine of an Impersonal God in its Effects on Morality and Religion*. Belfast: William Mullan, 1875.
McComb's Presbyterian Almanack, and Christian Remembrancer for 1875. Belfast: James Cleeland, 1875.
McCrady, John. "The Law of Development by Specialization: A Sketch of its Probable Universality." *Journal of the Elliott Society* 1 (1860): 101–14.
———. "The Study of Nature and the Arts of Civilized Life." *De Bow's Review*, new series 5 (1861): 579–606.
Moody, T. W. "The Irish University Question of the Nineteenth Century." *History* 43 (1958): 90–109.
Newman, Alfred K. "A Study of the Causes Leading to the Extinction of the Maori." *Transactions and Proceedings of the New Zealand Institute* 14 (1881): 459–77.
Noel, John. "Darwin, Religion, and Reconciliation: Cumberland University and the Teaching of Evolution in the Nineteenth Century." *Journal of Presbyterian History* (Fall 2004): 169–79.
Norton, Douglas. "Visions of Nature through Mathematical Lenses," in this volume.
Proctor, James. "Environment after Nature: Time for a New Vision," in this volume.
Pyenson, Lewis. "An End to National Science: The Meaning and the Extension of Local Knowledge." *History of Science* 40 (2002): 251–90.
Rankin, W. Duncan, and Stephen R. Berry. "The Woodrow Evolutionary Controversy."

In Joseph A. Pipa Jr. and David W. Hall, eds., *Did God Create in Six Days?* 53–100. Taylors, S.C.: Southern Presbyterian Press, 1999.
Rupke, Nicolaas. "A Geography of Enlightenment: The Critical Reception of Alexander von Humboldt's Mexico Work." In David N. Livingstone and Charles W. J. Withers, eds., *Geography and Enlightenment*, 319–39. Chicago: University of Chicago Press, 1999.
Rupke, Nicolaas. "Translation Studies in the History of Science: The Example of *Vestiges*." *British Journal for the History of Science* 33 (2000): 209–22.
———. "Nature as Culture: The Example of Animal Behavior and Human Morality," in this volume.
Said, Edward. *The World, the Text and the Critic*. London: Vintage, 1991.
Science and Revelation: A Series of Lectures in Reply to the Theories of Tyndall, Huxley, Darwin, Spencer, Etc. Belfast: William Mullan; New York: Scribner, Welford and Armstrong, 1875.
Secord, James A. *Victorian Sensation: The Extraordinary Publication, Reception, and Secret Authorship of* Vestiges of the Natural History of Creation. Chicago: University of Chicago Press, 2001.
Shapin, Steven. *The Scientific Revolution*. Chicago: University of Chicago Press, 1996.
Smith, Frank Joseph. "Calvinism, Science, and Dixie: The Philosophy of Science in Late Nineteenth Century Southern Presbyterianism." PhD thesis, City University of New York, 1992.
Stenhouse, John. "Darwin's Captain: F. W. Hutton and Nineteenth-Century Darwinian Debates." *Journal of the History of Biology* 23 (1990): 411–42.
———. "'A Disappearing Race Before We Came Here': Doctor Alfred Kingcome Newman, the Dying Maori, and Victorian Scientific Racism." *New Zealand Journal of History* 30 (1996): 124–40.
———. "The Rev. Dr James Copland and the Mind of New Zealand 'Fundamentalism.'" *Journal of Religious History* 30 (1996): 124–40.
———. "Darwinism in New Zealand, 1859–1900." In Ronald L. Numbers and John Stenhouse, eds., *Disseminating Darwinism: The Role of Place, Race, Religion, and Gender*, 61–89. Cambridge: Cambridge University Press, 1999.
Stephens, Lester D. *Science, Race, and Religion in the American South: John Bachman and the Charleston Circle of Naturalists, 1815–1895*. Chapel Hill: University of North Carolina Press, 2000.
Street, T. Watson. "The Evolution Controversy in the Southern Presbyterian Church with Attention to the Theological and Ecclesiastical Issues Raised." *Journal of the Presbyterian Historical Society* 37 (1959): 232–50.
Thijssen, J.M.M.H. "Between Apes and Angels: At the Borders of Human Nature," in this volume.
Todes, Daniel. *Darwin without Malthus: The Struggle for Existence in Russian Evolutionary Thought*. Oxford: Oxford University Press, 1989.
Topham, Jonathan R. "Scientific Publishing and the Reading of Science in Nineteenth-Century Britain: An Historiographical Survey and Guide to Sources." *Studies in History and Philosophy of Science* 31A (2000): 559–612.
Travers, William T. L. "On the Changes Effected in the Natural Features of a New Country by the Introduction of Civilized Races." *Transactions and Proceedings of the New Zealand Institute* 2 (1969): 299–313.

Tyndall, John. *Address Delivered Before the British Association Assembled at Belfast, with Addition*. London: Longmans, Green, and Co., 1874.

Watts, Robert. *The Reign of Causality: A Vindication of the Scientific Principle of Telic Causal Efficiency*. Edinburgh: T. & T. Clark, 1888.

Winsor, Mary. "Louis Agassiz and the Species Question." In William Coleman and Camille Limoges, eds., *Studies in History of Biology*, 89–117. Baltimore: Johns Hopkins University Press, 1979.

Woodrow, James. "An Examination of Certain Recent Assaults on Physical Science." *Southern Presbyterian Review* 24 (1873): 327–76.

———. "A Further Examination of Certain Recent Assaults on Physical Science." *Southern Presbyterian Review* 25 (1874): 246–91.

———. "Examination of Certain Recent Assaults." Reprinted in Marion W. Woodrow, ed., *Dr James Woodrow as Seen by his Friends: Character Sketches by His Former Pupils, Colleagues, and Associates*. Columbia, S.C.: R. L. Bryan, 1909.

5

ENDURING METAPHYSICAL IMPATIENCE?

Robert E. Ulanowicz

Entangled Nature

The foregoing chapters of this volume should be sufficient to convince the reader that both the concept of nature, as well as whatever visions of physical reality it might occasion, are exceedingly intricate. It follows, then, that virtually any particular description of the world is going to encompass elements that overlap those of parallel narratives. Furthermore, points of contact are likely to fall along a spectrum from agreement to dissonance, with any degree of nuance possible in between. The operative word here might be *entanglement* in the sense that, more often than not, it is impossible to parse out what belongs properly to one narrative and not to another. In the light of ineluctable entanglement, the two predominant discourses on reality, science and theology, are thus irreversibly entangled—so much so that, in retrospect, suggestions, such as that of Stephen Jay Gould, to dissect the two approaches into "nonoverlapping magisterial authorities"[1] now seem to ring hollow.

This nature of things notwithstanding, the complex world is replete with those who wish to behold reality through the lens of fundamentalism, be it of the scientific or religious ilk. On the religious side, one finds a spectrum of opinions that would place the authority of Scripture over that of scientific theory. These run the gamut from the "young earth creationists," who believe that the universe was created in six days several thousand years ago, and who seek physical evidence to refute evolutionary theory, to the neocreationists,[2] who couch their views on the origins of the world in mostly scientific terms, but who give priority to supernatural explanations over natural ones. On the scientific side, one finds militant naturalists like Richard Dawkins,[3] who ascribe ultimate agency in living systems to genetic material, or Daniel Dennett,[4] who underscores the mechanical and reductionist nature of living dynamics to the exclusion of any other natural forms of causality.

Fundamentalists, by their very nature, are impatient with such ambiguities as entanglement might entail and go Gould one further by insisting that their particular minimalist description trumps all others. With respect to the disciplines of science and religion, theologian John Haught has labeled this impulse to "seize the territory of the other" as "metaphysical impatience,"[5] and innumerable examples of such attempts have characterized the past three hundred years.

Enlightenment Naturalism

During the nineteenth century, enormous advances were made in the basic sciences of physics and chemistry, and the nascent fields of geology, paleontology, ecology, sociology, and anthropology came into their own. Much of what was being discovered, however, was deeply unsettling to those religious believers who held to a literal interpretation of Judaeo-Christian Scriptures: The Earth was old by all reckoning—millions and billions of years, rather than the six thousand or so indicated by Scripture. Humans did not appear suddenly out of clay, but most likely evolved from apelike ancestors. Bread could not appear *ex nihilio.* The literal interpretation of Scripture was in full retreat, and naturalists, such as Thomas Huxley, gave full pursuit.

The materialist view of nature had been given considerable momentum by the development of mechanics during the eighteenth century in the wake of Newton's *Principia*.[6] By the beginning of the nineteenth century, the ground rules for science had taken form as a tacit but widely held set of metaphysical assumptions. These assumptions were strictly materialist, possibly for two reasons:

1. Emerging scientists were eager to divorce themselves from anything that might encroach upon the transcendental, lest they appear heterodox and fall victim to the power to ostracize still wielded by clerics in many areas.
2. There had long existed an underground community of closet materialists who were eager to undermine clerical power by obviating the metaphysical assumptions that supported it—literally wanting to "seize" the clerical domain.

The metaphysic that evolved in the wake of Newton was thus heavily skewed toward the material and bore little resemblance to the beliefs of the man who initiated the revolution.[7] Five basic postulates have been identified that supported the mechanical/material approach to nature.[8]

1. Newtonian systems are causally *closed*. Only mechanical or material causes are legitimate, and they always co-occur. Other forms of action are proscribed, especially any reference to Aristotle's "final," or top-down causality.
2. Newtonian systems are *atomistic*. They are strongly decomposable into stable least units, which can be built up and taken apart again.
3. Newtonian systems are *reversible*. Laws governing behavior work the same in both temporal directions. This is a consequence of the symmetry of time in all Newtonian laws, but in addition Aemalie Noether[9] demonstrated that symmetry in time and the notion of conservation in general are virtually equivalent.
4. Newtonian systems are *deterministic*. Given precise initial conditions, the future (and past) states of a system can be specified with precision.
5. Physical laws are *universal*. They apply everywhere, at all times and scales.

Even those readers having only a passing familiarity with science will probably have noted that, since early in the nineteenth century, several of these five tenets have already faced serious challenges. Soon after Laplace[10] had exulted in the power of Newtonian laws, Sadi Carnot[11] was expositing the irreversible nature of many physical processes. Earlier, Georges Buffon[12] had suggested that earth developed over a series of epochs and that history has a place in science. Later, Georges Cuvier[13] would assert that some species had gone extinct over the ages, and Charles Lyell[14] would pronounce that observable contemporary processes were adequate to explain geological history. It remained, however, for Lyell's close friend and colleague, Charles Darwin,[15] to anchor history (irreversibility) into science through his abstract dynamic of descent with modification under natural selection. Then, early in the twentieth century, relativity and quantum theories called both universality and determinism into serious question.

The erosion of the Newtonian worldview notwithstanding, some of its postulates continue to hold sway in various fields of endeavor, and almost every contemporary scientist clings to at least one or more of the tenets. Thus it is that closure is strictly enforced in the neo-Darwinian scenario of evolution. As noted above in reference to Daniel Dennett, contemporary evolutionary theory remains scrupulous in making reference to only material and mechanical causes.[16] Atomism (reductionism) still dominates biology—witness the preponderance of molecular biology today. A substantial fraction of scientists even continue to deny the reality of chance in the world. If only the depth and precision of one's observation were not so limited, they maintain, one could, in principle, predict what appear to be random behaviors. Finally, science was obviously viewed as universal and exhaustive by Stephen Hawking[17] and his colleague Carl Sagan when they impatiently wrote that there is "nothing left for a Creator to do."

Intelligent Design?

For decades following Darwin, some thinking religious were content to reply to scientific challenges by affirming the deeper "mythical" truth of scriptural accounts of miracles. Then the 1960s ushered in many challenges directed against cultural institutions, and science was not exempted. Members of the burgeoning postmodernist movement, such as Thomas Kuhn,[18] and Paul Feyerabend[19] questioned the privileged position of science. Suddenly, science no longer seemed as absolute and free of belief as many had assumed. Data could no longer be considered independent of the normative presuppositions of the investigator acquiring them. The "disinterested observer" of Newtonian wisdom turned out to be a chimera. These challenges had little impact on the more literal believers in scientism, many of whom simply ignored the criticisms, hoping they would go away. E. O. Wilson,[20] for example, in his best-seller *Consilience*, takes but a few paragraphs to dismiss the entire postmodern critique out of hand.

Emboldened by the vulnerabilities exposed in the once-sacrosanct bastion of science, some theists have attempted to turn the tables and seize back some of the territory they have lost over the past three centuries. Prominent in the public eye today are the proponents of "intelligent design" (ID), who accept evolution through descent and most of the genetic theory that accompanies neo-Darwinism.[21] Key to the ID argument is the notion of "complex specified information" (CSI), which proponents claim is conserved and cannot be created via the mechanisms permitted under neo-Darwinism. Failing an explanation for such "irreducible complexity," advocates of ID ascribe the exquisite complexity in many biotic forms to the intelligence of a transcendental designer—a Creator.

Reactions to ID have been interesting. The consensus from the scientific community has been straightforward and, not too infrequently, derisive and vituperative. This is another "god-of-the-gaps" argument, it is proclaimed. Science is an ongoing and evolving enterprise; just because there are gaps in our ability to explain natural phenomena, there is no reason to believe that, given time, they will not be filled by lawful explanation. That is, ID is a prime example of metaphysical impatience.[22] The intensity of emotion that sometimes accompanies this declaration possibly derives from an insecurity on the part of the critics, owing to the circumstance that ID *does* appear to put its finger on a gap. The tendency when a community is under siege is to brook no dissent. Thus, some conscientious agnostics are currently the targets of enmity from fellow scientists of an orthodox bent for the dissidents' effrontery in claiming that neo-Darwinism is seriously incomplete.[23] Somehow, neo-Darwinism is deemed exempt from positivist scrutiny.

None of which is to imply that ID has been welcomed by most theologians, many of whom regard the notion of *design* as an inadequate metaphor for creation. Still, possibly out of a recognition that the proponents of ID have done more homework in philosophy than have their counterparts in the scientific community, the reaction from theologians has been a bit more nuanced. A major stumbling block for some theists is that following the ID argument in its entirety leads one too far down the strongly mechanical pathway of neo-Darwinism. Others worry that ascribing gaps entirely to the Divine opens further the Pandora's Box of Theodicy. Finally, some simply agree with the scientists that ID represents metaphysical impatience.[24]

And so the pendulum continues to swing, causing the more thoughtful or contemplative on either side of the debate to ask whether any relief is possible from having to endure episodic outbursts of impatience. Does any alternative middle metaphysical ground exist that could accommodate the aspirations of both the theist and the metaphysical naturalist without serving as a launching pad for seizing the intellectual domain of the other? Is there no perspective on nature that will allow both parties to abstain from derision and respect the intellectual position of those with whom they disagree? Perhaps, ironically, a more considered and thoughtful critique of ID could help point the way to such a common ground.

Information Arising

As noted above, a key element in the ID argument derives from information theory. William Dembski,[25] for example, cites a familiar result from information theory to the effect that the complexity inherent in any distribution, when compared to any reference distribution, can be cleanly parsed into two components—one representing an ordered complexity, and the other a residual, unorganized complexity. Dembski calls the first component "complex specified information" (CSI), and maintains that this term is conserved. Furthermore, ID holds that there is an irreducible component of CSI that conventional evolutionary mechanisms cannot explain, so that it must be specified by an outside intelligence—hence ID.

Leaving aside the specifics of Dembski's claim that CSI is conserved in ontogeny/phylogeny, it is important to note that the information-theoretic term representing CSI is *not* regarded as conserved in many other applications. One early description of how such order might increase in generic, self-organizing systems was Ilya Prigogine's "order through fluctuations" narrative.[26] Prigogine considered a metastable system that was poised at a "bifurcation" point between two possible states, at least one of which was tacitly assumed to be more ordered. Which state the system eventually came to occupy was considered to be determined by a simple, generic chance perturbation to the metastable configuration.

Perhaps a more didactic example of how information can increase by chance is provided by the concept of *autocatalytic selection*, which is believed to occur in a variety of living systems.[27] By *catalysis* is meant that one particular process tends to augment another. In ecology, for example, the growth of a submerged plant (first process) might provide more leaf area upon which more colonies of diatomaceous algae (commonly referred to as "periphyton") can grow (second process). By *autocatalysis* is meant a concatenation of catalytic processes that loops back upon itself. For example, the growth of periphyton just mentioned might provide more food for very small herbivorous aquatic animals, collectively called "zooplankton," that feed upon them (third process). In one common family of submerged aquatic plants, called Bladderworts (genus *Utricularia*), interspersed among the plant leaves are small "utricles" that function as traps for the zooplankton, providing extra nourishment for the plant that started the catalytic chain. Thus the growth of *Utricularia* augments itself indirectly.[28]

The key attribute of such "causal circuits" was expressed by Gregory Bateson[29] in his observation that random events impinging upon causal circuits result in nonrandom effects. The bias induced by causal circuits by way of autocatalytic action is easy to describe. If there is some chance change in the behavior of any participating process, and if that change either makes that process more sensitive to catalysis by its immediate antecedent in the loop, or a better catalyst of the subsequent one (or if both conditions pertain), then the catalysis will propagate around the cycle and the process that was changed will receive greater support from its antecedent neighbor. Conversely, if the change either makes the process less sensitive to catalysis by its antecedent or a poorer catalyst to the next member, the process in question will receive diminished catalysis from its immediate antecedent. The dynamics naturally provide a bias, an asymmetry or a "selection" that favors any changes that contribute to enhanced autocatalysis. Using the same information-theoretic decomposition cited by Dembski, I[30] showed how autocatalytic action serves to augment the "ordered complexity" term (CSI) as it applies to the structure of networks of feeding exchanges among ecosystem components.

The bottom line is that, even if the proponents of ID are correct in their assertion that neo-Darwinian dynamics cannot resolve all CSI, it remains possible (and even likely) that self-organization theory is capable of doing so. Dembski's assertion would then stand as an example of metaphysical impatience.

Irreducible Complexity Redux

Such criticism notwithstanding, Dembski possibly does science a service when he focuses upon the parsing of complexity into organized and disorganized components, if only because the term complementary to CSI (the one that Dembski neglects) is rarely emphasized in most scientific discourse.[31] That term may be called *residual chance*, and, appearances to the contrary, chance has never rested comfortably within science. In that context, it should be noted that Darwin's theory had atrophied significantly by the turn of the twentieth century, eclipsed at the time by developmentalist theories.[32] The evolutionary waters had been muddied by Gregor Mendel's observation that changes in the characteristics of succeeding generations of pea plants were discrete, chance events, rather than continuous. It was not until Ronald Fisher and Sewall Wright copied the earlier probabilistic arguments of Ludwig von Boltzmann and J. Willard Gibbs and demonstrated how chance could be incorporated into the evolutionary scheme that a renaissance in Darwinian thought occurred. Their "grand synthesis" effectively put the genie of chance back in its bottle and made the living world look regular and predictable—at least to a statistical degree.

The question few have bothered to ask is this: How far can probability theory go in resolving the complementary component that represents "disordered complexity"? Is it conceivable that Dembski focused on the wrong term, and that he might have better spent his time attending to any irreducible complexity within the second term of his decomposition? One notable physicist who pursued this line of inquiry was Walter Elsasser.[33] Elsasser argued that nature is replete with one-time events—events that happen once and never occur again. Accustomed as most investigators are to regarding chance as simplistic, Elsasser's claim sounds absurd. That chance is always simple, generic, and repeatable is, after all, the foundation of probability theory. Elsasser, however, used combinatorics to demonstrate the overwhelming likelihood of singular events. He reckoned that the known universe consists of somewhere on the order of 10^{85} simple particles. Furthermore, that universe is about 10^{25} nanoseconds in age. So at the outside, a maximum of 10^{110} simple events could possibly have transpired since the Big Bang. Any random event with a probability of less than 1 in 10^{110} of recurring simply won't happen. Its chances of happening again are not simply infinitesimally small, they are hyperinfinitesimally small. They are physically *unreal*.

That is all well and good, one might respond, but where is one going to find such complex chance? Those familiar with combinatorics are aware, however, that it doesn't take an enormous number of distinguishable components before the number of combinations among them grows hyperastronomically. As for Elsasser's threshold, it is reached somewhere in the neighborhood of seventy-five distinct components. Chance constellations of eighty or more distinct members will not recur in thousands of lifetimes of the universe. Now it happens that ecologists routinely deal with ecosystems that contain well over eighty distinct populations, each of which may consist of hundreds or thousands of identifiable individual organisms. One might say, therefore, that ecology is awash in singular events. They occur everywhere, all the time, and at all scales.

None of which is to imply that each singular event is significant. Most simply do not affect dynamics in any measurable way; otherwise, conventional science would have been impossible. A few might impact the system negatively, forcing the system to respond in some homeostatic fashion. A very rare few, however, might accord with the prevailing dynamics in just such a way as to prompt the system to behave very differently. These become incorporated into the material workings of the system as part of its history. The new behavior can be said to "emerge" in a radical but wholly natural way that defies explanation under conventional assumptions.[34]

Less Contentious Ground?

Because singular events do not recur, they elude treatment by conventional probability theory. They represent true gaps in the causal fabric of the natural world. No longer is it accurate to depict reality as an unbroken continuum. Nature is causally porous at all levels (and not just among the netherworld of quantum phenomena).[35] Furthermore, when radical chance is combined with autocatalytic selection, it becomes possible to identify inherently nonmechanical, natural phenomena.[36] Ecosystem dynamics arising from such a combination turns each of the five Newtonian postulates on its head. To wit:

1. Ecosystems are *open* to the influence of contingency and nonmechanical agency. Spontaneous events may occur at any level of the hierarchy at any time, and they may propagate either up or *down* the causal hierarchy.
2. Ecosystems are not deterministic machines; they are *contingent* by nature.
3. The realm of ecology is *granular,* rather than universal. Models of events at any one scale can explain matters at another scale only in inverse proportion to the remoteness between them. Obversely, selection at other levels circumscribes the domain within which irregularities and perturbations can damage a system. Chance does not necessarily unravel a system.
4. Ecosystems, like other biotic systems, are *historical.* Irregularities (simple or complex) often degrade predictability into the future and obscure hindcasting. Time takes a preferred direction, or *telos,* in ecosystems—that of increasing autocatalysis.
5. Ecosystems are *organic* in composition and behavior. Communication among elements of an organic system results in clusters of mutually reinforcing configurations within which components grow successively more interdependent. Hence, the observation of any component in isolation (if still possible) reveals regressively less about how it behaves within the ensemble.

While this alternative metaphysic might not be hailed as good news by some, it should be remembered that irreducible chance is a necessary and ready cofactor in natural creation. It represents nature at its most fecund—what Stuart Kauffman called the "expanded dimensionality of the adjacent possible."[37] Freedom and flexibility are expanding in nature faster than law can account for outcomes. Hitherto, development has been regarded as a monistic tendency, centered on organized complexity. The Darwinian narrative hinted at a different dynamic, which, under the ecological perspective, now reveals itself as a full-blown dialectic between agonistic tendencies. On the one hand is the inexorable drift toward disorder and decay. On the other is the anabolic drive toward ever more harmonious and efficient autocatalytic configurations.[38] The Hegelian nature of the confrontation becomes manifest as soon as one realizes that any system too weak in either tendency falls into jeopardy. Systems that are insufficiently coherent and robust risk displacement by others that are better organized; those that are too tightly constrained around their most efficient performance become "brittle,"[39] unreliable, and vulnerable to collapse.[40]

So it would seem that Dembski and colleagues have been looking in the wrong place for irreducible complexity. Not all gaps in nature are for want of lawful theories (new forms of which are continually arising). Rather, some gaps appear as necessary elements of the ontic landscape. To paraphrase John Polkinghorne, Ted Davis writes, "There are gaps and there are gaps."[41] By the former he meant phenomena that remain to be explained by natural science. By the latter, he indicated the *necessary* openings in nature, like those posited by Heisenberg with his Uncertainty Principle, or those indicated by Elsasser with his notion of radical chance. As Polkinghorne put it, "Those gaps must be intrinsic and ontological in character." The former likely will be filled in due time; it is simply *irrational,* however, to believe that all the latter will ever be closed.

A Necessary Patience

Properly considered, irreducible disordered complexity serves as an opaque epistemological "veil of ambiguity" that precludes metaphysical impatience. At first glance, those opposed to naturalism might object, noting that the scenario of autocatalytic selection bears a marked resemblance to Darwinism in that description remains entirely within the realm of the natural, and that a key role is played by blind chance. Because the ecological narrative does not stray from the natural, many agnostics can support it in good faith (despite the derision they might thereby incur from fellow materialists of a more fundamentalist mind-set). But the use of the word *blind* is a legacy of old habits. Simple, generic chance could be nothing else but blind; however, there is simply no reason whatsoever to expect that complex chance always will be isotropic and adirectional.

Even when the complex event in question exhibits no sign of asymmetry or order, it does not preclude the possibility of its corresponding in some significant way with an extant configuration of processes so as to give rise to a meaningful, irreversible change in dynamics. By way of analogy, the protein arrangements on the exterior of a microbe or the nucleotide sequence of an arbitrary strand of DNA, by themselves, exhibit no hint of order or direction, but when either the former comes into proximity with a microphage having a lock-in-key complementary surface arrangement or the latter encounters the network of cytoplasmic reactions, very significant and decisive consequences ensue.

And so the materialist can continue to assume that no agency whatsoever lies behind complex chance. It remains a complete dead end. (There are even many instances, as when the event in question leads to extremely deleterious consequences, in which the theist might also want to adopt this same position.) But the theist who does not preclude the possibility of epimaterial action finds more than ample "wiggle room" for agency to pass into the natural realm—call such agency immanent divine action, Design, Spirit, or whatever. The effects of a radical chance event cannot be distinguished from arbitrary action.[42] Nor must such action occur at a point-source opening, as would be characteristic of a "tinkerer" at work in the subatomic netherworld.[43] Complex ontic gaps exist everywhere, at all levels of the complex realm, and provide sufficient causal porosity to allow for compounded and coordinated actions broad in scope. Immanent divine action cannot be rationally proscribed.

Of course, the same veil of ambiguity that precludes the exclusion of further agency also frustrates any attempt to ascertain that numinous action is at work. Reality extant as a sample of one makes any conclusive test impossible. The veil of ambiguity renders metaphysical impatience otiose!

An Ecological Pathway

It was suggested above that the Newtonian metaphysic did not arise out of a neutral political and social context. As a methodological tool, however, it has served admirably in elucidating the workings of simple systems and has heralded a stream of impressive advances that have made life materially more bearable. At the same time, however, the bias toward minimalism has, according to Gregory Bateson[44] impeded access to the sacred. Going even further, there is reason to question whether the Newtonian strictures might be obscuring a fuller vision of purely natural phenomena. Hence, both the scientific and spiritual dimensions of humanity might be favored by the adoption of a less restrictive metaphysic, like the one suggested here by ecology.

It bears mention in passing how the ecological metaphysic mitigates some of the ostensible conflicts between science and theology that at times may have prompted the temptation of metaphysical impatience.[45] While *free will* was an outright impossibility in a Newtonian world, it presents no enigma within the ecological worldview, where features can emerge quite naturally. This metaphysic accommodates contingency, and in granularity it highlights the looseness among the several layers of phenomena that separate the firings of neural synapses from those higher-level, slower cognitive functions directly involved in decision making. In addition, top-down influence is not excluded from system dynamics.[46] The flexibility in the fabric of causality that could open it to immanent divine action provides hope for those who offer prayers of supplication. While the problem of *theodicy* will likely continue to haunt the believer, acknowledging the potential that petty evils can have for natural creative processes helps to mitigate somewhat the paradox of evil in a providential world.[47]

Ardent admirer of Newton though he was, Darwin envisioned nature as an "entangled bank,"[48] a notion that was truly prophetic in its pointing beyond Newtonian metaphysics. Unfortunately, it is little noted how Darwin's emphasis on process can open one's eyes to the fact that not all order and pattern in the world are the immediate consequence of physical laws. As Karl R. Popper has suggested,[49] no one should remain satisfied with the narrow conception of physical force, and science is compelled to entertain more general "propensities" that enfold radical chance into their operation. As a consequence, the world no longer appears as a rigid hegemony of physical laws. Against the background of a flexible, fecund, but organized reality, the fundamentalist/minimalist vision of nature, wherein science and religion remain fully separable and autonomous, takes on vanishing credibility.

Our ever-growing awareness of an entangled nature forces, in its turn, an ongoing conversation between naturalist and theist. In fact, the metaphor of the ecosystem dynamic as a dialectic[50] can provide a useful simile for human dialogue to imitate: Advances in science and theology often will be out of phase with each other, resulting in real or apparent conflicts. With sufficient patience, however, the hope remains that, at deeper levels, the two endeavors cannot help but richly inform each other.[51]

Notes

1. Stephen Jay Gould, "Nonoverlapping Magisteria," *Natural History* 106 (1997): 16–22.
2. E.g., Phillip E. Johnson, *Darwin on Trial* (Washington D.C.: Regnery Gateway, 1991).
3. Richard Dawkins, *The Selfish Gene*. (New York: Oxford University Press, 1976).
4. Daniel Dennett, *Darwin's Dangerous Idea: Evolution and the Meanings of Life* (New York: Simon & Schuster, 1995).
5. John F. Haught, *God after Darwin: A Theology of Evolution* (Boulder, Colo.: Westview Press, 2000).
6. Robert E. Schofield, *Mechanism and Materialism: British Natural Philosophy in an Age of Reason* (Princeton, N.J.: Princeton University Press, 1970), and Robert E. Ulanowicz, "Beyond the Material and the Mechanical: Occam's Razor Is a Double-Edged Blade," *Zygon: Journal of Religion and Science* 30 (1995): 249–66.
7. Richard S. Westfall, *The Life of Isaac Newton* (Cambridge: Cambridge University Press, 1993), and Robert E. Ulanowicz, "Beyond the Material and the Mechanical," 249–66.
8. David J. Depew and Bruce H. Weber. *Darwinism Evolving: Systems Dynamics and the Genealogy of Natural Selection* (Cambridge, Mass.: MIT Press, 1995), and Robert E. Ulanowicz, "The Organic in Ecology," *Ludus Vitalis* 9 (2001): 183–204.
9. Aemalie Noether, "Invariante Variationsprobleme." *Nachrichten von der Königlichen Gesellschaft der Wissenschaft zu Göttingen, Math-phys. Klasse* (1918): 235–57.
10. Pierre S. Laplace, *Essai philosophique sur les probabilités* (Paris: Mme. Ve. Courcier, 1819).
11. Sadi Carnot, *Reflections on the Motive Power of Heat* (New York: ASME, 1824; translated 1943).
12. Georges L. L. Buffon, *Histoire naturelle, général et particulièr*, Supplement, vol. 5 (Paris: De l'imprimerie royale, 1778).
13. Georges Cuvier, *Discours sur les révolutions de la surface du globe, et sur les changements qu'elles ont produits dans le règne animal*, 3rd ed. (Brussels: Culture et civilisation, 1969; orig. 1825).
14. Charles Lyell, *Principles of Geology, Being an Attempt to Explain the Former Changes of the Earth's Surface, by Reference to Causes Now in Operation* (London: J. Murray, 1830).
15. Charles Darwin, *On the Origin of Species by Means of Natural Selection, Or, The Preservation of Favoured Races in the Struggle for Life* (London: J. Murray, 1859).
16. Dawkins, *The Selfish Gene*, and Dennett, *Darwin's Dangerous Idea*.
17. Stephen W. Hawking, *A Brief History of Time* (New York: Bantam Books, 1998).

18. Thomas S. Kuhn, *The Structure of Scientific Revolutions*, 2nd ed. (Chicago: University of Chicago Press, 1970).
19. Paul K. Feyerabend, *Against Method: Outline of an Anarchistic Theory of Knowledge* (New York: Verso, 1978).
20. Edward O. Wilson, *Consilience: The Unity of Knowledge* (New York: Alfred A. Knopf, 1998).
21. William A. Dembski, *Intelligent Design as a Theory of Information*, Access Research Network, 1998. http://www.arn.org/docs/dembski/wd_idtheory.htm.
22. John F. Haught, *God after Darwin: A Theology of Evolution* (Boulder, Colo.: Westview Press, 2000).
23. Stanley N. Salthe, Personal Communication, and "Analysis and Critique of the Concept of Natural Selection," 2006, http://www.nbi.dk/~natphil/salthe/Critique_of_Natural_Select_.pdf.
24. John F. Haught, *Deeper Than Darwin: The Prospect for Religion in the Age of Evolution* (Boulder, Colo.: Westview Press, 2003).
25. Dembski, *Intelligent Design as a Theory of Information.*
26. Gregory Nicolis and Ilya Prigogine, *Self-Organization in Nonequilibrium Systems: From Dissipative Structures to Order through Fluctuations* (New York: John Wiley & Sons, 1977).
27. Humberto R. Maturana and Francisco J. Varela, *Autopoiesis and Cognition: The Realization of the Living* (Dordrecht, The Netherlands: D. Reidel, 1980); Robert E. Ulanowicz, *Ecology, the Ascendent Perspective* (New York: Columbia University Press, 1997); Stuart A. Kauffman, *Investigations* (Oxford: Oxford University Press, 2000); Terrance W. Deacon, "Reciprocal Linkage between Self-organizing Processes Is Sufficient for Self-reproduction and Evolvability," *Biological Theory* 1 no. 2 (2006): 136–49.
28. Robert E. Ulanowicz, "Utricularia's Secret: The Advantage of Positive Feedback in Oligotrophic Environments," *Ecological Modelling* 79 (1995): 49–57.
29. Gregory Bateson, *Steps to an Ecology of Mind* (New York: Ballantine Books, 1972).
30. Robert E. Ulanowicz, *Growth & Development: Ecosystems Phenomenology* (New York: Springer-Verlag, 1986).
31. But see Robert W. Rutledge, B. L. Basorre, and Robert J. Mulholland, "Ecological Stability: An Information Theory Viewpoint," *Journal of Theoretical Biology* 57 (1976): 355–71.
32. Depew and Weber, *Darwinism Evolving;* and Peter J. Bowler, *The Eclipse of Darwinism: Anti-Darwinian Evolution Theories in the Decades around 1900* (Baltimore: Johns Hopkins University Press, 1983). But see also Richard England, "Natural Selection, Teleology, and the Logos: From Darwin to the Oxford Neo-Darwinists, 1859–1909," *Osiris* 16 (2001): 270–87, for an account of how Anglican Christians had attempted to retrieve a theistic slant on Darwin.
33. Walter M. Elsasser, "Acausal Phenomena in Physics and Biology: A Case for Reconstruction," *American Scientist* 57 (1969): 502–16.
34. See Gregory Peterson, "Who Needs Emergence?" in this volume.
35. Charles Saunders Peirce, "The Doctrine of Necessity Examined," *The Monist* 2 (1892): 321–38; and Karl R. Popper, *The Open Universe: An Argument for Indeterminism* (Totowa, N.J.: Rowman and Littlefield, 1982).
36. Robert E. Ulanowicz, "Life after Newton: An Ecological Metaphysic," *BioSystems* 50 (1999): 127–42.

37. Kauffman, *Investigations.*
38. See also Antje Jackelén, "Creativity through Emergence: A Vision of Nature and God," in this volume.
39. Crawford S. Holling, "The Resilience of Terrestrial Ecosystems: Local Surprise and Global Change," in W. C. Clark and R. E. Munn, eds., *Sustainable Develoment of the Biosphere* (Cambridge: Cambridge University Press, 1986), 292–317.
40. Ulanowicz, *Ecology, the Ascendent Perspective.*
41. Edward B. Davis, "A God Who Does Not Itemize Versus a Science of the Sacred," *American Scientist* 86 (1998): 572–74.
42. Stanley N. Salthe, "Problems of Macroevolution (Molecular Evolution, Phenotype Definition, and Canalization) as Seen from a Hierarchical Viewpoint," *American Zoologist* 15 (1975): 295–314.
43. Cf. Roger Penrose, *Shadows of the Mind* (Oxford: Oxford University Press, 1994).
44. Bateson, *Steps to an Ecology of Mind.*
45. Robert E. Ulanowicz, "Ecosystem Dynamics: A Natural Middle," *Theology and Science* 2 (2004): 231–53.
46. Ulanowicz, *Ecology, the Ascendent Perspective;* and George F. R. Ellis, "Physics, Complexity and Causality," *Nature* 435 (2005): 743.
47. Ulanowicz, "Ecosystem Dynamics"; Catherine Keller, *God and Power: Counter-Apocalyptic Journeys* (Minneapolis: Fortress Press, 2005); Sidney Callahan, *Women Who Hear Voices* (New York: Paulist Press, 2003); and Antje Jackelén, "Creativity through Emergence: A Vision of Nature and God," in this volume.
48. Darwin, *On the Origin of Species.*
49. Karl R. Popper, *A World of Propensities* (Bristol, UK: Thoemmes, 1990).
50. Ulanowicz, "Ecosystem Dynamics."
51. Haught, *Deeper Than Darwin.*

Bibliography

Bateson, Gregory. *Steps to an Ecology of Mind.* New York: Ballantine Books, 1972.

Bowler, Peter J. *The Eclipse of Darwinism: Anti-Darwinian Evolution Theories in the Decades around 1900.* Baltimore: Johns Hopkins University Press, 1983.

Buffon, Georges L. L. *Histoire naturelle, général et particulièr,* Supplement, vol. 5. Paris: De l'imprimerie royale, 1778.

Callahan, Sidney. *Women Who Hear Voices.* New York: Paulist Press, 2003.

Carnot, Sadi. *Reflections on the Motive Power of Heat.* New York: ASME, 1824; translated 1943.

Cuvier, Georges. *Discours sur les révolutions de la surface du globe, et sur les changements qu'elles ont produits dans le règne animal,* 3rd ed. (orig.: 1825). Brussels: Culture et civilisation, 1969.

Darwin, Charles. *On the Origin of Species by Means of Natural Selection, or, The Preservation of Favoured Races in the Struggle for Life.* London: J. Murray, 1859.

Davis, Edward B. "A God Who Does Not Itemize Versus a Science of the Sacred." *American Scientist* 86 (1998): 572–74.

Dawkins, Richard. *The Selfish Gene.* New York: Oxford University Press, 1976.

Deacon, Terrance W. "Reciprocal Linkage between Self-organizing Processes Is Sufficient

for Self-reproduction and Evolvability." *Biological Theory* 1, no. 2 (2006): 136–49.
Dembski, William A. *Intelligent Design as a Theory of Information*. Access Research Network, 1998. http://www.arn.org/docs/dembski/wd_idtheory.htm (accessed September 2005).
Dennett, Daniel C. *Darwin's Dangerous Idea: Evolution and the Meanings of Life*. New York: Simon & Schuster, 1995.
Depew, David J., and Bruce H. Weber. *Darwinism Evolving: Systems Dynamics and the Genealogy of Natural Selection*. Cambridge, Mass.: MIT Press, 1995.
Ellis, George F. R. "Physics, Complexity and Causality." *Nature* 435 (2005): 743.
Elsasser, Walter M. "Acausal Phenomena in Physics And Biology: A Case for Reconstruction." *American Scientist* 57 (1969): 502–16.
———. "A Form of Logic Suited for Biology?" In Robert Rosen, ed., *Progress in Theoretical Biology*, vol. 6, 23–62. New York: Academic Press, 1981.
England, Richard. "Natural Selection, Teleology, and the Logos: From Darwin to the Oxford Neo-Darwinists, 1859–1909." *Osiris* 16 (2001): 270–87.
Feyerabend, Paul K. *Against Method: Outline of an Anarchistic Theory of Knowledge*. New York: Verso, 1978.
Gould, Stephen J. "Nonoverlapping Magisteria." *Natural History* 106 (1997): 16–22.
Haught, John F. *God after Darwin: A Theology of Evolution*. Boulder, Colo.: Westview Press, 2000.
———. *Deeper Than Darwin: The Prospect for Religion in the Age of Evolution*. Boulder, Colo.: Westview Press, 2003.
Hawking, Stephen W. *A Brief History of Time*. New York: Bantam Books, 1988.
Holling, Crawford S. "The Resilience of Terrestrial Ecosystems: Local Surprise and Global Change." In W. C. Clark and R. E. Munn, eds., *Sustainable Develoment of the Biosphere*, 292–317. Cambridge: Cambridge University Press, 1986.
Johnson, Phillip E. *Darwin on Trial*. Washington D.C.: Regnery Gateway, 1991.
Kauffman, Stuart A. *The Origins of Order: Self-Organization and Selection in Evolution*. Oxford: Oxford University Press, 1993.
———. *Investigations*. Oxford: Oxford University Press, 2000.
———. *Reinventing the Sacred*. New York: Basic Books, 2008.
Keller, Catherine. *God and Power: Counter-Apocalyptic Journeys*. Minneapolis: Fortress Press, 2005.
Kuhn, Thomas S. *The Structure of Scientific Revolutions*, 2nd ed. Chicago: University of Chicago Press, 1970.
Laplace, Pierre S. *Essai philosophique sur les probabilités*. Paris: Mme. Ve. Courcier, 1819.
Lyell, Charles. *Principles of Geology, Being an Attempt to Explain the Former Changes of the Earth's Surface, by Reference to Causes Now in Operation*. London: J. Murray, 1830.
Maturana, Humberto R., and Francisco J. Varela. *Autopoiesis and Cognition: The Realization of the Living*. Dordrecht, The Netherlands: D. Reidel, 1980.
Nicolis, Gregory, and Ilya Prigogine. *Self-Organization in Nonequilibrium Systems: From Dissipative Structures to Order through Fluctuations*. New York: John Wiley & Sons, 1977.
Noether, Aemalie. "Invariante Variationsprobleme." *Nachrichten von der Königlichen Gesellschaft der Wissenschaft zu Göttingen, Math-phys. Klasse* (1918): 235–57.
Peirce, Charles Saunders. "The Doctrine of Necessity Examined." *The Monist* 2 (1892): 321–38.

Penrose, Roger. *Shadows of the Mind*. Oxford: Oxford University Press, 1994.
Polkinghorne, J. C. *Belief in God in an Age of Science*. New Haven, Conn.: Yale University Press, 1998.
Popper, Karl R. *The Open Universe: An Argument for Indeterminism*. Totowa, N.J.: Rowman and Littlefield, 1982.
———. *A World of Propensities*. Bristol, UK: Thoemmes, 1990.
Rutledge, Robert W., B. L. Basorre, and Robert J. Mulholland. "Ecological Stability: An Information Theory Viewpoint." *Journal of Theoretical Biology* 57 (1976): 355–71.
Salthe, Stanley N. "Problems of Macroevolution (Molecular Evolution, Phenotype Definition, and Canalization) as Seen from a Hierarchical Viewpoint." *American Zoologist* 15 (1975): 295–314.
———. "Analysis and Critique of the Concept of Natural Selection." 2006. http://www.nbi.dk/~natphil/salthe/Critique_of_Natural_Select_.pdf.
Ulanowicz, Robert E. *Growth & Development: Ecosystems Phenomenology*. New York: Springer-Verlag, 1986.
———. "Beyond the Material and the Mechanical: Occam's Razor Is a Double-Edged Blade." *Zygon: Journal of Religion and Science* 30 (1995): 249–66.
———. "Utricularia's Secret: The Advantage of Positive Feedback in Oligotrophic Environments." *Ecological Modelling* 79 (1995): 49–57.
———. *Ecology, the Ascendent Perspective*. New York: Columbia University Press, 1997.
———. " Life after Newton: An Ecological Metaphysic." *BioSystems* 50 (1999): 127–42.
———. "The Organic in Ecology." *Ludus Vitalis* 9 (2001): 183–204.
———. "Ecosystem Dynamics: A Natural Middle." *Theology and Science* 2 (2004): 231–53.
Westfall, Richard S. *The Life of Isaac Newton*. Cambridge: Cambridge University Press, 1993.
Wilson, Edward O. *Consilience: The Unity of Knowledge*. New York: Alfred A. Knopf, 1998.

6

GOD FROM NATURE: EVOLUTION OR EMERGENCE?

Barbara J. King

Introduction

One quiet day at Arnhem Zoo in the Netherlands, adult female chimpanzee Krom became intrigued with some tires that zoo staff had sprayed with water and hung on a horizontal log extending out from the apes' climbing structure. Krom repeatedly pulled on a specific tire, in an attempt to free it. She "worked in vain on this problem for over ten minutes, ignored by everyone except Jakie, a seven-year-old whom Krom had taken care of as a juvenile." When Krom gave up her attempt, Jakie immediately walked to the same spot: "Without hesitation he pushed the tires one by one off the log. . . . When he reached the last tire, he carefully removed it so that no water was lost, carrying it straight to his aunt, and placing it upright in front of her."[1]

I love this story. Understood in the context of research on the intelligence of African apes, it tells us that Jakie is much more than a skilled tire-wrangler. Jakie showed the capacity for both empathy and compassion, a capacity also present in wild chimpanzees. When a leopard killed Tina, a ten-year-old chimpanzee living in Tai National Forest, Ivory Coast, dominant male Brutus kept watch over her body for nearly five hours. Brutus granted access to Tina's body to some apes but not others, and allowed only a single youngster near. This was Tina's five-year-old brother Tarzan. Tina and Tarzan had recently lost their mother, and now, Tarzan sat by his sister's body and gently pulled on her hand.[2]

Just as Jakie had recognized what his aunt wanted, Brutus recognized the close sibling relationship between Tina and Tarzan. Brutus' allowing Tarzan to sit with his sister in death is (again interpreting from the broad corpus of work in ape studies) an act of empathy and compassion akin to Jakie's.

Why open an essay on nature, science, and religion with Jakie and Brutus? In *The Emergence of Everything,* Harold J. Morowitz cites the great apes as one of twenty-eight examples of emergence from the earliest beginnings of the universe to humankind's future. *Everything,* for Morowitz, ranges from stars, cells, and reptiles to human ancestors, agriculture, and philosophy. Primates, collectively, are cited in his core twenty-eight, but of two-hundred-plus species of primates, only the great apes are plucked from taxonomic generality and discussed at length.

The great apes, Morowitz writes, have changed our planet. Jakie and Brutus help to clue us in as to why Morowitz makes such a startling claim. Through my own research, I have come to know the African great apes—chimpanzees, gorillas, and bonobos—as primates who live in emotional connection to their social partners. At times, the apes express this connection through empathy, imagination, and compassion; at other times, through selfish behavior and brutality (a fact that, I believe, only strengthens a claim for apes' behavioral similarity to humans).

In this chapter, I use African great ape behavior as a springboard to exploring a pair of questions intimately tied to new visions of nature, science, and religion: Did religion evolve? And can we trace, using the techniques of paleoanthropology, expression of the human religious imagination through prehistory in the same way that we can trace the evolution of language, culture, and technology? I offer some answers, then identify a second set of questions that I cannot answer fully, but nonetheless wish to discuss: Is the expression of the human religious imagination a predictable culmination of events in primate and human evolution? Alternatively, did religious behavior emerge? Or are processes of evolution and emergence both involved?

Although this chapter is unique in its focus on nonhuman primates (but see Johannes Thijssen's "Between Apes and Angels: At the Borders of Human Nature," in this volume), the questions it asks link it closely with some others. Gregory Peterson's unpacking of the concept of emergence and its various uses has rescued me from adopting a too-rigid definition. I share Antje Jackelén's sense that "exploring a concept that so far escapes deep understanding" leaves an author vulnerable and yet, simultaneously, invites a re-envisioning of aspects of the human religious imagination. When Robert Ulanowicz notes the possibility of spontaneous and contingent events in open ecosystems, he explores the very aspects of emergence that I wish to explore. And when James Proctor asserts the importance of connection itself—rather than of the search for a greater connection between people and the environment—he touches on an argument I wish to make: Humans and great apes *by their very nature* are in connection with each other.

Defining Matters

What is religion? What is emergence? A thornier pair of questions might be hard to find. The challenge in defining religion (see Willem Drees, "The Nature of Visions of Nature: Packages to Be Unpacked," in this volume) increases exponentially when the goal is to explore the *earliest* expression of humanity's connection with the sacred. Any definition charged with capturing the essence not only of present-day Buddhism and fundamentalist Christianity, but also the expression of spirituality in the prehistoric world, would be workhorse in nature indeed. A definition I have found useful for exploring prehistory involves people expressing an emotional connection with supernatural beings via social, symbolic ritual. Those familiar with the work of anthropologist Clifford Geertz will recognize my debt to his definition of religion,[3] although I follow anthropologist Roy Rappaport[4] in highlighting the importance of ritual.

The remainder of this chapter could easily be devoted to further levels of defining: What is a symbol, a ritual, the sacred world? I will be content to highlight just one aspect. *Practice,* not *belief,* is at the heart of my definition. Although expediency is a factor here (after all, how would we search for belief in prehistory?), my main impetus comes from theorists and theologians who envision *active relationship with God, god, or spirits* at the center of the human religious imagination. "[T]he longing for relation is primary," Martin Buber wrote. The being (the "form") that confronts a person demands participation: "Tested for its objectivity, the form is not 'there' at all; but what can equal its presence? And it is an actual relation; it acts on me as I act on it."[5] Prayer is action, too; as Philip and Carol Zaleski say in *Prayer: A History,* it is "action that communicates between human and divine realms."[6]

Matters get no easier when we move to emergence, a concept with both a long history and a new buzz, as Peterson ("Who Needs Emergence?" in this volume) makes clear. Whether the result of the buzz is cutting-edge thinking or blooming-buzzing confusion depends, of course, on how the term is used. When emergence is brought to bear too vaguely, questions of complex change over time are clarified not at all. Yet, as I noted in referring to Antje Jackelén's chapter in this volume ("Creativity through Emergence: A Vision of Nature and God"), the promise of emergence is considerable, as reflected in Ursula Goodenough and Terence Deacon's enthusiasm for its potential. Emergence, write Goodenough and Deacon, "more than any other concept we have encountered, puts Humpty Dumpty back together again in ways that are wonderfully resonant with our existential and religious yearnings."[7]

Any voyager into the sea of emergence theory must navigate the waters of competing temptations: the embrace of emergence as a catchall explanation in the absence of sufficient critical analysis, versus the rejection of emergence without pinpointing when and where it might serve as a useful explanation. What *is* emergence, though? Sobered by the array of options in selecting a working definition (see Peterson, "Who Needs Emergence?" in this volume), and while heeding Philip Clayton's caution that "simple definitions fail to satisfy," I will use Clayton's own definition: "the theory that cosmic evolution repeatedly includes unpredictable, irreducible, and novel appearances." A key idea that underpins this definition is that emergent properties "are irreducible to, and unpredictable from, the lower-level phenomena from which they emerge."[8]

Clayton's writing on emergence is cogent, yet still, how slippery are these notions! Willem Drees nicely captures how the defining terms of emergence can be turned back on themselves:

> Emergence stands for a vision that appreciates nature as a creative process, and seems to accept the scientific understanding of this process as well. However, at the same time, *emergence* serves as label for an axiological order that values higher over lower structures, complexity over simplicity, and often culture over nature. As a vision, it is opposed to reductionism or mechanicism or even materialism, even though it is hard to see the difference between materialism and emergence, since emergence tells us how complex and interesting behavior arises in and through material structures. (in this volume)

Emergence theorists—or at least some emergence theorists—would, I think, insist that emergence is no more about culture than nature, and that it is not so much about material structures as it is about *relationships between material structures*[9] *and/or processes.* So goes the point and counterpoint with emergence. While recognizing the real issues here, I want to push beyond them and focus on the question of whether—or, better yet, when—emergence adds strong explanatory power to an evolutionary framework for religion.

Certainly, as Drees has noted, a strong attractor within the concept of emergence is its power to oppose simplistic reductionism. When anthropologists read that geneticist Dean Hamer thinks "we [humans] follow the basic law of nature, which is that we're a bunch of chemical reactions running around in a bag,"[10] many of us wish to grab simplistic reductionism and shake the life out of it. When fellow anthropologist Pascal Boyer titles a book *Religion Explained* and claims that religion is a by-product of human minds that have been set, since the Paleolithic, to overdetect agency in the world, many of us wish to renounce an anthropology that cleaves so closely to a sterile brand of evolutionary psychology. But here's the rub: How does opposing simplistic reductionism with *emergence* improve upon opposing it with *more nuanced evolutionary accounts,* accounts that deeply root religion in collective and symbolic expression of emotion? Let's start with a look at evolutionary changes in primate behavior.

African Great Apes

African great apes are more highly intelligent and expressive emotionally than other primates and most other mammals. Broad though this generalization is, it fits with data from comparative primatology. Chimpanzees, gorillas, and bonobos may make and use tools, take the perspective of others (via "theory of mind"), and remember past interactions with social partners while planning for the future.[11] African apes *routinely* behave in ways that reflect emotional bonds with their social partners; *may*, under certain circumstances, follow rules enforced by the community; and *when raised in enriched environments,* communicate impressively through symbols.[12] No apes, of course, bring together in a coherent behavioral "package" all the components of religious expression. Apes do not use symbols in social ritual, for instance, and, unless we primatologists have great surprises in store for us in the future, they experience no emotional connection with beings beyond the here and now.

Renowned chimpanzee expert Jane Goodall does, however, veer toward the position that chimpanzees experience an incipient religiousness. When chimpanzees at Gombe, Tanzania, come upon a thundering waterfall, they may stand in the water for almost an hour, and stare—despite the fact that these apes normally dislike getting wet. "We find," Goodall says, "that chimps do have a sense of wonder, of awe. I think we can see the roots of some kind of religion from chimp behavior—[inhibited] by their lack of language."[13]

During many years' research on captive apes, I have never observed behavior that compelled me to think in Goodall's terms. Thus, I do not search for the roots of religious behavior in apes, but instead look at what apes do and how they do it during routine, everyday social interactions. I argue that these apes "read" and respond to each other's emotions in subtle ways, so that the family or social group can be best understood as an emotional system in which the members are highly attuned to each other's postures, facial expressions, bodily orientations, gestures, and vocalizations. The apes respond to unpredictable and contingent social events as they unfold, so that *meaning is created as social partners act together*. A shift—physical or emotional—in one member of the group causes a shift in another, because, as in any system, the "parts" are internally related to each other.[14]

Just as not all human social events are unpredictable and contingent—

think of cultural routines, such as basic greetings—not all ape social events are, either. On occasion, a dominant ape may threaten and a subordinate may submit, or a male ape with an erect penis may approach and mate with a willing female, in a fixed sequence of actions. I confess to doubting, however, that many such events, in humans or apes, are truly fixed and predictable (although this does need to be tested as a hypothesis). My point is that apes engage in thoughtful and strategic adjustment based on a close reading of the social partner in ways that differ qualitatively from what happens when two wolves circle each other or two cats hiss and posture.

For six years, I have observed and filmed the gestural communication and body language in a captive gorilla group at the National Zoological Park in Washington, D.C.[15] For part of this time, six gorillas lived together as a family: a large silverback male leader, Kuja; an adult female, Mandara; and Mandara's four male offspring, the subadult Baraka (whom she had adopted as an infant), Ktembe (born 1997), Kwame (1999), and Kojo (2001). One morning in 2000, a conflict took place between Kuja and Baraka—but it is far more accurate to say that the conflict took place within the family as an emotional system.

Viewed on film, this conflict is striking for two reasons. First, as the intensity heats up between the two males, the family members rush *toward* the conflict rather than away from it. Baraka is obviously frightened: His facial expression and body language tell us this, even apart from his piercing screams. A human observer of the fight understands intuitively why he would be scared: He is facing a much heavier and more powerful male who could, if he chose, injure Baraka badly. Yet the juvenile male, Ktembe, rushes into the fray in support of his adoptive brother. Apparently fearless, he tugs on the comparatively enormous Kuja, who simply knocks Ktembe aside. Ktembe persists, and is joined by his mother, Mandara, carrying infant Kwame on her ventrum.

Second, it becomes clear that Baraka, despite his evident fear, refuses to submit fully to Kuja. Although Baraka has the appropriate facial expression and leans away whenever Kuja approaches, he never crouches or lies flat in the position of submission to a dominant. In an embodied way, Kuja *tells* Baraka what to do: First, he tries to push Baraka down flat; when Baraka persists in staying upright, Kuja pulls on his legs in an iconic insistence that Baraka *get down*. Yet Baraka ignores the message.

My claim is that empathy and meaning-making are folded into this conflict. In a visible expression of emotional bonds, Baraka's adoptive mother and younger brother support him, and indicate through their persistence and concern that they have some sense of what Baraka is experiencing as the fight unfolds—just as Jakie understood what his aunt wanted (the tire) and Brutus understood that young Tarzan should be allowed close to his older sister's body. Baraka, if I am right, knew fully well what Kuja wanted him to do—he had grown up in a group where older-younger male conflict had occurred, and he had observed the process of appropriate submission to a dominant. Kuja, if I am right, recognized that Baraka needed a refresher course in proper gorilla submission and forcibly pushed him down, then pulled on his legs to indicate what should be done. Baraka's adoptive family, if I am right, supported him based, in part, on empathy for his fear. (Baraka and Ktembe were transferred together to another zoo later on, when conflicts with Kuja became too frequent and tense.)

Emotionally based support of this nature is common in African apes. Chimpanzee politics—a phrase used advisedly by Frans de Waal, whose book of the same name was assigned as reading homework in 1994 by Newt Gingrich to Republican members of the U.S. Congress—is rife with such alliances, which are often the mechanism by which alpha males are propelled to power. Primatologists know that among status-striving chimpanzees, favors are remembered, as are rejections and snubs.

It is fascinating to track the developmental roots of empathy and meaning-making by studying infant and juvenile African apes. From birth, all great ape infants and their mothers participate in a dance of contingent mutuality; led by the mother, these partners adjust to each other's actions minute by minute. Gradually, as the infants shift from uncoordinated body movements to postural and gestural communication, they come to learn how to be a full partner in the dynamic dance. One of Goodall's key points has long been that maternal style—confident and competent, uncertain and neglectful, or somewhere in between—and maternal personality have enormous effects on the developing child's personality and life outcomes. Research with apes living in enriched conditions (in close interaction with humans and objects from human life) tells us that rearing environment can unlock latent (and astonishing) capacities in African apes. Bonobos Kanzi and Panbanisha, now housed at the Great Ape Trust in Iowa, communicate about abstract concepts (what "good" and "bad" behavior means, for example) through use of symbols called lexigrams, and comprehend human speech to a degree that has surprised many. These bonobos participate in, and indeed cocreate, cultural routines in their chimpanzee-human community, and follow rules for appropriate behavior established by that community.[16]

The behavior of African great apes reviewed in this section establishes an evolutionary platform for the later development of the human religious imagination. Evolutionarily speaking, what happened next?

Human Evolution

Humans and the African great apes shared a common ancestor. About six million years ago, the two lineages, ape and human, split apart and began evolving in different directions. The humanlike species just on the other side of that split, the earliest hominids, are shrouded in mystery. We know something of their anatomy; they were bipedal creatures with ape-sized brains. About their behavior and culture, we know almost nothing. Yet the behavior of African great apes reviewed in the previous section may serve as a useful model for early hominid behavior. Not only are we dealing with our closest living relatives in the African great apes, but, more significantly, with species that show very well-developed emotional expression as well as cognitive and communicative skills (see note 12).

Naturally, some risks exist with projecting present-day ape behavior back into prehistory, but most biological anthropologists accept them as small. Certainly, the risks are smaller than those associated with an alternative that was popular in the earlier days of anthropology, using modern-day forager societies to model the early hominids. *All* modern humans, including foragers (hunter-gatherers), have evolved emotional, cognitive, and linguistic skills far in excess of anything that early hominids might have possessed; mapping of modern forager lifeways directly onto the lives of extinct hominids is a comparison not only troubling and potentially offensive, but certain to be inaccurate.

By contrast, it is overwhelmingly likely, given the robust distribution of behaviors across ape species and environments, that early hominids would have shown *at least* the same level of emotional attachment, meaning-making, empathy, and compassion as today's African apes. The idea, then, is to use African great apes to model the *minimal* capacities of early hominids. From that platform, and keeping in mind a definition of religion as practice, a gradual development of the human religious imagination throughout the six-million-year period of human evolution can be traced.

For decades, anthropologists thought that the famous "Lucy" fossil, termed *Australopithecus afarensis*, represented our earliest human ancestor. Discovered in Ethiopia in 1974 by Don Johanson, Lucy was dated to 3.2 million years and declared to be at the very root of our hominid family tree. Now, over thirty years later, the time span of human ancestry has nearly doubled. The oldest fossil hominid is now taken to be *Sahelanthropus tchadensis*, found in Chad and dated to 6–7 million years ago. Even skeptics who contest the hominid status of *Sahelanthropus* recognize that good evidence exists for hominids back to 5 or 6 million years ago.

For the first half of this sweeping period—until about 2.5 million years ago—our sole window on hominid social life is through the ape evolutionary platform. Extremely few tangible artifacts inform our understanding of this early period. Then, Lucy and her kind gave way to hominids classified into our own genus *Homo*, when individuals began to make stone tools regularly and use them to process animal carcasses. The brain expanded beyond the ape level and, gradually, toolkits became more complex and reveal to us today a hint of underlying social processes.

Hand axes, for example, were fashioned by hominids called *Homo ergaster* who lived in Africa starting around 1.8 million years ago. Teardrop-shaped and splendidly symmetrical, hand axes were made with an imposed form. Although it is difficult to ascribe a specific symbolic meaning to this symmetry, it does seem clear that, for hominids, form mattered over and above function—form *symbolized* something important so that individuals communicated in some way about preserving it over time.

Indeed, the form of hand axes is so preserved as to appear static over a very long period. In parallel fashion, hominid brain size remained stable (except where body size increased). Yet it is important not to think of early *Homo* as somehow primitive or stuck in time. This is the species that first migrated out of Africa and into Asia. Small groups began to move year by year, a little at a time and perhaps without any purposeful planning, but heading inexorably north. As they traveled, early *Homo* faced dangerous predators, a harsh climate, and all the other environmental challenges a fierce world presents. What would make us think that these hominids inhabited a social world any less emotionally complex—full of empathy, meaning-making, and compassion, as well as violence and brutality—than that of today's African apes?

At around 600,000, brain size leaps upward[17] and cultural changes begin to accrue more rapidly. Hominids controlled fire, and fashioned sophisticated weapons, such as spears that were effective aids in hunting big game like elephants and mammoth. Here and there can be found a few hints of recognition of self, as well as an appreciation of aesthetics, in stones or rocks that were altered and curated by hominid groups.

By the latter period of the Neandertals (who lived 125,0000–27,000 years ago) and early *Homo sapiens* (who originated at 200,000 years ago), a turning point had been reached in the expression of a spiritual sense. Despite popular myth, Neandertals were not shambling cavepeople. Coexisting with early human populations, rather than giving rise to them, Neandertals, at certain locations throughout Asia or Europe, buried their dead in symbolic graveside rituals and adorned themselves with jewelry in an expression of personal identity. Archaeological evidence and theory are invaluable here: Steven Mithen paints a fascinating picture of Neandertals as steeped in music-making and other artistic emotional expression, and Brian Hayden argues provocatively that Neandertal rituals are spiritual because they involve some contemplation of an afterlife.[18]

Despite the complexity of their behavior, the Neandertals went extinct, outcompeted by *Homo sapiens:* ourselves. We see in *Homo sapiens,* starting at about 35,000 years ago, a vivid expression of emotion- and symbol-laden ritual relating to otherworldly realms. The double-child burial in Russia at about 28,000 years ago is a superb example. One child was a girl, age nine, the other a boy about thirteen. Each was covered, in death, with red ochre and buried with spectacular grave goods. Most startling are the thousands of ivory beads, probably sewn onto clothes; also included were bracelets, ivory animal carvings, ivory pins, and disc-shaped pendants.[19] Archaeologist Randall White estimates that the two children's beadworks alone would have taken many thousands of hours of labor to complete.[20] No individual, indeed no single family, could have orchestrated the double-child burial. A social network of some sort, infused with emotional connection between its members, must have been in place.

The tricky part, of course, is to connect the symbol usage in early *Homo sapiens* ritual with a sacred realm. Evidence for spiritual awareness, and even shamanic activity, seems to me very strong by the time of the cave painters at Lascaux, Chauvet, and Altamira in the France-Spain region. Images were sometimes painted in "deep caves," hard-to-access and dark areas, and quite likely to induce an emotional response in observers. At Lascaux, the so-called "shaft drawing" may be interpreted convincingly in shamanic terms: a crudely rendered male human with a birdlike head and prominent erection leans back at an angle, next to a bird perched on some kind of pole or stick. The figure's crude quality separates it from other, far more sophisticated animal images; the figure may well be a shaman depicted in an ecstatic trance (birds and polelike staffs are both often associated with shamans).[21] Other prehistoric artwork, too, illustrates a tantalizing, blended human-animal theme.[22] From Chauvet Cave, also in France, comes a striking image of a painted half-man/half-bison, and from Hohle Fels Cave in Germany, a half-human/half-lion figurine made of mammoth ivory.

Evolution and Emergence

This whirlwind tour of six million years of prehistory[23] suggests that our species' past can be described by a gradual evolution toward a full flourishing of collective, symbolic ritual in emotional connection with the sacred. In recognizing this, it helps to understand hominids not just as bipedal walkers and club-wielding mammoth-hunters, but also as feeling persons with intense emotional ties: mothers who sang to their babies; brothers who play-chased across the savanna; men who grieved when their kin or their friends died. Daily life was, in prehistory as it is now, emotional life, even if the exact nature of the emotions involved cannot be known. (I would not wish, for instance, to project onto Neandertals the notion of love; love is an emotion with meaning created in a specific time and place, and is not safely propelled into the past.)

My key claim in this chapter is that prehistoric spirituality is rooted in an ever-deepening capacity for hominids to relate to each other and to the larger world, first through routine daily behaviors and eventually through symbolic ritual. As hominid minds—or more accurately, the minds of *some* hominid individuals—became better able to project themselves beyond everyday survival, perhaps at graveside rituals or in dark caves, a tendency resulted to seek connection with supernatural beings or forces suspected to influence aspects of life, including what was not understood or what was feared.

As Barbara Smuts puts it in another context, "Minds with a degree of conscious awareness may respond to changes in their environments in different ways and/or in more innovative ways than less conscious minds. If some of these creative responses are adaptive . . . they can become more fixed in the population. Individual creativity could thus influence selection, and selection, in turn, could favor increased capacities for awareness and innovation."[24] Innovation of this sort, over the sweep of human evolutionary history, clearly included a turn toward the spiritual—although I prefer to talk about *mutual* creativity, meant to include embodied and emotional creativity (as I believe Smuts intends as well), in keeping with a focus on mutual meaning-making in primate evolutionary history.

My key question is whether these changes over time can be understood through a conventional (insightful) evolutionary account alone, or whether emergence adds significant explanatory power. Is the origin of religion predictable from the changes over time traced in this chapter?

Perhaps, as outlined in Stanley Greenspan and Stuart Shanker's book *The First Idea,* as brains evolved, emotional communication within nurturing contexts gradually became more nuanced and complex. According to this view, emotions *pull along, make possible,* or *unlock* creative cognitive advances. Greenspan and Shanker explain how emotional signaling in the context of nurturing care drives the development of ever-higher levels of thinking. This idea puts a different twist on a conclusion reached by Goodenough and Deacon: "Symbolic cognition . . . precipitates a cascade of reorganizational cognitive and coevolutionary events that eventually produce a brain with a capacity for the kind of mindfulness, intersubjective projection, aesthetic sensibility, and empathy that is now possible."[25] Whatever is considered to be the precipitating factor, it is clear that cognition and emotion must be considered together through the sweep of human evolutionary history.

Indeed, it is the creative power of sophisticated thinking and emotional nurturing that exposes the fatal flaws in simplistic reductionist thinking. Some theorists insist that there's no need even to consider unpredictable emergents, because lower-level structures adequately explain what happens at higher levels. Dean Hamer claims to have uncovered a specific genetic allele that contributes to spiritual tendencies in modern human populations. Though he backpedals from any outright genetic determinism, Hamer packaged his ideas as *The God Gene*. And in *Religion Explained*, Pascal Boyer suggests that religion is a mere by-product of the way our brains evolved. Boyer understands emotion to be created and experienced within each individual's mind; feelings are "the outcome of complex calculations that specialized systems in our minds carry out in precise terms." In turn, though these feelings may be *expressed* in social contexts, including religious contexts, religion itself is a "mere consequence or side effect" of the human brain.[26]

Yet, as we have seen, the human religious imagination cannot be divorced from the realms of social expressions of emotion. The emotion connected with religious experience is not just *felt* between individuals but *created* between individuals; it is about love and longing (or fear and anxiety, or all four of these together) as *enveloping experiences*. We humans have evolved to create, through interaction, these emotions in ever-more-complex ways over time.

Here we encounter the second set of questions, the set for which I cannot confidently offer complete answers. What sort of tension exists in coming at the origins of religion via evolution or via emergence? Can the push factor of creative mutuality, and its effect on symbolic practice in ritual, be said to somehow predict a shift from the relating between individuals to the relating between people and supernatural beings? Is this another way of characterizing a conceptual divide that Antje Jackelén ("Creativity through Emergence: A Vision of Nature and God" in this volume) describes in terms of "nature is enough" versus "nature is not enough" to explain all aspects of our world, including religious awe?[27]

Working to understand *how* religion has evolved throughout time is of signal importance in grappling with issues involving emergence. Adopting an evolutionary perspective that insists on the importance of shifts in nurturing and relating as well as in genes and brains allows us to think critically about emergence in a new light. It is less helpful merely to assert a role for emergence in the development of religion than to try to distinguish which expressions of the religious imagination meet Clayton's definition of emergent properties as unpredictable, irreducible, and novel.

Is awe in the face of the universe's mysteries, or love for God or gods or spirits, unpredictable, irreducible, and novel in the sweep of human evolutionary history? Novel, yes, certainly; but unpredictable and irreducible? I don't know. I do know that many of the building blocks for religion were in place very early on: emotional connection and communication with the social partner together with some degree of meaning-making, compassion, empathy, and a release from immediate here-and-now thinking. As brain size increased and, indeed, as the brain reorganized, what had previously been islands of behavior now came together into a coherent and complex whole. That hominids engaged in graveside and deep-cave rituals seems to me to flow naturally from, indeed to be created by, the trajectory of primate and human emotional evolution. Does it tell us anything new to call these *emergent?*

Let's ask the question more positively: When is it most useful to speak of emergent behaviors in the prehistory of religion? Perhaps when Neandertals or early *Homo sapiens* began to engage with other worlds, fully apart from the here and now; or when shamans began to communicate between the everyday world and a spirit world; or when humans began to build huge temples (eleven thousand years ago in Anatolia, present-day Turkey) or monumental pyramids intimately connected to visions of an afterlife (fifty thousand years ago in Egypt); or when the major organized religions began to codify not only practice but also belief?

Or perhaps it is enough to say, as I wish to do, that significant aspects of the human religious imagination developed from nature. Rooting religion in nature is the opposite of a simplistic reductionism. The nature of which I speak is deeply social, emotional, and creative, and I see this as no paradox at all.

Notes

1. F. de Waal, *Our Inner Ape*, 182–83.
2. For a full description of this and other complex behaviors observed in West African chimpanzees—still less well known than Jane Goodall's East African variety—see Boesch and Boesch-Achermann, *The Chimpanzees of the Tai Forest*; event involving Tina reported on 248–49.
3. See C. Geertz, *The Interpretation of Cultures*, 90.
4. R. Rappaport, *Ritual and Religion in the Making of Humanity.*
5. M. Buber, *I and Thou*, 78, 61.

6. Zaleski and C. Zaleski, *Prayer: A History*, 5.
7. U. Goodenough and T. W. Deacon, "The Sacred Emergence of Nature," 854.
8. Clayton distinguishes between strong and weak emergence; see his *Mind and Emergence*; quoted material is from 39.
9. My thanks to Ursula Goodenough for making this point to me about the relationships between material structures.
10. From an interview in *Time* magazine, "Is God in Our Genes?" (October 25, 2004), http://www.time.com/time/magazine/article/0,9171,995465,00.html.
11. For a compelling look at bonobo behavior with an emergence framework, see B. Smuts, "Emergence in Social Evolution: A Great Ape Example."
12. Readers interested in ape behavior and cognition may consult works noted in the text by Christophe Boesch and Frans de Waal, as well as my own. Good edited volumes include Anne Russon and David Begun, eds., *The Evolution of Thought* (Cambridge: Cambridge University Press, 2004), and Sue Taylor Parker et al., eds, *The Mentalities of Gorillas and Orangutans* (Cambridge: Cambridge University Press, 1999).
13. Quoted in A. Powell, "Science and Spirituality: Good Chemistry?" *Harvard University Gazette* (October 25, 2001), http://www.hno.harvard.edu/gazette/2001/10.25/16-science.html.
14. B. J. King, *The Dynamic Dance*.
15. I am grateful to Lisa Stevens and the staff at the National Zoological Park's Great Ape House for making this research possible, and to William and Mary undergraduates Ann Hagan, Christy Hoffman, Margie Robinson, Rebecca Simmons, and Kendra Weber for research assistance.
16. Segerdahl et al., *Kanzi's Primal Language*.
17. According to Steven Mithen in *The Singing Neanderthals*.
18. See Mithen, ibid., and B. Hayden, *Shamans, Sorcerers, and Saints*.
19. V. Formicola and A. Buzhilova, "Double Child Burial from Sunghir (Russia)."
20. R. White, posted on Institute for Ice Age Studies Website, http://www.insticeagestudies.com/.
21. See Hayden, *Shamans, Sorcerers, and Saints*, 149.
22. B. King, *Being with Animals*.
23. For a fuller account, see King, *Evolving God*.
24. B. Smuts, "Emergence in Social Evolution: A Great Ape Example," in Clayton and Davies, eds., *The Re-Emergence of Emergence* (Oxford: Oxford University Press, 2006).
25. U. Goodenough and T. W. Deacon, "The Sacred Emergence of Nature," 863.
26. Boyer, *Religion Explained*, 128, 330.
27. See also J. Haught, *Is Nature Enough?*

References

Boesch, Christophe, and Hedwige Boesch-Achermann. *The Chimpanzees of the Tai Forest*. Oxford: Oxford University Press, 2000.

Boyer, Pascal. *Religion Explained*. New York: Basic Books, 2001.

Buber, Martin. *I and Thou*. New York: Touchstone, 1970.

Clayton, Philip. *Mind and Emergence*. Oxford: Oxford University Press, 2004.

De Waal, F.B.M. *Our Inner Ape*. New York: Riverhead Books, 2005.
Formicola, V., and A. Buzhilova. "Double Child Burial from Sunghir (Russia)." *American Journal of Physical Anthropology* 124 (2004): 189–98.
Geertz, Clifford. *The Interpretation of Cultures*. New York: Basic Books, 1973.
Goodenough, Ursula, and T. W. Deacon. "The Sacred Emergence of Nature: Reduction and Emergence." In P. Clayton and Z. Simpson, eds. *The Oxford Handbook of Religion and Science*, 853–71. Oxford: Oxford University Press, 2006.
Greenspan, S. I., and S. G. Shanker. *The First Idea*. New York: Da Capo Press, 2004.
Hamer, Dean. *The God Gene*. New York: Doubleday, 2004.
Haught, John F. *Is Nature Enough?* Cambridge: Cambridge University Press, 2006.
Hayden, Brian. *Shamans, Sorcerers, and Saints*. Washington, D.C.: Smithsonian Institution Press, 2003.
King, Barbara J. *Being with Animals*. New York: Doubleday, in press.
———. *The Dynamic Dance*. Cambridge, Mass.: Harvard University Press, 2004.
———. *Evolving God*. New York: Doubleday, 2007.
Mithen, Steven. *The Singing Neanderthals*. London: Weidenfeld and Nicolson, 2005.
Morowitz, Harold J. *The Emergence of Everything*. Oxford: Oxford University Press, 2002.
Rappaport, Roy. *Ritual and Religion in the Making of Humanity*. Cambridge: Cambridge University Press, 2002.
Segerdahl, E., S. Savage-Rumbaugh, and William Fields. *Kanzi's Primal Language*. New York: Palgrave McMillan, 2005.
Smuts, B. B. "Emergence in Social Evolution: A Great Ape Example." In Philip Clayton and Paul Davies, eds., *The Re-Emergence of Emergence*. Oxford: Oxford University Press, 2006.
Zaleski, Philip, and Carol Zaleski. *Prayer: A History*. Boston: Houghton Mifflin, 2005.

7

WHO NEEDS EMERGENCE?

Gregory Peterson

In the past decade, there has been a renaissance of scholarship on emergence.[1] The reason is straightforward enough. On the one hand, the theories and data of contemporary science seem to leave little room for the traditional dualisms that once formed our vision of things, with biochemistry seeming to dissolve the difference between life and nonlife and neuroscience dissolving the difference between mind and brain. On the other hand, holding that "nothing but" atoms exist seems to quickly make nonsense of the world as we experience it on the everyday level and, taken to its extreme, arguably hinders as much as helps the scientific endeavor that is understood to support the *nothing but* thesis. Emergence, then, provides a middle ground, getting "something more" from "nothing but," proclaiming, in Aristotle's famous phrase, that the "whole is more than the sum of its parts."[2] A good deal of recent effort has been devoted to explicating what exactly these slogans might legitimately be taken to mean, with the result that there is more clarity about what is being talked about than when the usage of the concept first started to seriously spread.[3] This refining and explication have significantly aided an understanding of what is at stake in the debate, even though significant disagreement remains. Part of this disagreement, I would suggest, stems from the disciplinary perspective involved. The implication of this is that the answer to the question of whether or not emergence is important is, "It depends." Specifically, the importance of emergence varies, depending on the area of inquiry that we are speaking of; the form of emergence that is utilized also varies correspondingly.

Speaking of Emergence

The modern discourse on emergence has roots in the writings of G. H. Lewes and John Stuart Mill in the late nineteenth century, influencing the blossoming of the British emergentists of the early twentieth century. This blossoming proved to be short-lived, probably due both to changes that occurred in the sciences and to the nascent winds of positivism arriving on the scene.[4] Although scholars supporting emergence language persisted in the middle twentieth century, interest was reinvigorated in the 1970s, which saw the publication of several volumes relating to the topic and which acquired eloquent advocates in Nobel Prize–winning neuroscientist Roger Sperry and in biochemist turned theologian Arthur Peacocke.[5] It is notable that much of the interest in emergence has been spurred by reflection on issues that arise in the life sciences and, especially, on the problem of understanding the relation of mind-brain-body, with the primary concern being categories of explanation, ontology, and causality. This renewed interest has only accelerated in the 1990s and the early 2000s, and the enthusiasm on the part of some is probably best suggested by the title of Harold Morowitz's book, *The Emergence of Everything: How the World Became Complex*, which claims that there are at least twenty-eight levels of emergence in the cosmos, explaining virtually all the major transitions of natural history and complexity, from simple molecules to complex human societies. As usage of the term *emergence* spread, so too did its meaning and its purported implications. At the same time, criticisms have started to appear, suggesting either that the emergence thesis is mistaken or that the implications taken to be most important do not actually follow logically from the concept.[6]

The claim that the whole is more than the sum of its parts admits of a number of interpretations. From the perspective of scientific explanation and description, the emergence intuition arises out of the simple need to explain the physical world in a way that is both useful and intellectually satisfying. While it may be true that foxes and rabbits are both composed of atoms, this tells us almost nothing interesting about either foxes or rabbits or what makes them different, and it does not suggest anything about their potentially complex relations as predator and prey. "Fox-ness" and "rabbit-ness" involve something more from an explanatory perspective, and to understand both foxes and rabbits in a deep way involves knowledge of biochemistry, anatomy, evolutionary history, and ecological niche. Similarly, a scientific understanding of the human person involves not only knowledge of the ten trillion or so cells that compose the human body, but also how they are arranged and organized, and the way that lower levels of organization and complexity give rise to higher levels (thoughts, desires, emotions) that, in turn, seem to loop down on the very physical structures that gave rise to them, suggesting an understanding of causation that is not only "bottom up" but also "top down." This sort of analysis easily lends itself to a hierarchical way of viewing the world, with the more complex arising out of the less, like a set of Russian *matryoshka* dolls. The interactions of individual subatomic particles give rise to atoms, which, in turn, conjoin with one another to form molecules, making possible the basic structures of life—subcellular structures, cells, multicellular organisms, complex multicellular organisms, and eventually human beings, who, in turn, form complex societies, produce cultures and political structures, and engage in economic enterprises, each activity producing patterns that may themselves be understood to be emergent. This hierarchy of the sciences, both understood as a relation of disciplines and of the realities that they describe, has sometimes been put forth as a key feature of the emergentist perspective, suggesting the seamless links between stages of evolution and the connections between the physical and the complex structures that the physical gives rise to.[7]

It is at this point that differences start to develop, as the import given to such phrases as *top-down causation* and *nonreducible wholes* varies considerably. Categorizing these differences is not so easy, but there seems to be some consensus for a threefold division in usage, and Philip Clayton has suggested the labels *façon de parler*, weak, and strong to categorize these three forms.[8] As the label of the first form of emergence suggests, *façon de parler* emergence is not really emergence at all. Rather, it is the reductionist's shorthand way of speaking of complex phenomena. By this account, wholes really are nothing but the sum of their parts, understood either to be identical to the parts or eliminable in favor of them. Daniel Dennett may be taken as an advocate of this kind of view. Although Dennett is careful to distinguish between *good reduction* and *greedy reduction*, he is also insistent in the claim that there is nothing else but the physical and that, when push comes to shove, lower levels get explanatory priority over higher levels. Although one may speak of "real patterns," these patterns are best understood as a kind of useful fiction. Only the bottom level carries ontological and causal weight in the final analysis.[9]

By contrast, weak emergentists claim some ontological and causal significance to the whole, often under the aegis of nonreductive physicalism (NRP). A general feature of NRP is to accept that physics and chemistry are essentially complete, and that the world is causally closed. Despite this, the advocate of NRP will argue that most efforts at reduction fail, and that one can speak of wholes as real and having top-down powers that influence the behavior of the constituents of which the whole is made. These issues have been of central concern to the philosophy of mind where the stakes have been highest. Clearly, the argument asserts, the mind arises out of the activity of the neurons of which it is composed, but this should not be taken to suggest that the mind is a causally inert epiphenomenon. Rather, we might think of the causal force of mental activity as analogous to that of the software of a computer: Although the software does not exist independently of its instantiation in a computer or set of computers, it nevertheless plays a clear and obvious causal role in the computer's operations. Software is just as real as hardware, mind is just as real as brain, and both exert a real and significant causal force.

Although the intuitions behind this view are strongly attractive, there has been considerable difficulty in making the logical relations work, as implied in NRP. If mind just supervenes on the brain, it is not clear that it can have any causal role except in a merely metaphorical sense.[10] Since the laws of physics cannot be contradicted and the law of conservation of energy must be obeyed, it is not clear how the mind's causality is anything more than the sum of the causes of the neurons of which it is composed. Nevertheless, if some successful account were able to be given, it would provide a kind of very interesting middle ground that recognizes the reality and significance of complex objects (including the human mind) without having to add anything extra or mysterious to the physical world.[11]

This leaves the third option, strong or radical emergence. Strong emergence suggests that, at least in some cases and under certain circumstances, new realities and causal powers emerge that are not simply the sum of their constituent parts. Ultimately, strong emergence rejects the claims of causal closure and completeness: There are realities beyond those of the lowest level of physical matter, and these cannot be simply explained by or derived from these lowest levels. What makes strong emergence a form of emergence rather than simply an ontological dualism or pluralism is the assertion of a dependency relationship—the physical lower level is still required for the higher-level complex entity to exist; it is just not sufficient for the higher-level complex entity's existence. Something more—some new law, causal force, or category of being—emerges as a result of the complex interactions of lower-level particles and causal laws.

It is important to note that strong emergence is indeed radical, and might be better labeled as such. Strong emergence not only rejects a simplistic atomism, it also rejects a narrow physicalism, at least when physicalism is understood to entail the claim that only those things that have already been discovered by science can be said to exist. One would expect examples of strong/radical emergence to be rare, and it is noteworthy that few such examples are given in the literature, one being the Pauli exclusion principle that governs the behavior of electrons as part of the structure of the atom; the other being the human mind, especially the property of consciousness.[12] In the case of the Pauli exclusion principle, the evidence for strong/radical emergence stems from the well-established nature of the principle and its inderivability from current, more fundamental physical laws. Similarly, the evidence for the human mind as an instance of strong emergence stems from the evidence provided by introspective experience and the apparent lack of derivability from existing principles of neuroscience. As these examples might suggest, one potential danger of the position of strong/radical emergence is that it becomes an "emergence of the gaps," invoked to explain any unexplained phenomenon, in which case it would be guilty of simply being an argument from ignorance. At its best, however, strong/radical emergence would be something very different, a reality posited not just because of an existing gap, but because of positive evidence that suggests the difficulty or even impossibility of deriving from lower-level entities and laws.

These characterizations are very brief, but they should give some indication of the ideas at play when emergence is discussed, and it is easy to see why the language of emergence is so attractive to many who are engaged in thoughtful reflection on the world as viewed by the sciences. On the one hand, the past centuries have seen the tremendous and ever-expanding success of the scientific endeavor. Physical principles clearly underlie all of life, and the scope of physical explanation can inspire a kind of awe. On the other hand, the reductive drive of science can be threatening, perhaps not so much when we say that a cell is just an arrangement of molecules, but certainly when we say that a human person is just a bundle of cells, for we are inclined to say that there is more to being human than to speak of just the cells we are composed of, as this does little to distinguish one organism from another. Emergence satisfies both of these urges, acknowledging the importance of scientific discovery, while suggesting that the reductive story is not all there is.

Acknowledging this, however, leaves out a crucial point, which is that the importance of emergence and the importance of the different kinds of emergence may well vary considerably due to the given task at hand, and it is this variability of the intellectual and social context that may explain some of the important diversity of usage of the concept.

Emergence and Scientific Practice

While many scientists have prominently advocated an emergentist perspective on the world, much of this is done in the context of reflecting on the meaning and significance of science, not in the context of actual scientific practice. In the case of Sperry and Peacocke, both engaged in their reflections on emergence in the latter stage of their careers, and did so in no small part to make sense of the scientific work that they had participated in. This is a different matter, however, from claiming that emergence is important for the practice of science and active scientific theorizing. Such a claim is not terribly difficult to make. Scientists regularly engage in analysis of wholes and parts, and while sometimes analysis solely in terms of the parts is important, a good deal of scientific work involves understanding the very complex ways that wholes and parts interact and form complex, tangled networks. I take this to be at least a partial point of the contribution in this volume by Robert Ulanowicz ("Enduring Metaphysical Impatience?"), who suggests that proper scientific theorizing requires the positing of dynamic, interlevel networks that can be characterized by the importance of structure, boundary constraints, and feedback loops. The satisfactoriness of such multilevel theorizing is connected to categories of predictive power and empirical fruitfulness. A clear example of this is the rise of cognitive approaches in psychology, which have largely replaced old-school behaviorism, a research program that came to be seen as too drastically limited precisely because it was unwilling to posit emergent phenomena (mind, emotions, representation) that were plainly useful and relevant for furthering psychological research.

Although this analysis suggests that practicing scientists are often engaged in an emergentist framework as part of their work, it is much less clear that scientists need to be invested in a way that has philosophical significance and that distinguishes between the three primary senses of emergence outlined above. In terms of the practice of science, scientists may easily endorse either a *façon de parler* or a weak account of emergence for any given scientific theory involving interlevel explanation (e.g., between a cell as a whole and the parts of which it is composed) with little impact on the outcome of their work. This indifference to the philosophical implications is perhaps exemplified by Willem Drees' article in this volume ("The Nature of Visions of Nature: Packages to Be Unpacked"), which relegates emergence to a "useful notion within science," but one with limited applicability outside of that context.[13] It is also perhaps illustrated in the several articles by Terrence Deacon and Ursula Goodenough on the subject.[14] Deacon and Goodenough suggest three stages of emergence, beginning with thermodynamics, from which can arise morphodynamics (shaping interactions involving complex structures) and teleodynamics (involving semiotics and meaning). What is interesting about their approach is, at least by comparison, the relative lack of interest in those categories that most occupy philosophers when engaging the topic, whether and how emergent entities are real, and whether and how they can be said to have top-down causal force. Rather, these researchers' interest is in what might be understood as a more obviously scientific one, informed by a desire to understand how it is that such emergent systems come to be and how they might work.

Furthermore, there is at least one usage of emergence language that is, from the scientific perspective, potentially problematic, and that is when emergence language is itself used without any supporting justification. Assertions about human consciousness as emergent sometimes take this form when the invocation of emergence is used as an explanation in a way that suggests that nothing more needs to be said. In this case, the problem is not the claim that consciousness is emergent, even strongly emergent, but the implied accompanying suggestion that no further scientific research need be done on the matter. In this instance, emergence simply becomes an emergence of the gaps.

Put another way, I would suggest that the practice of science typically employs categories of emergence primarily because they are useful and only secondarily because they might be true. This is not to say that the sciences do not imply specific conceptions of emergence (think of the Pauli exclusion principle as a case of strong emergence), but that, from the scientific perspective, the ontological distinctions are less important. What is important is the success, testability, and fertility of the scientific theories. The sticky details of ontology can be safely left to others.[15]

Emergence, Philosophy, and Theology

By contrast, philosophers are keenly interested in both the ontological and causal questions, and typically see the two as linked: To be is to do. In the philosophy of mind, this has been significantly manifested in the literature on supervenience, which has tried to explicate how weak emergence in the guise of nonreductive physicalism can preserve the reality and causal efficacy of mind without denying causal closure and ontological completeness at the level of physics and chemistry. An acceptable solution has proved elusive, and it has been argued by some longtime participants in the debate that a satisfactory solution cannot be found.[16] The amount of effort expended by philosophers on supervenience and emergence in the context of philosophy of mind also suggests the different interests and concerns at play. Whereas scientists are much more inclined to seek to identify the processes that give rise to emergence and to develop models that can produce testable results, the philosopher is focused on implications that reductionist and emergentist claims have for a broader worldview and in terms of ethical import. As a result, the category of emergence becomes centrally important for philosophy of mind, for if physicalism holds, it is philosophically crucial to find some way to speak of the reality and causal relevance of thoughts, motives, and conscious reflection in a way that is compatible with the implications of a physicalist perspective. It is much less crucial from a philosophical perspective to ask how nature produces such emergent complexity or how the philosophical account can lead to testable results.

Correspondingly, the distinction between weak and strong emergence looms large in the philosophical debate and is strongly tied to intuitions about the satisfactoriness of a physicalist worldview and, in particular, the satisfactoriness of computational models for explaining human consciousness and free will. Indeed, in the philosophical discussion the emphasis on the mind-body problem dwarfs more general reflections on emergence, and is perhaps indicated by David Chalmers' claim that there is only one instance of strong emergence, and that is the human mind.[17]

The interest of theology in the emergence debate is somewhat different, as the focus of theology is typically different from that of philosophy. There is certainly overlap: Philosophy, as traditionally defined, includes epistemology, metaphysics, and ethics, disciplines also of concern to theology. Theology as a discipline takes the perspective of the whole, attempting to chart a picture of the world that relates to central questions of meaning and purpose, a task that contemporary philosophy, at least, has become less interested in. In the case of the Western monotheistic traditions, theology is centrally concerned with understanding and explicating the reality of God and God's relation to human beings and the world. Encompassed by this is some account of theological anthropology, an understanding of the place and significance of human beings and prospects for salvation, redemption, and the good life. This would suggest that theology has a great deal at stake in the emergence debate, at least insofar as theological anthropology is concerned with a more profound understanding of creation and of human nature, and theologians have expressed considerable interest in philosophies of emergence, as witnessed by the many theologians who have written on the subject.[18]

In the specific context of Christian theology, there seem to be two likely points of further interest. The first concerns the overall relation of God and the world, which was a topic of interest to C. Lloyd Morgan, one of the early British emergentists.[19] If the relation of God to the world is analogous to the relation of human mind to body, then it can be argued that just as the mind emerges from the activities of body and brain, so, too, can God be understood as an emergent reality of the world. Although such an analogy has been employed, it is significantly in tension with much of the theological tradition. Why? Because God would seem to be either an epiphenomena (if the relation were characterized by weak emergence) or possibly a causal agent or force, but one that is dependent on the world and not the reverse, in which case God could not be creator and perhaps not redeemer, either. Indeed, if the concept of emergence were used to explain the God-world relation at all, it would seem that reversing the relation would be the more interesting move, to understand the world as emergent from (and therefore dependent on) God. This approach would be much more consonant with the panentheism held by many in the science-theology dialogue, although it makes the connection between God and the world too tight for some, as we would all be part of/emergent from the Divine.

A second concern would focus on the significance of emergence for understanding nature and whether emergence might function as a vision of nature. Here, the theological interest is potentially greater than the philosophical one, as a natural world that is itself capable of producing emergent novelty is different from one that requires divine action for each step of novelty and complexity. Acceptance of the capacity of nature to produce emergent novelty is, in a significant sense, what separates advocates of intelligent design from those who accept the broad sufficiency of the evolutionary account, as the former specifically deny the capacity of nature to produce the many and varied forms of complex life forms that we see around us. In this context, emergence speaks to the sufficiency of nature and the role that God plays in natural processes, and indeed what kind of God we are speaking of—one who constantly intervenes or one who gifts the natural world with some power of creative novelty, impacting how theists view the goodness of creation and matters of environmental care. Interestingly, the theological interest here is perhaps more similar to the scientific one than the philosophical, emphasizing not so much the fact of emergence but how emergent structures may be said to come about through natural processes alone.

The essay in this volume by Martha Henderson ("Rereading a Landscape of Atonement on an Aegean Island") hints at this kind of approach, although she is writing as a geographer, not as a theologian. While her thesis is informed by the science of self-organizing complexity, and so by the scientist's concern with emergence, her approach is clearly suggestive of the theologian's, to re-envision the relation of land and community in a way that is different from her predecessors' views. Like the theologian's approach, Henderson's has clear valuational implications, implying that an emergentist view allows us to see the relation of people and land as other than a story of degradation and environmental harm.

Conclusion: An Emergent Vision of Nature?

Emergent complexity runs through the natural world, and the history of our planet, indeed the cosmos, may be understood as the evolution of emergent and complex structures and life forms, including not least the human species. The category of emergence may be understood to provide a vision of nature, a frame through which to view the world, albeit not the only one, and perhaps not the best. Emergent nature requires an evolutionary nature, and both, in turn, may be seen as contributing to malleable nature. Furthermore, there is not one kind of emergence, but many, and the implications and significance of theories of emergence depend on the task at hand. This realization might be taken to suggest that theories of emergence are not so much end points as beginnings. To label something *emergent* by itself tells us very little. It is the explicating of emergence that is the real task, and one that depends strongly on the disciplinary concerns at play.

Notes

1. Mark A. Bedau and Paul Humphreyes, eds., *Emergence: Contemporary Readings in Philosophy and Science* (Cambridge, Mass.: Bradford Books, 2008); Philip Clayton and Paul Davies, eds., *The Re-Emergence of Emergence: The Emergentist Hypothesis from Science to Religion* (New York: Oxford University Press, 2006); William Stoeger and Nancey Mur-

phy, eds., *Evolution and Emergence: Systems, Organisms, Persons* (New York: Oxford University Press, 2007).

2. Terrence Deacon and Ursula Goodenough, "From Biology to Consciousness to Morality," *Zygon: Journal of Religion and Science* 38 (2004): 801–19; Aristotle, *Metaphysics*, 8.4.
3. Cf. Philip Clayton, *Mind and Emergence: From Quantum to Consciousness* (New York: Oxford University Press, 2004); Gregory R. Peterson, "Species of Emergence," *Zygon: Journal of Religion and Science* 41 (2006): 689–712.
4. Much of this history has been recounted in Clayton, 2004.
5. Roger Sperry, "Mental Phenomena as Causal Determinants in Brain Functions," in Gordon G. Globus, Grover Maxwell, and Irwin Savodink, eds., *Consciousness and the Brain: A Scientific and Philosophical Inquiry* (New York: Plenum, 1976); Arthur Peacocke, "Reductionism: A Review of the Epistemological Issues and Their Relevance to Biology and the Problem of Consciousness," *Zygon: Journal of Religion and Science* 11 (1976): 307–66.
6. E.g., Donald H. Wacome, "Reductionism's Demise: Cold Comfort," *Zygon: Journal of Religion and Science* 39, no. 2 (2004): 321–37; Jaegwon Kim, "Being Realistic about Emergence," in Clayton and Davies, eds., *The Re-Emergence of Emergence*, 321–37.
7. For a classic explication of the hierarchical view implied by theories of emergence, see Arthur Peacocke, *Theology for a Scientific Age: Being and Becoming—Natural, Divine, and Human*, enlarged ed. (Minneapolis: Fortress Press, 1993). Antje Jackelén ("Creativity through Emergence: A Vision of Nature and God," in this volume) raises an important issue with a too easy embracing of the language of hierarchy, given its association with power and sex/gender hierarchies in broader arenas.
8. Clayton, *Mind and Emergence*. I have elsewhere used a similar division in terms of reductive, nonreductive, and radical emergence. See Peterson, "Species of Emergence."
9. Daniel Dennett, *Darwin's Dangerous Idea: Evolution and the Meanings of Life* (New York: Touchstone, 1995). Emergentists are sometimes accused of setting up reductionism as a "straw man" argument that no one really holds. It may be the case that no one, or at least very few, hold the view consistently, but a survey of the literature very quickly finds rhetorical passages that strongly imply such a reductionism, and the logical implication of a view like Dennett's seems to lead directly to such a reductionism.
10. Cf. Jaegwon Kim, *Mind in a Physical World: An Essay on the Mind-Body Problem and Mental Causation* (Cambridge, Mass.: MIT Press, 2000).
11. Nancey Murphy and Warren Brown have recently made just such an attempt, appealing to a dynamic systems theory showing that the mind cannot be understood simply as being instantiated in the brain but arises out of the interaction of the brain and body with its environment. It is not clear to me that this solves the problem of NRP as much as it widens the scope of what it is that the mind supervenes on. See Nancey Murphy and Warren Brown, *Did My Neurons Make Me Do It? Philosophical and Neurobiological Perspectives on Moral Responsibility and Free Will* (New York: Oxford University Press, 2007).
12. Paul Humphreys, "How Properties Emerge," *Philosophy of Science* 64 (1997): 1–17.
13. Willem Drees, "The Nature of Visions of Nature: Packages to Be Unpacked," in this volume.
14. See Deacon and Goodenough, 2003; Terrence Deacon, "Emergence: The Hole at the Wheel's Hub," in Clayton and Davies, eds., *The Re-Emergence of Emergence*, 111–50.
15. Robert Ulanowicz's essay in this volume ("Enduring Metaphysical Impatience?") suggests that this might not be true for all scientists or all scientific approaches, as he implies that

an emergentist view, in an ontological and causal sense, is integral to how ecologists go about studying the world.

16. Kim, *Mind in a Physical World;* David Chalmers, *The Conscious Mind: In Search of a Fundamental Theory* (New York: Oxford University Press, 1997).
17. David Chalmers, "Strong and Weak Emergence," in Clayton and Davies, eds., *The Re-Emergence of Emergence,* 244–56.
18. Peacocke, *Theology for a Scientific Age;* Murphy and Brown, *Did My Neurons Make Me Do It?;* among others.
19. C. Lloyd Morgan, *Emergent Evolution* (London: Williams and Norgate, 1923).

Bibliography

Bedau, Mark A., and Paul Humphreys, eds. *Emergence: Contemporary Readings in Philosophy and Science.* Cambridge, Mass.: Bradford Books, 2008.

Chalmers, David. *The Conscious Mind: In Search of a Fundamental Theory.* New York: Oxford University Press, 1997.

———. "Strong and Weak Emergence." In Philip Clayton and Paul Davies, eds., *The Re-Emergence of Emergence: The Emergentist Hypothesis from Science to Religion,* 244–56. New York: Oxford University Press, 2006.

Clayton, Philip. *Mind and Emergence: From Quantum to Consciousness.* New York: Oxford University Press, 2004.

Clayton, Philip, and Paul Davies, eds. *The Re-Emergence of Emergence: The Emergentist Hypothesis from Science to Religion.* New York: Oxford University Press, 2006.

Deacon, Terrence, and Ursula Goodenough. "From Biology to Consciousness to Morality." *Zygon: Journal of Religion and Science* 38, no. 4 (2003): 801–19.

Dennett, Daniel. *Darwin's Dangerous Idea: Evolution and the Meanings of Life.* New York: Touchstone, 1995.

Humphreys, Paul. "How Properties Emerge." *Philosophy of Science* 64, no. 1 (1997): 1–17.

Kim, Jaegwon. *Mind in a Physical World: An Essay on the Mind-Body Problem and Mental Causation.* Cambridge, Mass.: MIT Press, 2000.

———. "Being Realistic about Emergence." In Clayton and Paul, eds., *The Re-Emergence of Emergence,* 189–202.

Lewes, G. H. *Problems of Life and Mind,* 2 vols. London: Kegan Paul, Trench, Turbner, & Co., 1875.

Morgan, C. Lloyd. *Emergent Evolution.* London: Williams and Norgate, 1923.

Morowitz, Harold. *The Emergence of Everything: How the World Became Complex.* New York: Oxford University Press, 2002.

Murphy, Nancey, and Warren Brown. *Did My Neurons Make Me Do It? Philosophical and Neurobiological Perspectives on Moral Responsibility and Free Will.* New York: Oxford University Press, 2007.

Peacocke, Arthur. "Reductionism: A Review of the Epistemological Issues and Their Relevance to Biology and the Problem of Consciousness." *Zygon: Journal of Religion and Science* 11 (1976): 307–36.

———. *Theology for a Scientific Age: Being and Becoming—Natural, Divine, and Human,* enlarged ed. Minneapolis: Fortress Press, 1993.

Peterson, Gregory R. "Species of Emergence." *Zygon: Journal of Religion and Science* 41 (2006): 689–712.
Sperry, Roger. "Mental Phenomena as Causal Determinants in Brain Functions." In Gordon G. Globus, Grover Maxwell and Irwin Savodink, eds., *Consciousness and the Brain: A Scientific and Philosophical Inquiry*. New York: Plenum, 1976.
Stoeger, William, and Nancey Murphy, eds. *Evolution and Emergence: Systems, Organisms, Persons*. New York: Oxford University Press, 2007.
Wacome, Donald H. "Reductionism's Demise: Cold Comfort." *Zygon: Journal of Religion and Science* 39, no. 2 (2004): 321–37.

8

CREATIVITY THROUGH EMERGENCE: A VISION OF NATURE AND GOD

Antje Jackelén

Had this book been written some 250 years ago, the most fascinating theme to talk about in terms of a vision of nature would have been nature's conformity to law. The unchanging character of the laws of nature would have been hailed as the gateway to secure knowledge of nature and as the key to the forces of nature so as to make them amenable to human purposes.

The profile of this volume is quite different: A significant number of contributions focus on nature as irregularly dynamic. Thus, this research project reflects major shifts and developments in philosophy, science, and theology throughout the last century. Whereas the philosophy of earlier modernity has been interested in continuity and linearity—expressed in the temporal sense in terms of continuous progress as the ideal, and in the spatial sense in colonization of the rest of nature and the world—the philosophy of later modernity has turned its attention to ruptures and diversity instead. Whereas classical physics, enchanted by the discovery of the universality of natural law, celebrated the continuity of absolute time and space, twentieth-century science has turned much of its interest to discontinuities—be it the significance of quantum leaps, the role of disturbances for the formation of galaxies, the part of bifurcation points in chaotic systems or processes of emergence in complex systems. While not questioning the validity of the laws of nature or the merits of modern Western philosophy in general, the focus of inquiry has moved more toward understanding the processes that bring about discontinuity, change, and novelty. This has led researchers in many fields to ask how such features as relatedness and interactivity bring about the emergence of highly complex systems.

Can emergence talk help us as we today are groping for new visions of nature that include the best of scientific and theological knowledge, that are attentive to the groaning of an environmentally stressed nature, and that can tell the story of becoming we see unfolding in nature? In this essay, I will explore a vision of nature as creation that arises from the concepts of complexity and emergence. The appealing plasticity of the concept of emergence has its counterpart in some obvious difficulties, such as the absence of clear definitions, the lack of the predictability of outcomes and the risk of conflating description and value judgments. In light of my discussion of the possibilities and pitfalls of emergence, I will nevertheless conclude that the concept of emergence, in its concrete as well as in its metaphorical sense, can contribute some specific suggestions as to how to speak well of nature, God, and ourselves. Such talk, in turn, is a prerequisite for responsible action.

The Emergence of Complexity: God or Nature?

How then does complexity emerge—is it nature that does it or is it divine creativity that does it? Framing the question this way implies an antagonism that on and off has kept many good minds busy. Yet, as I will argue by way of discussion of complexity and emergence, this opposition is quite unnecessary. The work of creativity may be adequately described in ways that are both immanent to nature and transcend nature. In fact, views that attribute creativity to both nature and divine energy can be traced back to the early centuries of the Common Era.

In the late fourth century, the church father Basil of Caesarea, one of the so-called three Cappadocians, delivered a series of nine sermons on Genesis 1:1–25. In these homilies, known as the *Hexaemeron*, Basil included a lot of information about botany, zoology, geography, and astronomy, most of which reflects very well the level of scientific knowledge of his time.[1] With amazing ease, he moves between *God* and *nature* (*physis*) as actors in creation. It is nature that "encloses the costly pearl in the most insignificant animal, the oyster";[2] nature has placed the grain of the wheat "in a sheath so as not to be easily snatched by grain-picking birds";[3] nature has placed such powerful organs of voice in the lion "that frequently many animals that surpass him in swiftness are overcome by his mere roaring."[4] Animals follow "the law of nature strongly established and showing what must be done,"[5] and so do humans: We have got "natural reason which teaches us an attraction for the good and an aversion for the harmful . . . implanted in us,"[6] and we have "natural virtues toward which there is an attraction . . . from nature itself."[7] Teachings about social order are not introducing anything new, according to Basil; they are merely a continuation of natural order. When Paul gives directions regarding the relationship between parents and children, he recommends nothing new; he just "binds more tightly the bonds of nature."[8] In this regard, if he were alive today, Basil might even go so far as to actually agree with Barbara King's argument for the development of religious imagination from nature.[9] Nevertheless, there is a point where King and Basil would part company. For him, immanence alone will not do. In spite of the active causal role Basil attributes to nature and the law of nature, he has no problem whatsoever seeing God in the same things. Basil praises the sea urchin for its capacity of forecasting calm or rough waters by its behavior. By this, Basil concludes, "the Lord of the Sea and the winds placed in the small animal a clear sign of [God's] own wisdom."[10] Hence: "There is nothing unpremeditated, nothing neglected by God. . . . [God] is present to all, providing means of preservation for each."[11] For Basil, God apparently acts "in, with and through" nature and there is no contradiction in that. What he observes is both natural and divine; it cannot be reduced to the dualism of either God or nature. Stated in more philosophical terms: God's creativity works in both immanent and transcendent ways. According to Basil's view, processes of complexification can be as natural as they are divine and as divine as they are natural.

Complexity and Emergence: A Paradox Made Plausible?

It is a truism that the notion of complexity is far too complex to be caught in a simple definition.[12] Emergence is part and parcel of complexity: Complexity theory is "an incentive for an emergentist worldview."[13] As programmatic as this statement by physicist Paul Davies may sound, there are problems and risks with an emergentist worldview. A problem is that complexity and emergence are notions that at present nobody fully understands. Both concepts may be fairly well-defined in specific scientific and philosophical contexts, but as soon as they migrate into the realm of general understanding they become increasingly fuzzy.[14] Gregory Peterson's contribution in this volume ("Who Needs Emergence?") addresses this problem by discussing the emergence of emergence and different kinds of emergence.[15] Exploring a concept that until now has escaped full understanding and daring to go so far as to build some concrete suggestions on it, as I do in this essay, has its risks and leaves the author vulnerable. Yet, this vulnerability is a necessary part of a vision: Only time will tell whether this is a vision or whether it was a delusion. I am convinced, however, that, at the very least, here is some valuable material for further re-envisioning the creative relationship of nature (including humans) and God.

Apart from many differences in definitions, it is a fundamental insight of complexity research that complexity is ontological; that is, it is inscribed in the order of being and is not a feature in the eye of the beholder. This means that, in spite of the elusiveness of its definition, there is something deeply objective about complexity. The fact that complexity research has strengthened the position of mathematical language, if at all possible, bears witness to this. Douglas Norton's essay in this volume ("Visions of Nature through Mathematical Lenses") illustrates this poignantly: The chaotic necessity that drives the idea of nature as emergent is firmly grounded in the language of mathematics.[16] Ultimately, complexity research is about the attempt to show that systems as disparate as sand piles and anthills, earthquakes, immune systems, economies, and ecosystems conform to common mathematical principles. There is a growing view that complexity is evident at all scales. If that is right, then the assumption that macroscopic complexity always is the result of simplicity at the microscale is mistaken.

Complex systems are analyzed in terms of *levels* of complexity. In purely physical and philosophical terms, hierarchy seems to be the big gain of the deal. Precisely the idea that the whole order of being (and maybe even becoming) can be described in terms of a conclusive hierarchy of levels of complexity constitutes the core of the practical and aesthetic appeal of this concept.[17] In that, however, it runs counter to some recent developments in philosophy and theology that have pointed out that the concept of hierarchy is charged with so many problems that it needs to be abandoned or at least submitted to radical critique. Feminist and liberation theologians are not the only ones to have analyzed hierarchical thought structures that result in hierarchical social structures, which, in turn, produce oppression of human and nonhuman nature. Half a century ago, theologian Paul Tillich pointed to the problems inherent in hierarchical descriptions. Dismissing the concept of levels, he works with dimensions, realms, and degrees instead.[18] Whereas levels can hardly be imagined other than as ladder-type structures, dimensions, realms, and degrees allow for the conceptualization of more complex patterns of relationship. One dimension—here understood in the colloquial sense of "scope" or "aspect," rather than in its precise mathematical sense—can certainly govern and override the other in the same way as levels do. Yet, dimensions and realms can do more than that: They can succeed and precede each other, they can interact and be independent of each other, and they can overlap and complement each other. One can be superior to the other in some respects and inferior in others.[19] Such a concept oscillates between continuity and discontinuity and comprises both. It may even be more in line with how hierarchy is understood in ecology: Ecologists tend to recognize that any top-down influence that may be exerted in natural systems is not as absolute as often assumed in the philosophical and sociological use of the term *hierarchy*. The neatness of philosophical and physical hierarchies does not have a one-to-one correspondence with the levels of order we see in the world of living systems. Nature in that sense is messier than many of its descriptions. Evidently, here lurks the risk of a disconnect between emergence as an abstract philosophical concept and emergence as the description of concrete natural processes.

Phrasing the problem as a question: Can emergence focus on dynamics and potentiality in such a way that the notion of levels is both validated and relativized? The answer seems to be yes, if one understands emergence as the coming into being of new modes and levels of (self-)organization and (co-)operation that transcend the limits of a system's inherent causality. In that sense, emergence transcends the rigidity of the physical origins of life, which, of course, implies neither the absence of causality nor an understanding of causal chains limited to compoundity or complicatedness. As Paul Davies has noted, "Complexity reaches a threshold at which the system is liberated from the strictures of physics and chemistry while still remaining subject to their laws. Although the nature of this transition is elusive . . . , its implications . . . are obvious."[20] In this sense, complexity is something like a paradox made plausible! Radical indeterminacy is understood as a very natural transition. Emergence is radically surprising, yet not totally enigmatic.

Einstein, in his day, was very puzzled by the uncertainty implied in quantum physics. In fact, he found this indeterminacy repellent. Later interpretations have accepted this indeterminacy, however, and given it a positive spin, as it were, by understanding it as potentiality. Seen in this light, emergence can be interpreted as introducing—or better—accounting for, potentiality on every scale from the subatomic to the macroscopic. In my view, it is this particular feature that constitutes the radical character of emergence.

Culture and Nature: Abandoning Dualisms

Much of traditional reasoning about nature, including human nature, has been anthropocentric and individualistic. This perspective has been noticeably challenged by bio- and ecocentric models of thought. Anthropology in general, and the question of human uniqueness in particular, have gained new theological, philosophical, and scientific actuality in many respects—genetics, primatology, evolutionary psychology/behavioral ecology, and artificial intelligence research being the most prominent on the scientific side. In this process, the chasm between nature and culture suggested by modernity has been unmasked as illusory, requiring new visions of connectedness, as James Proctor argues in this volume ("Environment after Nature: Time for a New Vision").[21]

Modern science has had a twofold input in framing the understanding of the relationship between nature and culture. On the one hand, it has strikingly contributed to the objectification of nature. The spirit of Baconian science offers graphic expressions in this regard. "The Beautiful

Bosom of *Nature* will be Expos'd to our view: we shall enter into its *Garden*, and taste of its *Fruits*, and satisfy our selves with its *plenty*."[22] These are powerful metaphors that, as we know today, have deeply and often fatefully influenced modern views of nature and of human beings as "*maîtres et possesseurs*"[23] of this same nature. Metaphors and concepts like these have served to justify the domination, the exploitation, and the rape of nature.[24] Newtonian physics, with its concepts of absolute space and time, fostered an understanding of nature as the solid stage on which the drama of culture is performed. It seemed that nature is dominated by a cyclic order. Driven by repetitious cycles, nature forms merely the passive backdrop to the dynamic events in a culture that is developing linearly in history.

On the other hand, both Darwinian and Einsteinian science has contributed to the abandonment of this dualism. With the theory of evolution through natural selection, nature gained part in linearity and historicity. With the theories of relativity, the polarity between passive nature and active culture was rescinded. Nature is not an object in a huge container called absolute time, but time is in nature. History is not the account of a universal now moving inexorably and uniformly through time. It is the account of a space-time continuum of crisscrossing light cones curved around fields of gravity.[25]

When this problematization of the subject-object relationship by nineteenth- and twentieth-century science is taken seriously, the way is open to think in terms of a differentiated relationality that blurs many clear-cut borders between nature and culture. Only a careless thinker will interpret this as a lowering of the standards of rationality and scientific accuracy. Accounting for complexity—that is, for processes of becoming, multidimensionality, and relationality—clearly requires more than descriptions that limit themselves to states of being and one-dimensionality. It is precisely this development that has set the stage for much of the interest in understanding processes in terms of emergence. In its wake, a number of concepts crave clarification. Design and order are two of those.

Revisiting Design and Order

Theology has always depended on nontheological models of thought in order to frame its discourse about nature and God. For ages, Christian theologians have drawn on philosophers, especially Plato and Aristotle. When philosophy of nature turned into science, it was science that contributed to the shaping of theological thought about nature. Generally, the assumption was not that science would lend objective truth to theological statements. More often, scientific theories would provide inspiring metaphors for the articulation of a theological language that matches contemporary contexts. Both areas of knowledge have in various ways contributed to the shaping of worldviews throughout the centuries, which is shown by the historical contributions to this volume.[26] The following exploration of what it may mean to speak of nature as creation in light of emergence thought will serve as a current example of the interaction of theological thought with scientific concepts.

Within a framework based on the fundamental distinction between binaries, such as matter and form or matter and spirit, the doctrine of a creation out of nothing (*creatio ex nihilo*) makes a lot of sense. It safeguards the sovereignty of God by allowing for nothing beside God at the moment of creation. It also emphasizes the goodness of all creation: If everything comes from the word of a good Creator, nothing can fall outside, in the domain of a potentially evil force. But the doctrine also has its downsides. In the end, divine goodness tends to be overpowered by the idea of divine omnipotence. The doctrine also leaves Genesis 1:2—about the Earth being a formless void and God's spirit hovering over the face of the waters—without any intelligible interpretation.

Some theologians have pointed out that the notion of creation out of chaos is closer to the biblical sources than creation out of nothing. This, of course, does not decide the case, as the history of Christian thought knows of many doctrines that lack a clear scriptural foundation; but it provides, at least, motivations for considering alternatives. Mythically, chaos has tended to be understood as evil. Creation, then, is basically synonymous with the slaughter of the chaos beast. Theologian Catherine Keller identifies this understanding, which she calls *tehomophobia* (from the Greek *phobos*, meaning "fear," and the Hebrew *tehom*, meaning the "deep, the sea, or the chaos") as harmful.[27] The creative potential of the *tehom* fell victim to a tradition demonizing it as evil disobedience.[28] Order came to be understood as fully good and disorder as totally evil. By contrast, Keller points out that the biblical material also contains an often-neglected *tehomophilic* (from the Greek *philia,* meaning "friendship, love") strand, which is less interested in hegemonic and linear order and that interprets creation as cocreation.[29] Waters and the earth do their own creation (Genesis 1:20–24), and God takes delight in the play of Leviathan, the chaos monster (Psalm 104:26). The so-called wisdom literature, in particular, expresses views that are *tehomophilic* rather than *tehomophobic* and that are much less wedded to the dualism of order and disorder.

This observation calls for a radical change of perspective, from understanding chaos as enemy only to an understanding of chaos as potentiality. In light of this shift, Keller suggests that *creatio ex nihilo* be complemented by *creatio ex profundis,* out of the profundity and womb of God, which is understood as the multidimensional continuum of all relations.[30] Drying up the sea (*tehom*) of potentiality is fatal—as fatal as the emptying of the earth's aquifers. Christian repression of the transitional and wild is not only bad for the environment, as Keller opines;[31] it also eliminates the possibility of understanding complexity and emergence as significant features of the natural world, I would add. If this issue remains unsettled, we tend to build in yet another ostensible conflict between scientific and religious views of the world.

It can be argued on theological, philosophical, and scientific grounds that the dualism of matter and form, or order and disorder can no longer constitute a sufficient framework for understanding nature, creation, and creativity. The door seems open for a liaison between emergence and *tehomophilic* understandings of creation and creativity. It must not be forgotten, though, that such understandings come at a cost. They give up something of the clarity of distinction between matter and form or spirit, good and evil, order and disorder. Creation is a risk for everybody involved, including God; its story needs to be read as a narrative of transformation and of metamorphosis, as philosopher John Caputo claims, and not as a neat onto-theological metaphysics.[32]

Keller is not the only theologian to choose messiness over clarity. Elizabeth Johnson has suggested that it was the fear of chaos that motivated an obsession with order in God, coming along with a support of hierarchical and oppressive structures.[33] Or in the words of Ruth Page: "The axiom of Christian faith that God is a God of order and not of disorder has meant in practice that disorder has been ignored, or, explained away, or written off as sin. . . . But that has left Christianity speechless in the face of much disorder. . . . The emphasis on order has never reflected the dual experience of stability and change, the disequilibrium inherent in present order in open systems. . . ."[34] A discourse obsessed with order has not been able to account adequately for development and creativity as they unfold in, with, and through the interrelatedness that marks nature and culture.

As the work of Keller, Johnson, and Page shows, theology has resources to develop other ways of talking about creation than to focus on design and order. Although the replacement of "God the designer" with God the "infinitely liberating source of new possibilities and new life,"[35] God as "serendipitous creativity,"[36] or God "the networker"[37] is not without problems, it is in accord with important elements of Christian theology, such as the concept of freedom, certain aspects of eschatology, the primacy of the possible before the real, and the notion of novelty.[38] It has the additional benefit of demonstrating that the question of "intelligent design" does not deserve the place in the limelight of religion and science that it so often assumes.

This reflection on design and order versus chaos and disorder suggests that there are good theological reasons to call into question some of the binaries and dualisms that have set the tone in much discourse for centuries. Moreover, it is striking how this theological development is paralleled by scientific and philosophical understandings of emergence. With varying enthusiasm, scientists and philosophers interpret emergence as a way of letting go of binaries and transcending dualisms that seem to belong to a bygone era of intellectual history.[39] In this case, developments in theological thought are in consonance with sciences that describe the natural world by using the terminology of emergence and self-organization. If emergence and self-organization are true marks of nature, then *tehomophilic* strands do indeed provide a more comprehensive understanding of creation and creativity than *tehomophobic* ones.

Creation in Light of Emergence: Possibilities and Pitfalls

When Catherine Keller states that "the wounds inflicted by certainty . . . will be better healed by a discourse of uncertainty than by just another sure truth,"[40] this resonates perfectly well with the insight of one of the leading figures in complexity research. Contemplating the fact that we cannot even predict the motions of three coupled pendula, Stuart Kauffman exclaims: "Bacon, you were brilliant, but the world is more complex than your philosophy."[41]

Natural selection is not enough to account for the development from cell to organism to ecosystem, according to Kauffman's theory of complexity. Kauffmann claims the insufficiency of natural selection for diametrically opposite reasons than the intelligent design movement, though. In his work, the energy that drives creative processes is called "self-organization," instead of design. He concludes that we need both science and story to make sense of the universe.[42] Evolutionary theory must be rebuilt as "a marriage of two sources of order in biology—self-organization and selection,"[43] suggesting that science in general should be regarded in terms of an "intermarriage of law and history."[44] This, he muses, may be the starting point of a general biology that can formulate laws for all biospheres.[45] Kauffman is not alone: He draws heavily on Per Bak's concept of self-organized criticality as a general mechanism to generate complexity. The Brussels school, under Ilya Prigogine, would substantiate Kauffman's claim about history. We have reached a description of physics that brings a narrative element into play on all levels, says Prigogine.[46] Systems biology provides yet another example; it takes on many of these insights and is currently gaining influence in both research and teaching.

Nevertheless, complexity theory and emergence are not undisputed. Critical issues can be raised in several respects. First of all, as already noted, there is a lack of clear definitions. The absence of consensus in this regard leads to a lack in clarity as to how to assess the potential of emergence. Second, the emphasis on the impossibility of predicting the development of complex systems is itself at odds with the traditional criteria for good science; namely, the ability to make testable predictions. Complexity theorists insist that at the poised stage between order and chaos, the unfolding consequences of the next step cannot be foretold. Will the next grain of sand falling on a sand pile evoke a trickle or a landslide? Nobody can tell. We can only be locally wise, not globally wise, as Kauffman puts it.[47] The theory of complexity is of necessity abstract and statistical;[48] furthermore, it appears to be insufficient.

Both Bak and Kauffman draw support from Stephen Jay Gould's theory of punctuated equilibrium and his emphasis on contingency in the evolutionary process. Other views, such as those embodied in Simon Conway Morris' convergence thesis, seem to accord less weight to the contingency of evolutionary processes.[49] This indicates a possible third difficulty: The conviction that large avalanches and not gradual change make the link between quantitative and qualitative behavior and thus form the basis of emergent phenomena is central to complexity theory.[50] It seems, however, that the final word on whether evolution should be understood in terms of revolution, as Bak suggests,[51] has not yet been spoken.

A fourth critical issue pertains to the role of the "exactly right" level of criticality. The idea that supercritical, chaotic rules will wash out any complex phenomenon that might arise and that subcritical rules will freeze into boring, simple structures, while only the critical state will allow complexity, sounds plausible, if not seductive: Ecology must be posited precisely at the critical state separating the extremes, or rather at the phase transition between those extremes. The conclusion sounds appealing. "A frozen state cannot evolve. A chaotic state cannot remember the past. This leaves the critical state as the only alternative."[52] However, caution may be called for. Cosmology provides an example of how fascination with just the right critical level (in this case, the exactly right level of matter density to slow down the expansion of the universe indefinitely) did not prove to be the right road to travel. The nonexpert craves an explanation that clarifies the distinction between the desire to detect teleology and the state of facts in this regard. The window of possibility for viable structures may be much wider than the fascination with the edge of chaos suggests.

A fifth issue is, in my opinion, the most problematic one; namely, the conflation of description and implicit value judgments, which seems to come very easily with emergence, as also pointed out by Willem B. Drees in this volume.[53] Careful and critical interpretation is called for, especially when emergence is used in order to justify social norms based on what is perceived as a universal, natural, and hierarchical order. In one breath, the editorial description of a recent book states that the emergence of new order and structure in nature and society is explained by physical, chemical, biological, social, and economic self-organization, according to the laws of nonlinear dynamics.[54] The scope of this list is quite breathtaking. The author of the book, philosopher of science Klaus Mainzer, suggests that symmetry and complexity are not only useful models of science, but that they are universals of reality: "In the beginning there was a dynamical symmetry expanding to the complex diversity of broken symmetries,"[55] which leads to the emergence of new phenomena on all levels from atoms to art. On the basis of his understanding of phase transitions, Mainzer argues that, in order to meet the challenges of globalization, "We should deregulate and support self-regulating autonomy," because "The sociodiversity of people is the human capital for a sustainable progress . . . in the evolutionary process of globalization."[56] Mainzer derives social and political norms directly from the scientific and philosophical study of emergence. A leap of such dimensions requires a careful and critical analysis; this need must not be hidden under the cover of emergence as an all-embracing theory.

In light of these five issues, what is the theological relevance of emergence? Where is its place in theological reasoning? Does emergence argue for the existence of God? The answer is no. Even though Kauffman expresses the hope that the new science of complexity may help us to recover our sense of the sacred,[57] it is as feeble a proof of the existence of God as Thomas Aquinas' five ways that build on the principle of simplicity rather than complexity. The laws of complexity do not allow for a deus ex machina; they build solely on dynamic interactions among elements of a system—the principle called self-organization. No intervention from outside is needed. Complexity requires long processes of evolution, but it "can and will emerge 'for free' without any watchmaker tuning the world."[58]

Neither can emergence be claimed as a proof for the failure of materialistic accounts. Quite the contrary, the concept of emergence has gained considerable appeal just because it seems to underwrite a materialistic worldview.[59] The least to be said is that there is ample wiggle room for interpretation here. For example, both Ursula Goodenough and Terrence Deacon on the one hand, and Philip Clayton on the other, argue for strong forms of emergence. Yet, there is a fundamental difference between their proposals. Goodenough and Deacon use emergence in order to argue that everything is perfectly intelligible within a naturalist framework, thus making any theist notion superfluous.[60] Religious feelings like awe can be fully accounted for within the realm of the natural. They are not dependent on a God-relationship; hence the possibility of a nontheistic religious naturalism. According to this view, nature is enough. Philip Clayton, on the contrary, uses emergence precisely to break open such a naturalist system by exploring how emergence may suggest transcendence. In his proposal, nature is not self-enclosed but, in principle, is upwardly open to divine influence on various parts of the natural world.[61] According to him, nature is not enough.

The conclusion following from this theological twilight is that a hermeneutical approach of methodological naturalism fits this area of science as well as any other. The theological relevance of emergence is not to be sought in the historical area of proofs for the existence of God. It is not in the field of apologetics. Rather, theological reflection on emergence has a heuristic function. It encourages a fresh look at old things by discussing the ways in which emergence thought can help to respond to the call for visions of nature that fulfill the criteria stated at the beginning of this essay: visions that include the best of scientific and theological knowledge, that are mindful of the groaning of an environmentally stressed nature, and that can tell the story of becoming that we see unfolding in nature.

As already mentioned, there are exciting parallels with regard to understanding how nature works and how we speak about nature as creation. There are points of contact between Keller's talk of creation as cocreation and Kauffman's and Bak's terminology of coconstruction[62] and coevolution of interacting species,[63] that is, the coordinated evolution of entire ecosystems.[64] Emergence theorists talk about interacting dancing fitness landscapes and life as a global, collective, cooperative phenomenon. There is a direct affinity between such talk and the language of much of recent theology that often favors metaphors of dance and concepts of relationality.[65]

My appraisal of creation out of nothing in light of emergence differs from the critique of process theologians who tend to see *creatio ex nihilo* as the most disastrous distortion of Christian faith.[66] Instead, I argue for maintaining the concept for both its anti-Manichaean merits and its affirmation that everything created has an implicit God-relationship. However, I also argue that the *ex nihilo* needs to be released from its metaphysical restriction to the dualism of form and matter so that it can be used as a lens for a theological understanding of processes of emergence. As cosmology insists, nothing is not nothing (at least with regard to quantum fluctuations): The metaphysical short-circuiting of the *nihilo* can be overcome on rational grounds. Creation understood as the emergence of "*something more* from *nothing but*"[67] can be a legitimate interpretation of creation out of nothing. Such a reading comes with the additional benefit of lessening the gap between *creatio originalis* (original creation) and *creatio continua* (continuous creation).

In sum, I agree with Peterson[68] that emergence entails a critique of claims of completeness and closure. I am careful, however, to distinguish these claims from attempts of sneaking in a variation of a god-of-the-gaps argument and to avoid an overemphasis on the hierarchical levels of emergence. The latter neglects horizontal relationships at the expense of vertical ones and has a propensity to conflate description and values. Both these risks appear to be more imminent, when emergence thought is based predominantly on physics and philosophy. Essays by Henderson, Proctor, and Ulanowicz in this volume lead me to the conclusion that the ecological scale seems to be the most appropriate one with which to gain an understanding of the scope of interrelatedness that is the hotbed of emergence. At the ecological scale, it seems easier to avoid the tyranny of the ladder metaphor that often comes with emergence talk, because ecology has a tendency to relativize distinctions between higher and lower levels. The ecological scale may also help to address the intricate relationships between facts and values, because ecology always needs to ask the question: What is a value for whom?

As I now move on to sketch out some specific elements of a vision of human and nonhuman nature and God, I deliberately enter the gray area between a concrete and metaphorical use of emergence—a methodological move that in itself may count as emergent.

Elements of a New Vision in Light of Emergence

In my view, the following elements are conducive to an appropriate vision of nature in light of emergence—a vision that is informed by both science and a theologically reflected understanding of nature and God. I will sketch these elements as brief responses to six questions. Rather than providing definitive answers, these short statements are meant to provide material for further reflection.

1. *How can we talk about* nature?

In light of emergence, nature presents itself as shaped by two seemingly opposing tendencies. On the one hand, it is marked by an openness that facilitates evolution and complexification; on the other hand, it bears the mark of a restraint that imposes order. Consequently, we see a powerful inherent creativity in nature, provided by the laws of nature. Again, we see the paradox made plausible—yet not domesticated. This vision of nature suggests that the opposing tendencies are linked together; openness and restraint are one in nature.[69] Theologically speaking, this would mean that Manichaeism has rightly been debunked as wrong teaching. Concepts that work with the unity of the hidden and revealed God are better suited to express this vision. Nature is not a chain of sand grains or beings trickling from the hand of a supposedly almighty creator. It is better understood as the story of becoming and complexification. The mix of catastrophism and creativity has its correspondence in a creator who is present both immanently and transcendently and for whom creation also is a process of *kenosis* (Greek for "emptying") and vulnerability. Creation is a "dicey business" for everybody, including God.[70]

Concepts of complexity count on a phase space of potentiality linked to natural phenomena. This is an image for the idea that every event is "surrounded by a ghostly halo of nearby events that didn't happen, but could have."[71] In terms of theological analogy, this could mean that a field of transcendence is coupled with factual reality. The "adjacent possible"[72] has a role in processes of actualization; it can, in fact, be seen as a part of actuality. It is in this sense that I think Paul Tillich understood the *eschaton* ("the last, the ultimate"), when he spoke of the *eschaton* as the "transcendent meaning of events."[73] This concept of a space of potentiality implies a beneficial disruption of simple notions of intervention. It entrusts to the rubbish heap of history the equation that identifies any divine action with a violation of the laws of nature. Searches for expanded concepts of causation are well justified and called for.[74]

2. *How can we talk about* human nature?

The concept of emergence is potentially helpful in addressing the question of human nature, and specifically the question of human uniqueness within a vision of nature that emphasizes the continuity between nature and culture. Traditional theological understandings of human uniqueness (*imago dei*) have often focused on cognitive traits, like human rationality and intelligence, that have set humans apart from the rest of nature. On the contrary, an understanding of human uniqueness in terms of emergence, such as that suggested by Wentzel van Huyssteen, emphasizes both continuity and discontinuity with the rest of nature: It accounts for our close ties with the animal world as well as for the uniqueness in which symbolic and cognitively fluent minds bring about language, art, technology, religion, and science.[75] It is a prerequisite for a comprehensive vision of nature to understand humans as the part of nature they are, while at the same time articulating the specifics of human potential and responsibility.

3. How can we speak about God?

God is not the designer of outcomes; rather, God is the wellspring of the frameworks within which complexification can occur. The watchmaker image of God has given way to a networker image of God.[76] This is the definite end of any deistic concept of a God who winds up a cosmic clock and then retires to watch the process of mechanical unwinding. In this vision, God is the transcendent creator as well as the immanent creative energy. This concept acknowledges both *creatio originalis* and *creatio continua*. The idea that God has created the world as self-productive or self-organizing seems to offer a possibility of modifying concepts of God as a designer, so that they include evolutionary concepts, allowing for freedom and genuine novelty. God as the wellspring of complex autopoietic systems is Godself living a complex life, implying change, having freedom, and granting freedom. In light of this, problematic divine attributes, such as immutability and impassibility, can be revisited in a substantive way. Grace and freedom can be conceptualized without ruling out the notion of God's transformative power.

4. How can we speak about natural evil?

Theories of complexity are relevant to the question of natural evil. Why do earthquakes happen if creation is meant to be good? As Bak remarks, self-organized criticality (the state of maximum slope in the sand pile) can be conceived of as the theoretical underpinning for catastrophism, that is, the opposite philosophy to gradualism.[77] It implies that catastrophes happen and need to happen, and that they happen as a consequence of very small events. This thought has its theological counterpart in apocalypticism, which tends not to be a favorite subject of theologians. Often, its well-behaved cousin, eschatology, has been tremendously more popular than this unruly enfant terrible. Yet, as Keller rightly points out, from an ecotheological perspective, an antiapocalyptic stance that joins the *tehomophobic* strand and metaphorically and literally empties the dark sea (and thus creativity), as tempting as it may seem, colludes with a conservative triumphalism so often detrimental to the environment.[78]

Along these lines, the concept of emergence adds sophistication to one of the traditional ways of engagement with the unsolved problem of theodicy. It supports the pedagogical approach by suggesting that nature works so that there is a price to be paid for complexification, because complexification needs both order and disorder. Nature displays criticality and catastrophes as well as creativity and stability. This does not diminish the role of pain and evil in the world, and does not explain the magnitude of evil. Even pain that is understood within a framework of emergence is no less painful, but not being able to feel and articulate pain would be an even greater evil. This is why lament is a vital element of religious practice.

5. *What can emergence contribute to an understanding of* sacramentality?

The concept of emergence has a theological parallel in the concept of sacramentality. Both emergence and sacramentality can be understood as having the capacity of expressing the continuity between the physical, mental, and spiritual in terms of a differentiated relationality. Both express the fact that the less complex can birth the more complex. Bread and wine emerge into shared communion with Christ; out of water and word emerges a new life in Christ.

Properly understood, sacramentality is the radicalization of the idea that a phenomenon is more than it presents itself to be. It goes at least one step further than the emergentist understanding referred to with regard to human uniqueness and its emphasis on continuity and discontinuity with the rest of nature. It also moves beyond a general acknowledgment of the significance of potentiality. Rather than focusing exclusively on the actual, emergence encourages a view of reality as a blend of the actual and the potential. Sacramentality radicalizes this by declaring the potential to be part of the actual: For the human eye and tongue, bread is bread, and wine is wine; a sacramental view claims that the reality of communion in

Christ surpasses the apparent actuality by turning that which according to human perspective is (merely) potentiality into reality. Bringing emergence thought together with the theology of the sacraments seems fruitful both for Western theology as well as the theology of the Eastern churches and their understanding of sacramentality and divine energies.

It is here, in the context of sacramentality, that I stretch the use of *emergence* toward a metaphorical maximum. The following statement about theological method reverts to a more concrete understanding of emergence.

6. *What bearing does a heuristic understanding of emergence have on* theological method *in general?*

Emergence presupposes the existence of clusters of networks. A word of caution may be in order, though. The human mind with its seemingly insatiable desire to recognize patterns has a tendency to imagine networks and clusters of networks as a state of order. However, radical interconnectedness implies much more disorder than a sanitized concept like clusters of networks suggests—especially to the layperson. Mathematically, one can clearly distinguish the ordered structure from the disordered—and both are there![79] Bringing emergence and theology together may therefore in praxis be much riskier than the theory would indicate.

One of the less dramatic implications of viewing the practice of theology in light of emergence is the requirement that theology as a discipline needs to increase its attention to communal, ecumenical, and interfaith approaches. Developments in the religious landscape cannot be understood adequately by focusing on the religious experience of single individuals or the content of one specific religious tradition or geographic region alone. One has to take into account relations with the rest of nature, as well as a full societal scope. Global wisdom cannot be attained. Yet local wisdom cannot be attained without seeking global wisdom.

Conclusion

Many current intellectual pursuits across varied disciplines tend to be driven by the will to understand nature, science, and religion in terms of dynamic systems, interrelatedness, discontinuities, and processes of complexification. In this situation, emergence serves as a fruitful concept that seems applicable over the entire spectrum of knowledge. Emergence is not easily defined, however, and its concomitant interpretations can be as flawed as any. In the scientific realm, ecology appears to be especially well-suited to enhance the understanding of emergence by describing the interplay of opposing tendencies in nature and the variety of interactions across levels and networks.

In philosophy and theology, emergence contributes to the critique of ontological metaphysical statements. Narrative understanding is always necessary: Metaphysics cannot do without myth. Even here, opposing tendencies and binaries are seen in a perspective that is different from what usually goes by the name of Cartesian dualism. Reconsidering the role of the potential and the real allows for an understanding of binaries, such as immanent/transcendent, order/disorder, and nature/culture along the lines of what I have called a differentiated relationality. This, in turn, invites us to envision nature, God, evil, sacramentality, and theological method in the directions shown in this essay.

Notes

1. Except for geography, where Basil seems to be behind the standards of his own time. Saint Basil, *Exegetic Homilies*, translated by Sister Agnes Clare Way, C.D.P. (Washington, D.C.: The Catholic University of America Press, 1963), x.
2. Ibid., VII 6, 115.
3. Ibid., V 3, 71.
4. Ibid., IX 3, 139.
5. Ibid., VII 4, 112.
6. Ibid., VII 5, 113.
7. Ibid., IX 3, 141.
8. Ibid., IX 3, 141.
9. Barbara King, "God from Nature: Evolution or Emergence?" in this volume.
10. Saint Basil, *Exegetic Homilies*, VII 5, 114.
11. Ibid.
12. According to Danish theologian Niels Henrik Gregersen, for example, complex systems come in seven varieties; they can be descriptively, constitutionally, organizationally, causally, functionally, algorithmically, and effectively complex. Niels Henrik Gregersen, "Complexity: What Is at Stake for Religious Reflection?" in Kees van Kooten Niekerk and Hans Buhl, eds., *The Significance of Complexity* (Aldershot, UK: Ashgate, 2004), 135–65, particularly 136–41.
13. Paul Davies, "Introduction: Toward an Emergentist Worldview," in Niels Henrik Gregersen, ed., *From Complexity to Life: On the Emergence of Life and Meaning* (Oxford: Oxford University Press, 2003), 13.

14. On the migration of concepts and its hermeneutical implications, see Antje Jackelén, *The Dialogue between Science and Religion: Challenges and Future Directions*, ed. Carl S. Helrich (Kitchener, Ontario: Pandora Press, 2004), esp. 15–81.
15. Gregory Peterson, "Who Needs Emergence?" in this volume.
16. Douglas E. Norton, "Visions of Nature through Mathematical Lenses," in this volume.
17. See, e.g., Philip Clayton, *Mind and Emergence: From Quantum to Consciousness* (Oxford: Oxford University Press, 2004), and my discussion of emergence and hierarchy in Antje Jackelén, "Emergence Everywhere?! Reflections on Philip Clayton's *Mind and Emergence*," *Zygon* 41 (September 2006): 623–32.
18. Paul Tillich, *Systematic Theology*, vol. 3 (Chicago: University of Chicago Press, 1963), 15ff.
19. Cf. Robert Ulanowicz, "The Organic in Ecology," *Ludus Vitalis* 9 (2001): 183–204, who argues that higher-order composite systems, like ecosystems and socioeconomic systems are generally both simpler and longer-lived than their components.
20. Paul Davies, "Introduction: Toward an Emergentist Worldview," in Gregersen, ed., *From Complexity to Life*, 8.
21. James Proctor, "Environment after Nature: Time for a New Vision," in this volume.
22. Thomas Sprat, *History of the Royal Society*, Jackson I. Cope and Harold Whitmore Jones, eds. (St. Louis: Washington University Press, 1958), 327. Spelling adapted by the author.
23. "Masters and possessors" in René Descartes, *Discours de la méthode*, Introduction and notes by E. Gilson (Paris: Librairie Philosophique J. Vrin, 1989), 128.
24. Carolyn Merchant, *The Death of Nature: Women, Ecology and the Scientific Revolution* (San Francisco: Harper & Row, 1989).
25. Cf. Antje Jackelén, *Time and Eternity: The Question of Time in Church, Science, and Theology*, translated by Barbara Harshaw (Philadelphia and London: Templeton Foundation Press, 2005), 121–81.
26. See the essays by John Hedley Brooke, David N. Livingstone, Nicolaas A. Rupke, and Johannes M.M.H. Thijssen in this volume.
27. Cf. the Babylonian *Tiamat*, which can be translated as "watery chaos."
28. Catherine Keller, "No More Sea: The Lost Chaos of the Eschaton," in Dieter T. Hessel and Rosemary Radford Ruether, eds., *Christianity and Ecology: Seeking the Well-Being of Earth and Humans* (Cambridge, Mass.: Harvard University Press, 2000), 183.
29. Catherine Keller, *God and Power: Counter-Apocalyptic Journeys* (Minneapolis: Fortress Press, 2005), 137–45.
30. Ibid., 146–47.
31. Keller, "No More Sea," in *Christianity and Ecology*, 196.
32. John D. Caputo, *The Weakness of God: A Theology of the Event* (Bloomington and Indianapolis: Indiana University Press, 2006), 55–83.
33. Elizabeth Johnson, *She Who Is: The Mystery of God in Feminist Theological Discourse* (New York: Crossroad, 1994), 196–97.
34. Ruth Page, *God and the Web of Creation* (London: SCM Press, 1996), 37.
35. John F. Haught, *God after Darwin: A Theology of Evolution* (Boulder, Colo.: Westview Press, 2000), 120.
36. Gordon D. Kaufman, *In the Beginning . . . Creativity* (Minneapolis: Fortress Press, 2004).
37. Gregersen, "Complexity: What Is at Stake for Religious Reflection?" in van Kooten Niekerk and Buhl, eds., *The Significance of Complexity*, 156.

38. On the priority of the possible over the actual, see, e.g., Eberhard Jüngel, *God as the Mystery of the World,* translated by Darrell L. Guder (Grand Rapids, Mich.: Eerdmans, 1983), 216–17. On eschatology and novelty, see Jackelén, *Time and Eternity*, 93–97, 208–14.
39. Cf. Proctor in this volume, "Environment after Nature: Time for a New Vision," and Peterson in this volume, "Who Needs Emergence?"
40. Keller, *God and Power,* 150.
41. Stuart Kauffman, *At Home in the Universe: The Search for Laws of Self-Organization and Complexity* (Oxford: Oxford University Press, 1995), 303.
42. Stuart Kauffman, *Investigations* (Oxford: Oxford University Press, 2000), 119.
43. Ibid., xi.
44. Ibid., 267.
45. Ibid., 157.
46. Ilya Prigogine, "*Zeit, Chaos und Naturgesetze,*" in A. Gimmler, M. Sandbothe and W. Ch. Zimmerli, eds., *Die Wiederentdeckung der Zeit* (Darmstadt, Germany: Wissenschaftliche Buchgesellschaft, 1997), 91.
47. Kauffman, *At Home in the Universe,* 29.
48. Per Bak, *How Nature Works: The Science of Self-Organized Criticality* (New York: Copernicus, 1996), 9–10.
49. Simon Conway Morris, *Life's Solution: Inevitable Humans in a Lonely Universe* (Cambridge: Cambridge University Press, 2003). Convergence means that similar trends are found repeatedly in evolutionary history. The role of contingency would then be less significant than claimed by Gould.
50. Cf. Bak, *How Nature Works,* 32.
51. Ibid., 60.
52. Ibid., 127.
53. Drees' essay in this volume, "The Nature of Visions of Nature: Packages to Be Unpacked."
54. Klaus Mainzer, *Symmetry and Complexity: The Spirit and Beauty of Nonlinear Science* (Singapore and Hackensack, N.J.: World Scientific, 2005).
55. Ibid., 22.
56. Ibid., 272.
57. Kauffman, *At Home in the Universe,* 4–5.
58. Bak, *How Nature Works,* 48.
59. For example on the shape of the "culturological" approach, see Leslie A. White, *The Science of Culture* (New York: Farrar, Straus and Cudahy, 1949).
60. Ursula Goodenough and Terrence W. Deacon, "From Biology to Consciousness to Morality," *Zygon: Journal of Religion and Science* 38 (December 2003): 801–19. Ursula Goodenough, "Reductionism and Holism, Chance and Selection, Mechanism and Mind," *Zygon: Journal of Religion and Science* 40 (June 2005): 369–80.
61. Philip Clayton, *Mind and Emergence: From Quantum to Consciousness* (Oxford: Oxford University Press, 2004), 156–213, especially 193.
62. See, for example, chapter 8 in Kauffman's *Investigations,* "Candidate Laws for the Coconstruction of a Biosphere," and chapter 10, "A Coconstructing Cosmos?"
63. E.g., Bak, *How Nature Works,* 122.
64. Cf. Robert E. Ulanowicz, *Ecology, the Ascendent Perspective* (New York: Columbia University Press, 1997).

65. The interpretation of *dance* may differ between biology and theology, though. In biology, it is close to being a euphemism for the struggle for survival, whereas in theology it tends to have aesthetic and liturgical connotations.
66. Cf. David Ray Griffin, *Two Great Truths: A New Synthesis of Scientific Naturalism and Christian Faith* (Louisville and London: Westminster John Knox Press, 2004).
67. Goodenough and Deacon, "From Biology to Consciousness to Morality," 802.
68. Peterson, "Who Needs Emergence?" in this volume.
69. Robert E. Ulanowicz speaks of a "two-tendency universe" in *Ecology, the Ascendent Perspective*, 93–95. He discusses the significance of two opposing trends in ecosystem development in his "Process Ecology: A Transactional Worldview," in *International Journal of Ecodynamics* 1 (2006): 103–14.
70. Caputo, *The Weakness of God*, 64.
71. J. Wentzel van Huyssteen, "Evolution and Human Uniqueness: A Theological Perspective on the Emergence of Human Complexity," in van Kooten Niekerk and Buhl, eds., *The Significance of Complexity*, 199, drawing on the phase space concept as developed by Ian Stewart in his *Life's Other Secret: The New Mathematics of the Living World* (London: Penguin, 1998).
72. I have borrowed this term from Kauffman's *Investigations* without necessarily following his definition.
73. Paul Tillich, "*Eschatologie und Geschichte*," in R. Albrecht, ed., *Der Widerstreit von Raum und Zeit: Schriften zur Geschichtsphilosophie, Gesammelte Werke*, vol. 6 (Stuttgart: Evangelisches Verlagswerk, 1963), 77 ("*jedes beliebig kleine oder beliebig grosse Geschehen nimmt Teil am Eschaton, am transzendenten Geschehenssinn*").
74. Cf. Philip Clayton, "Natural Law and Divine Causation: The Search for an Expanded Theory of Causation," *Zygon: Journal of Religion and Science* 39 (2004): 615–36. See also Niels Henrik Gregersen, "The Idea of Creation and the Theory of Autopoietic Processes," *Zygon: Journal of Religion and Science* 33 (1998): 333–67. Gregersen suggests a distinction between structuring and triggering causes to the understanding of complexity and emergence. While a triggering cause always has a direct relation to an effect, a structuring cause has no one-to-one relationship to a particular effect.
75. Van Huyssteen, "Evolution and Human Uniqueness," 211–12. See also van Huyssteen, *Alone in the World? Human Uniqueness in Science and Theology* (Grand Rapids, Mich.: Eerdmans, 2006).
76. As pointed out, for example, in Gregersen's writings on the subject, e.g., in "Complexity: What Is at Stake for Religious Reflection?" 156.
77. Bak, *How Nature Works*, 131.
78. Keller, "No More Sea," 185.
79. Cf. Ulanowicz, *Ecology, the Ascendent Perspective*, 77–80.

Bibliography

Bak, Per. *How Nature Works: The Science of Self-Organized Criticality*. New York: Copernicus, 1996.

Saint Basil. *Exegetic Homilies*. Translated by Sister Agnes Clare Way, C.D.P. Washington, D.C.: The Catholic University of America Press, 1963.

Brooke, John Hedley. "Should the Word Nature Be Eliminated?" in this volume.
Caputo, John D. *The Weakness of God: A Theology of the Event*. Bloomington and Indianapolis: Indiana University Press, 2006.
Clayton, Philip. *Mind and Emergence: From Quantum to Consciousness*. Oxford: Oxford University Press, 2004.
———. "Natural Law and Divine Causation: The Search for an Expanded Theory of Causation." *Zygon: Journal of Religion and Science* 39 (2004): 615–36.
Davies, Paul. "Introduction: Toward an Emergentist Worldview." In Niels Henrik Gregersen, ed., *From Complexity to Life: On the Emergence of Life and Meaning*. Oxford: Oxford University Press, 2003.
Descartes, René. *Discours de la méthode*. Introduction and notes by E. Gilson. Paris: Librairie Philosophique J. Vrin, 1989.
Drees, Willem B. "The Nature of Visions of Nature: Packages to Be Unpacked," in this volume.
Goodenough, Ursula. "Reductionism and Holism, Chance and Selection, Mechanism and Mind." *Zygon: Journal of Religion and Science* 40 (June 2005): 369–80.
Goodenough, Ursula, and Terrence W. Deacon. "From Biology to Consciousness to Morality." *Zygon: Journal of Religion and Science* 38 (December 2003): 801–19.
Gregersen, Niels Henrik. "The Idea of Creation and the Theory of Autopoietic Processes." *Zygon: Journal of Religion and Science* 33 (1998): 333–67.
———. "Complexity: What Is at Stake for Religious Reflection?" In Kees van Kooten Niekerk and Hans Buhl, eds., *The Significance of Complexity*, 135–65. Aldershot, UK: Ashgate, 2004.
Griffin, David Ray. *Two Great Truths: A New Synthesis of Scientific Naturalism and Christian Faith*. Louisville and London: Westminster John Knox Press, 2004.
Haught, John F. *God after Darwin: A Theology of Evolution*. Boulder, Colo.: Westview Press, 2000.
Henderson, Martha L. "Rereading a Landscape of Atonement on an Aegean Island," in this volume.
Jackelén, Antje. *The Dialogue between Science and Religion: Challenges and Future Directions*, edited by Carl S. Helrich. Kitchener, Ontario: Pandora Press, 2004.
———. *Time and Eternity: The Question of Time in Church, Science, and Theology*. Translated by Barbara Harshaw. Philadelphia and London: Templeton Foundation Press, 2005.
———. "Emergence Everywhere?! Reflections on Philip Clayton's Mind and Emergence." *Zygon: Journal of Religion and Science* 41 (September 2006): 623–32.
Johnson, Elizabeth A. *She Who Is: The Mystery of God in Feminist Theological Discourse*. New York: Crossroad, 1994.
Jüngel, Eberhard. *God as the Mystery of the World*. Translated by Darrell L. Guder. Grand Rapids, Mich.: Eerdmans, 1983.
Kauffman, Stuart A. *At Home in the Universe: The Search for Laws of Self-Organization and Complexity*. Oxford: Oxford University Press, 1995.
———. *Investigations*. Oxford: Oxford University Press, 2000.
Kaufman, Gordon D. *In the Beginning . . . Creativity*. Minneapolis: Fortress Press, 2004.
Keller, Catherine. "No More Sea: The Lost Chaos of the Eschaton." In Dieter T. Hessel and Rosemary Radford Ruether, eds., *Christianity and Ecology: Seeking the Well-Being of*

Earth and Humans, 183–98. Cambridge, Mass.: Harvard University Press, 2000.
———. *God and Power: Counter-Apocalyptic Journeys*. Minneapolis: Fortress Press, 2005.
King, Barbara. "God from Nature: Evolution or Emergence?" in this volume.
Livingstone, David N. "Locating New Visions," in this volume.
Mainzer, Klaus. *Symmetry and Complexity: The Spirit and Beauty of Nonlinear Science*. Singapore and Hackensack, N.J.: World Scientific, 2005.
Merchant, Carolyn. *The Death of Nature: Women, Ecology and the Scientific Revolution*. San Francisco: Harper & Row, 1989.
Morris, Simon Conway. *Life's Solution: Inevitable Humans in a Lonely Universe*. Cambridge: Cambridge University Press, 2003.
Norton, Douglas E. "Visions of Nature through Mathematical Lenses," in this volume.
Page, Ruth. *God and the Web of Creation*. London: SCM Press, 1996.
Peterson, Gregory R. "Who Needs Emergence?" in this volume.
Prigogine, Ilya. "Zeit, Chaos und Naturgesetze." In A. Gimmler, M. Sandbothe and W. Ch. Zimmerli, eds., *Die Wiederentdeckung der Zeit*, 79–94 . Darmstadt, Germany: Wissenschaftliche Buchgesellschaft, 1997.
Proctor, James. "Environment after Nature: Time for a New Vision," in this volume.
Rupke, Nicolaas A. "Nature as Culture: The Example of Animal Behavior and Human Morality," in this volume.
Sprat, Thomas. *History of the Royal Society*, Jackson I. Cope and Harold Whitmore Jones, eds., 327. St. Louis: Washington University Press, 1958.
Thijssen, Johannes M.M.H. "Between Apes and Angels: At the Borders of Human Nature," in this volume.
Tillich, Paul. "Eschatologie und Geschichte," In R. Albrecht, ed., *Der Widerstreit von Raum und Zeit: Schriften zur Geschichtsphilosophie*, Gesammelte Werke, vol. 6, 72–82. Stuttgart: Evangelisches Verlagswerk, 1963.
———. *Systematic Theology*, vol. 3. Chicago: University of Chicago Press, 1963.
Ulanowicz, Robert E. *Ecology, the Ascendent Perspective*. New York: Columbia University Press, 1997.
———. "The Organic in Ecology." *Ludus Vitalis* 9 (2001): 183–204.
——— . "Process Ecology: A Transactional Worldview." *International Journal of Ecodynamics* 1 (2006): 103–14.
———. "Enduring Metaphysical Impatience?" in this volume.
van Huyssteen, Wentzel J. "Evolution and Human Uniqueness: A Theological Perspective on the Emergence of Human Complexity." In van Kooten Niekerk and Buhl, eds., *The Significance of Complexity*, 195–215.
———. *Alone in the World? Human Uniqueness in Science and Theology*. Grand Rapids, Mich.: Eerdmans, 2006.
White, Leslie. *The Science of Culture: A Study of Man and Civilization*. New York: Farrar, Straus and Cudahy, 1949.

9

REREADING A LANDSCAPE OF ATONEMENT ON AN AEGEAN ISLAND

Martha L. Henderson

Introduction

We were bumping along a winding gravel track on the ridge that separates the bays of Yera and Kalloni on Lesvos Island. Our guide was the local fire officer, our goal to see recent fire scars on the landscape. The road took us over a crest of a long ridge where ancient pines were charred and the annual grasses now thick following the previous year's wildfires. The road was gullied and without tire tracks. We were the rare visitor to this abandoned land. Driving along the top of the ridge, we wandered back and forth across the crest until we rounded a corner in the road. Before us was a gray granite wall of rock with small, dark, twisted trees entangled in the sheer rock face.

"Look at the erosion!" gasped my research partner. The landscape resembled the oft-repeated interpretation of Greek island landscapes as the result of centuries of misuse and overgrazing.

"Not erosion," said our guide, and put the truck into reverse.

"If not erosion, then what is it?" I asked myself. Like most geographers trained during the twentieth century, my colleague and I assumed that George Perkins Marsh had correctly identified the origins and problems of eastern Mediterranean landscapes as stemming from overgrazing.[1]

Working in the nineteenth century, Marsh, hailed as the originator of the American conservation movement, sounded the alarm to alert Americans about the potential impacts of deforestation, which he believed had caused vast soil erosion in the Aegean region. In his time, Marsh was speaking from a social desire to promote land and natural resource conservation. In his 1864 classic, *Man and Nature: Physical Geography as Modified by Human Action,* Marsh writes that "Man has too long forgotten that the Earth was given to him for usufruct alone, not for consumption, still less for profligate waste."[2] The concept of Earth being given to humanity underlies Marsh's thinking on conservation. He assumes that Earth is a gift from God and must be treated in ways that respect this religious connection. In Marsh's eyes, the world was to be conserved and maintained as a green Earth, an Edenic landscape.

"Our guide sees another vision of a legitimate religious landscape," I thought. His vision of the landscape did not include misuse or a lack of conservation of soils. He did not observe a lack of devotion to an angry god or the need to re-create a green Eden. To our guide, the landscape was Edenic by some other definition of godliness. I wondered what he envisioned that allowed him to see the landscape as acceptable.

The next day I went back to my research projects at the Geography Department at the University of the Aegean in Mytilini, where I was spending the winter as a Fulbright scholar. What did our guide see that my American, university-trained eyes did not? I began to reread the landscape as a record of local and regional relationships between social pressures, nature, and religion. Owing to the domination of culture and society by the Greek Orthodox church, I decided to examine church doctrine and practices to see what the church doctrine was relative to the physical environment. The story I found became the basis for rereading the landscape not as the broken relationship between nature, science, and religion, but as a symbol of an interwoven set of land uses informed by cultural and religious practices.

In this essay I attempt to reconnect the relationship between these separated bodies of Western knowledge by rereading the cultural landscape of Lesvos Island. Instead of the Roman-ruined landscape, I propose that the Aegean landscape is representative of customary celebratory and weekly consumption of lamb associated with religious beliefs. Over time, the demand for lamb has led to the coevolution of landscape and identity that is not driven by an Edenic struggle but, instead, reflects a vision of reality based on the local chaos of lived life in a state of atonement as envisioned by the Greeks and early Christians.[3] Today, the increasing economic pressures from the European Union and urban populations of Eastern Orthodox Greeks make high demands on rural ecologies for the production of lamb. The environmental messages from the ecumenical patriarch, Bartholomew I, may indicate a need to revitalize the current ecological setting and reduce the current rates of erosion within a self-organizing system.

My efforts to reread the landscape of Lesvos Island from the perspective of emergence and self-organizing systems touch on the work of other authors in this text. This essay is akin to the essays of John Hedley Brooke ("Should the Word *Nature* Be Eliminated?") in his clarification of nature as a subset of cultural and social ideas, Robert Ulanowicz ("Enduring Metaphysical Impatience?") and the processes of emergence in ecosystems, and Barbara King and her work on the emergence of religious practices ("God from Nature: Evolution or Emergence?"). My research as presented here offers an example of emergence theory, self-organizing systems, ecological conservation, religious landscapes, and locality.

Emergence and Self-Organizing Systems

Questions about the relationship between nature, science, and religion with regard to landscape interpretation follow similar historical debates about creation, evolution, and human intervention found in other cultural and natural sciences. Are landscapes the result of natural forces and human response? Are the landscapes we see today the result of continual human destruction of nature? Or, are historic, local landscapes the result of coevolutionary adaptive processes over time?

One of the most influential observers of the late nineteenth century was George Perkins Marsh. Marsh, the U.S. ambassador to Turkey (1849–1854) and Italy (1864), wrote in the mid-1800s about the destruction of Aegean lands due to poor agricultural practices, erosion, and loss of productivity. In his book, *Man and Nature,* Marsh's greatest fear was that poor land use practices in the United States would overcut New England forests in a way similar to what he believed had happened in Anatolia and the Greek islands. Deforestation would lead to soil erosion and, as in Turkey and Greece, would cause destruction of the land and the impossibility of potential agricultural productivity. Marsh's 2003 biographer, David Lowenthal, writes, "Marsh was the first to recognize that man's environmental impacts were not only enormous and fearsome, but even cataclysmic and irreversible."[4] According to Lowenthal, Marsh's underlying premise was that human agency leads to unavoidable impacts.

While Marsh observed that impacts had the potential of being Earth-transforming, he did not hesitate to seek certain kinds of transformations when he thought they would enhance economic productivity. Marsh observed soil erosion due to forest destruction and the misuse of land—in other words, land use practices that precluded sound agricultural practices. He viewed land use practices and the landscape of the Mediterranean as in a state of destruction and social decline.[5] Marsh's works set into motion modern land conservation studies and the modern environmentalist movement. Conservation efforts assume the ability of humans to modify and transform landscapes with the desired goal of resource protection for use over time. The focus is on human action over time and protection of potential productivity through the lens of ecological decline.

The tendency to base American thinking about conservation primarily on Marsh's contention—that society tends to destroy nature and therefore act disrespectfully within a religious ethos of Western Christian beliefs—is a powerful paradigm. The paradigm functions well in offering explanations for conservation of land and resources, economic benefits of conservation, and maintenance of socially responsible behavior to the point of equating conservation with religion. The paradigm is limiting in its ability to explain and engender beliefs that are not environmentally deterministic. The possibilities of creative adaptation, adjustment, and continual renewal are not possible in the paradigm. As symbols of religious identity, redemption and renewal are better read in a landscape paradigm based on self-organizing systems.

The concept of self-organizing systems was first introduced in general systems theory in the 1960s. Since then, the concept has received various levels of support in biology and physics. Several authors have chronicled an expanding set of meanings, typologies, and uses in attempting to understand complex systems. Most recently, Evelyn Fox Keller has examined the meaning of the words that comprise the concept.[6] The physical geographer Jonathan D. Phillips classifies self-organizing systems into eleven types and suggests that each type may be a starting point for understanding landscape histories.[7] Thomas S. Smith and Gregory T. Stevens probe the potentials of understanding large-scale social systems as self-organizing.[8] They suggest that the concept of self-organizing systems shows direct linkages between biological response and psychological stress. I extend the relationship between biological response to include religious ideas that inform human interaction and the social construction of historical landscapes. Self-organized criticality, one of Phillips' "types," offers an integrative model that remembers when human and ecological systems reach points of renewal. These memories of landscape formation are renewed as human religious beliefs are recalled and acted upon in the contemporary moment.

Landscape formation and interpretation must, by definition, include both physical and cultural components over time. Landscape formation incorporates nature, those characteristics that are from the natural world, and culture. For example, fire, which is a natural phenomenon, becomes an integral part of a landscape when it is used to form a particular set of natural features that are preferenced by a cultural group. Integral theorist Erich Jantsch proposed reuniting humans and the natural world by observing that evolutionary processes are not easily understood when measured in linear space or strict disciplinary boundaries. Instead, Jantsch suggests that evolution is an interdisciplinary and synthetic system where humans and nature are engaged in an adaptive and self-organizing process. This process, in which human mental, emotional, and spiritual capabilities are engaged in human evolution, is best described by the concept of *noosphere*, a term derived from the Greek word for mind-sphere.[9]

Jantsch, relying on the concept of the noosphere, proposes a human-nature evolutionary process that is not linear or progressive, but, rather, self-organizing. A self-organizing universe, then, is able to make sudden and multidirectional changes to accommodate shifts in physical conditions, such as climatic or human conditions, or changes in technology. Humans increasingly affect the natural world in the current technologically driven era. Of vital concern are catastrophic changes to natural systems that may require significant shifts in human systems without loss of life, yet the immediacy of life-threatening situations may be required by a self-organizing system to regain abundance. Some anthropologists conclude that religion explains fluctuations between catastrophe and abundance. Jantsch would argue that it is religious practices that help to create the natural world and may, in fact, be a successful mechanism for limiting natural catastrophe and sustaining culturally appropriate uses of available resources at the local scale.

Reading Aegean Landscapes

Lesvos Island, located in the northeastern Aegean Sea, is a place of common, historical land use practices that have evolved over time. Scholarship on Lesvos Island landscapes is available in English, with much greater research published in Greek. The University of the Aegean's Geography and Environmental Studies Departments are located on the island. I am thankful for the insights of a number of scholars in these departments, although this interpretation of the island's landscape is mine alone.

The islands of the Aegean and the larger region of the eastern Mediterranean biosphere have been described by a number of botanists and ecologists. Robert Sallares provides an exhaustive interpretation of the ecology of the ancient Greek world in his text, *The Ecology of the Ancient Greek World*. Francesco Di Castri, in a number of volumes and research papers, also describes the biogeography of the region. Di Castri summarizes the region as "shrub lands found primarily within the xero-thermic range of Mediterranean climates, characterized by the dominance of wood shrubby plants with evergreen, broad and small, stiff and thick (schlerophyll) leaves, an over story of small trees sometimes being present, and with or without an under story of annuals and herbaceous perennials."[10]

Peregrine Horden and Nicholas Purcell, in their 2000 study of the Mediterranean region, titled *The Corrupting Sea*, embrace the Mediterranean Sea as a unifying force that has created human-nature relationships over time. The authors devote an entire chapter to religion, not as a series of social-belief systems supporting elite power, but as a set of practices that has empowered the survival of local peoples in local areas for periods of time

> . . . beyond recoverable history as to belong in the domain of myth. Millennial processes of interaction have made all Mediterranean landscapes essentially *anthropogene*.[11]

Horden and Purcell are careful to specify that they do not believe the physical environment is determinant in creating a sacred geography. The survival of religious sites and practices is preferenced by human choice and historical events, thus contributing to a philosophical stance that embraces spontaneity, chaos, and locality.

Like Horden and Purcell, the geographer A. T. Grove and the ecologist Oliver Rackham write in their volume, *The Nature of Mediterranean Europe: An Ecological History*, that the continual belief in the misuse and eventual demise of the Aegean is wrongheaded.[12] Published in 2001, the volume is the consequence of a critical study of historical landscape and ecological conditions in the eastern Mediterranean. The volume follows two previous volumes by Rackham that evaluate landscape and ecological changes on the island of Crete and the islands of the eastern Mediterranean. These earlier volumes evaluated the impact of climate changes on the environment. The authors' research on the islands indicated that historical climatic conditions have remained stable and have not adversely affected landscapes and land uses. Climate over the last five thousand years has been unstable with high-magnitude events that have not been adequately recorded. Thus, the evidence is incomplete as to the impact of climate on current landscapes and ecological mosaics.

Recently, some grasslands scientists have begun to explore the possibility that overgrazing and associated land degradation is not occurring. Writing in 1998, Ave Perevolotsky and No'am Seligman propose that the assumption that the region is overgrazed is false.[13] They argue that the region has been grazed for a very long period. They suggest that heavy grazing is an ecologically sound and efficient land use. Further, they suggest that it is grazing over long periods that has created the biosphere of today. This composition of plants and related nondomestic animals is maintained by heavy grazing pressure and fire. In their article "Role of Grazing in Mediterranean Rangeland Ecosystems," the authors point out that disturbances like grazing and fire are necessary to generate a diverse and healthy biosphere. Anthropogenic uses of the land are not only possible, they constitute a necessity.

In *The Nature of Mediterranean Europe: An Ecological History,* Grove and Rackham challenge the long-standing belief that landscape change and desertification across Southern Europe are due to historical misuse and abuse of the landscape and its soils, flora, and fauna. They identify this belief as the "ruined landscape theory."[14] The theory of a ruined landscape is based on the belief that Southern Europe had greater vegetation and biomass in previous times and that human activity—primarily overgrazing and deforestation—has led to land degradation and desertification. The authors' conclusions are that modern land use policies are the cause of current erosion. They assert that current land use policies that limit traditional grazing patterns have contributed to catastrophic wildfires. Erosion studies, according to the authors, fall short because they are not longitudinal studies that can adequately measure extreme events.

Groves and Rackham propose that it is not a ruined Edenic landscape, but a landscape that has come to its current status as the result of a corelationship between traditional pastoralism and soil conservation measures, such as terracing. Using powers of observation, knowledge of local agricultural and social practices, and a difference in interpretation of changes over time, they propose that until recently, eastern Mediterranean landscapes flourished with prescribed fire regimes, pasturage, and agricultural uses. Contemporary erosion and imminent desertification are more likely the result of natural climatic and geological phenomena that in the past were brought on by traditional cultivation and grazing. For example, past soil loss following an intense downpour was prevented by century-old terraces. Now, however, global and local economic forces do not encourage the repair of terraces and intense downpours of rain are more likely the cause of soil loss.

By clarifying the level of engagement between humans and the biological communities of the region, Grove and Rackham reject the need for conservation as if to reach a perfect, Edenic landscape in favor of a coevolving present landscape that is highly productive and efficient. A degraded landscape is, according to Grove and Rackham, only in the vision of the claimant. The Aegean landscapes that pushed George Perkins Marsh to observe a degraded landscape tell us more about Marsh's conservation and philosophical beliefs than about the actual processes at work. According to Grove and Rackham, "Most Mediterranean landscapes are blends of the natural and man-made."[15] The evidence for this "blend," the noosphere, is the cultural landscape itself, including ancient and historic religious practices.

The Historical and Contemporary Religious Landscape

The religious landscape of the Aegean is as diverse in time as it is in area. Ancient Greeks laid claim to their spiritual identity through myth. These myths were incorporated into Hellenic Greek culture, together with myths and stories from the Levant. Changing belief systems spread across the region, sometimes unifying the human populations and sometimes brutally dividing them. More often, the ideas of one system were incorporated into the next over time and place. The monotheistic religions that dominate the region—Judaism, Islam, and Christianity—share a common origin story that branches out into different beliefs over time and place. The history of the three groups stems from the same sense of Yahweh, but has fractured into the three sectarian beliefs. The dominant Christian sect of the region is Eastern Orthodoxy, which is often described as the most "innocent" or "pure" form of Christianity. Its authenticity is difficult to argue, given its constant presence in the region since 380. Centuries of similar religious beliefs and practices formed a major part of the noosphere and are represented in the regional landscape.

The Christian story of the ultimate sacrifice, Jesus as the lamb, relies on earlier myths and religious practices of the region that space and time do not allow for exploration in this chapter. As a reinvention of an earlier cosmology, Jesus Christ represents the sacrificial lamb. The necessity of meat to sacrifice is essential to a human relationship with the gods. This tradition of sacrificing meat, grounded in the Old Testament and symbolic in the New, becomes a significant part of the Aegean noosphere. The contemporary landscapes of the region are representative of centuries of a religious need for a burnt-animal sacrifice, especially lamb.

The Apostle Paul's travels throughout Anatolia and the Greek islands to Rome spread religious identity with Jesus' life and teachings. Greek people, the Hellenes, migrated throughout the region. Their language, cultural identity, intellectual history, and folk practices provided a singular means of communication across the region. Slowly, this Greek world incorporated Christian ideas into its culture. John Tomkinson calls this merger of Hellenic culture with Christian ideas the "Great Synthesis."[16] The acceptance of Christianity in the region was completed with the Roman Emperor Constantine's move to the ancient Greek colony at Byzantium on the Black Sea in 330. By 381, Byzantium was elevated in status to equal Rome, and the Eastern Orthodox Church took on the mantle of religious leadership for the eastern Mediterranean.

Constantinople, like Rome, was both a political and a religious center. The Byzantine Empire encircled the Aegean from 610 until 1204 when the Fourth Crusade sacked the city. The Ottomans took Constantinople in 1453, bringing an end to the Byzantine Empire. The political and economic power of the old empire was largely destroyed. The Ottomans did not fully silence Orthodox Christianity in the region, however. Urban and rural populations of the old empire continued to follow the liturgy and worship in ways consistent with traditional practices of Orthodoxy. Land use practices and agricultural production were maintained in ways consistent with the historical culture and ecology of Orthodox Byzantium.

Contemporary Orthodox Christians may not be as devout as they once were, but Great Week and its many services and celebrations still demand the full attention of Christians across the Aegean. The final celebration on Paschal Sunday is observed at home with family and friends. Preparations for the feast to end the forty-day Great Lent fast progress throughout the week. By tradition, men must secure a lamb for roasting. In urban areas, the meat shops that have suffered for business during the fast are crowded. Butchered and skinless lambs, eyes protruding, hang in shop windows ready to be taken home and roasted on charcoal grills, on driveways, or on front lawns, if necessary. These customs are evident all over Lesvos Island.

Lesvos Island's Religious Landscape

Lesvos Island, also known as Lesbos and the home of the ancient poetess Sappho, is located in the northeastern Aegean Sea. Only six kilometers off the Turkish coast, the island is more Anatolian in culture and physiography than Greek. The chevron-shaped island, with its two large southern bays, has the appearance of a jigsaw puzzle piece, perhaps an apt metaphor for its significance in the intellectual debate about creation, evolution, logic, and religious belief. The tower of St. Athanasios Cathedral in Mytilini, the island's primary city, faces Asia Minor, as does the old Venetian castle and the historic port. The island was historically self-sufficient in terms of food and energy.

The vegetation of the island has been altered over thousands of years, due to terracing, fires, and grazing of goats and sheep. Terracing was an early adaptation to the region's rugged topography, shallow soils, and sudden storms. Olive groves have traditionally occupied the terraces. Olive trees require little watering and have extensive root systems that hold the fragile Mediterranean soils in place on the terraces, preventing soil erosion. The synergy between terraced soils used for olive production and seasonal grazing on open lands conserved a dynamic and efficient use of available land and resources. The land use pattern on Lesvos allowed for cultivation of lowlands, terraced olive groves across the western half of the island, and grazing of sheep and goats at higher elevations on the eastern and consistently across the western side of the island. This pattern produced sufficient amounts of foodstuffs for the island, as well as export goods, such as olive oil for over seven thousand years.

Sedentary, urbanizing, agricultural practices in flatlands with available water sources for irrigation and agropasturalism, with its associated firepower, dominated Lesvos and other islands in the Aegean Basin. Nomad and sedentary lifestyles created a productive landscape, equally dependent on each other, but suspicious of one another. The nomadic, shepherding lifestyle was suspect because of its fleeting, periodic appearance in domestic, urbanized centers. A difference in apparent wealth and material goods also separated the two populations. But the two populations relied on each other for specialized goods and shared technologies. Most importantly, the urbanized populations counted on the pastoralists to provide one of the most significant necessities for cultural maintenance, the Paschal lamb and the charcoal with which to roast the symbolic meat.

The historical relationships between urban dwellers and rural shepherds have been in place for centuries. Only recently have Greek agricultural policies affected traditional land use patterns. Andreas Troumbis, an ecologist at the University of the Aegean, Lesvos, writes about the ecological role of cattle, sheep, and goats.[17] He suggests that humans and domesticated ruminants have interacted over long periods in ways that have transformed both the animals and human societies. Members of nomadic tribes, who once followed herds of animals, learned to domesticate animals and, consequently, altered the scale of human residence, regional biota, and the role of herded animals in maintaining ecosystems. Shepherds invigorated pasturelands and open space by periodically burning the land. Low-temperature fires would remove woody debris and encourage new shoots on trees and shrubs. The use of fire by shepherds to promote growth has a very long history in the region.

The geographers Theodoros Iosifides and Theodoros Politidis' recent article, "Socio-Economic Dynamics, Local Development and Desertification in Western Lesvos, Greece," suggests that deeply rooted social practices are the causes of the current rate of erosion on the island.[18] Iosifides and Politidis offer a unique look at local beliefs and practices of rural pastoralists on Lesvos Island. Based on intensive interviews with herders in the most arid region of the island, these geographers analyze their findings and provide an alternative explanation to the official Greek view that desertification can be prevented with better technology. Instead, Iosifides and Politidis suggest that the causes of current rates of desertification are due to governmental policies that ignore historical patterns of land use. The relationship between land use and society is a "web of relationships and interconnectedness between the local development trajectory and the overall environmental performance and problems in the area."[19] In light of their conclusions, I argue that grazing has become a necessity and that the system is currently experiencing a temporary period of excess on its way to a period of readjustment within a self-organizing system. Placed within this context, the landscapes of the Aegean require rereading from a post-Marsh perspective.

Rereading the Aegean Landscape

Recent ideas about human-land relationships coalesce around the concept that human agency is the most important factor and that Earth processes in landscape formation are at the whim of human need or desire.

Throughout their volume, Grove and Rackham suggest that degradation of the region is not occurring. They maintain that the assertions of land degradation are unfounded and that there are systems in place that have protected the land base from disruption, erosion, overgrazing, and reduction in productivity. They suggest that a better understanding of the relationship between herbaceous cover, fire, grazing, and human demands is needed to explain the state of Aegean landscapes. Conservation, then, would be seen as an integral part of the status quo. If conservation is required, it would have to be done with better scientific observations of past landscapes and current problems.

Evidence for the long use of fire as a tool for creating and maintaining specific mosaics of herbaceous cover is mounting. Giovannini and colleagues have shown that without regularly occurring fires, erosion tends to increase. Periodic fires remove enough debris to prevent high temperatures associated with catastrophic fire events. The ongoing removal of debris in low-temperature fires enables root systems and larger shrubs to prevent soil losses. Grazing has an equally powerful ability to prevent catastrophic events, such as high-temperature fires. Grazing invigorates grass and shrub growth. While grazed landscapes have often been described as degraded, grazing forces new growth on a seasonal basis. The interpretation of grazed landscapes as degraded occurs in the event of severe overgrazing due to changes in traditional land use practices.[20]

Between these poles of thought is a third theory: coevolution. Usually defined by ecologists as symbiotic relationships between organisms, coevolution becomes a more sensitive topic when it is used to describe the relationship between humans and Earth. To place humans in a position of having evolved with Earth calls into question where, when, and how humans are related to Earth as equal to others.

The work of these scientists stresses an underlying relationship between humans and Earth processes that may, in fact, support the rewriting of modern interpretations of religious and scientific texts. If the landscapes, with the use of fire and grazing, are actually the most efficient from the perspective of conservation of resources, then humans are acting in a way that protects Earth processes. Traditional religious practices created and conserved Earth for everlasting as a self-perpetuating system. The noosphere was fully integrated into the biosphere, and the result was a functioning and self-organizing landscape. Conservation measures were needed at the local scale in immediate situations of warfare, earthquakes, or storms, but the basic infrastructure and demands could withstand years to centuries of continual use.

Horden and Purcell offer further reflection on the ways in which religion has affected the Mediterranean landscape. While much of their work is focused on the material landscape, they also observe that the practice of Christianity was a unifying force. The local knowledge that a given set of topographic characteristics are continuous, contiguous, historically written, and practiced daily creates "territories of grace."[21] According to Horden and Purcell, cultural and physical survival is dependent not on a specific set of land use practices (conservation), but on changing human initiative to meet relatively few significantly catastrophic events. Horden and Purcell underscore the significance of hierarchical Byzantium's inclusion of local, ancient religious practices. The ancient practice of meat sacrifice, which is also embraced by Orthodoxy, has the same effect on the grazed rural landscape. The seemingly eroded landscape of the Aegean is not a landscape of degradation; it is a coevolved religious landscape that has for centuries offered its human inhabitants mechanisms that reinforce a specific noosphere.

The practice of sacrificing lambs is an ancient and significant part of religion in the Aegean. Herding sheep for religious purposes, as well as serving as an efficient form of food production, has been a part of the region's social and ecological evolution for centuries. The noosphere of meat production and consumption on specific Orthodox feasts maintained the self-organizing system of nature and religion. The noosphere that has evolved in the region created an interwoven set of relationships between humans, animals, and biota. These interrelationships created the landscape of the Aegean and the larger eastern Mediterranean region, which, until recently, was able to handle catastrophic natural and human events and maintain a functioning and self-renewing environment.

The web of relationships identified by Iosifides and Politidis describes a noosphere with an ethos of Orthodoxy, the ecological conditions of Lesvos Island, and urbanized populations dependent on rural pastoralists. Since 1800, rapid urbanization and globalization have led to catastrophic events and land use patterns that the old traditional system cannot accommodate. It is now that the bell pealed by Marsh is starting to attract conservation plans and pronouncements. The degradation noted by Marsh is now occurring, not due to traditional patterns that he misinterpreted as overgrazing and severe erosion, but as a result of national and international agricultural policies of the late twentieth century. Increasing numbers of goats and sheep pastured on declining terraced olive groves without the advantage of cyclical fires is radically changing the self-organized and adaptive landscape of Greece.

A New Vision

William Yeats' poem, "Sailing to Byzantium," captures the possibilities of the older traditional world of Eastern Europe as an antidote to the rapidly changing Western Europe. Yeats' imagined realm of old Byzantium is visible in the historical landscape of the Aegean. Urban centers surrounded by terraced lands, pastures, and pine forests still hold the basic mosaic of social, cultural, and physical components of the old Orthodox empire. The old traditional culture of Byzantium is kept alive by the practice and order of the Orthodox Church. Today, old Byzantium is revealed in the contemporary nations of the Aegean. Contemporary governments and membership in the European Union are having an impact on local economies, especially when agricultural policies and land use policies are favoring urban centers over the more remote and rural populations. At the beginning of the twenty-first century, when agricultural and economic policies favor a globalized economy, the landscape of the Aegean is much closer to becoming what George Perkins Marsh misinterpreted in the nineteenth century.

Rethinking the relationship of nature, science, and religion requires an understanding of the role of landscape in representing belief systems. Thomas Maxwell suggests that a rereading of landscape texts from a deep ecological perspective requires a redefinition of local landscapes as integrated ecological and spiritual wholes.[22] Landscapes are the result of the interaction between humans, the noosphere, and Earth processes over long periods. Conservation over long periods has evolved in ways that protect both human and natural settings. Karl S. Zimmerer suggests that the capacity of humans to engage critically and carefully with nature could lead to a wiser form of environmental conservation that provides mechanisms for reinforcing local, rural geographies that can accommodate catastrophic stress.[23]

Aegean landscapes have proven to be critical for interpreting significant questions about human-nature relationships. It is through the theoretical and philosophical dimensions of landscape that a holistic view of humanity is possible. Landscape analysis sidesteps the problems of reductionism, linear science in favor of "nonlinear and partly chaotic mutual-causal and reciprocal processes of the coevolution of Mediterranean people and their landscapes."[24] Instead of smaller and smaller bits of reality, landscape analysis creates an avenue for interpreting and understanding "wholeness, connectedness, integration, synthesis, and complementarities of ordered complexity, and from non- and multidisciplinarity to inter- and transdisciplinarity."[25] Similarily, the theologian Kimberly Whitney believes that the moral and ethical necessity of making just choices about local landscapes protects an unfolding, not determined, landscape.[26]

The landscape is beginning to change, due to expanding urbanization at the cost of rural life. Bartholomew I, ecumenical patriarch of the Orthodox Church, has chosen to speak out about the issues of pollution, the necessity to restore marine life, and the rekindling of Orthodox beliefs about nature and ecology. Bartholomew, known as the green patriarch, has spoken on a number of occasions about the necessity for Orthodox Christians to consider their heritage and tradition with respect to nature. Other Orthodox theologians have continued the discourse set in motion by Bartholomew I. The Very Reverend Stanley S. Harakas, speaking before the Orthodox Summit on the Environment, sponsored by the National Council of Churches in the United States, the Standing Conference of Canonical Orthodox Bishops in America, and the Oriental Orthodox Churches in the United States, said, "We as Orthodox Christians have our answers. We, too, must describe the Orthodox Christian theological vision as the larger context to ecological reflection."[27] Harakas' vision can be summarized as (1) God is the ultimate reality; (2) creation (nature) depends on God; (3) sacrifice is necessary for redemption and resurrection; (4) God maintains community with humans through the Holy Spirit; (5) misuse of nature is a sin related to the lack of humility; and (6) humans must live in a state of sacrifice in order to protect nature. These principles of Orthodoxy and its relationship with nature are to be lived out by every Orthodox Christian, not by governments, business, or international treaties. Harakas, like Bartholomew I, turns to the local, to individual beliefs and practices, as shepherded by the church, to protect nature. "So, as the microcosm of the universe, we are called to serve as the priests of creation for all beings, we are called to live in a God-like communion with our Creator, with our neighbor, with the material world, because precisely, as the Psalmist (Psalm 41:1) says: 'The Earth is the Lord's and the fullness thereof, the world and those who dwell therein.'"[28]

Bartholomew I has taken a strong stand as the green patriarch to reunite religion with science and nature. When speaking at the Symposium on the Sacredness of the Environment, held at the Saint Barbara Greek Orthodox Church in Santa Barbara, he stated, "The entire universe participates in celebration of life, which St. Maximos the Confessor described as a 'cosmic Liturgy.'" We see this cosmic liturgy in the symbiosis of life's rich biological complexities. As human beings, created "In the image and likeness of God" we are called to recognize this interdependence between our environment and ourselves.[29] Bartholomew I instructs Orthodox Christians to conserve ocean and land resources not for potential economic gain, but because these are God's creations. In his vision, Bartholomew speaks to the need for local and rural wisdom, rather than global economic and political logic. Asking individuals to practice a systemic conservation reinvigorates the web of relationships between nature, science, and humans, or the noosphere. Bartholomew I reminds humans of their creative power in an adaptive natural world. Humans, then, are part of the self-organizing system at the local and regional level.

The concept of the noosphere allows for a reading of Aegean islands as a self-organizing system where religious practices, traditional land use practices, and local economies created a sustainable environment that embraced humanity. The self-organizing system included natural human modifications to the land, such as terracing and controlled fires. These land use activities prevented catastrophic events, such as major wildland fires, water erosion, and soil depletion. The religious practices that demanded grazing prevented catastrophic events and promoted abundance. In this dynamic landscape, the noosphere on the Aegean islands integrated the human and natural world in a coevolutionary process at the local scale. By rethinking Marsh's powerful paradigm that humans have a tendency to destroy the gift of nature, and envisioning nature as a self-organizing system that includes humans, the connections between humans, nature, and society can be redefined.

Notes

1. George Perkins Marsh, *Man and Nature: Physical Geography as Modified by Human Action*, reprint ed. Edited by David Lowenthal (Seattle: University of Washington Press, 2003).
2. As cited in John Opie, *Nature's Nation: An Environmental History of the United States* (Orlando: Harcourt Brace & Company, 1998), 372.
3. I am thankful to Antje Jackelén for her guidance on theological terms and their meanings.
4. Marsh, *Man and Nature*, xvi.
5. Ibid., 1–52.
6. Evelyn Fox Keller, "Ecosystems, Organisms, and Machines," *BioScience* 55 (December 2005): 1069.
7. Jonathan D. Phillips, "Divergence, Convergence, and Self-Organization in Landscapes," *Annals of the Association of American Geographers* 89 (September 1999): 468.
8. Thomas S. Smith and Gregory T. Stevens, "Emergence, Self-Organization, and Social Interaction: Arousal-Dependent Structure in Social Systems," *Sociological Theory* 14 (July 1996): 133.
9. The term *noosphere* is the English version of the Greek word *nous* meaning "mind." The term *noosphere* first appears in Valdimir Vernadsky's theory of human cognition as fundamental to transformation of nature. The term appears later in the works of Pierre Teilhard de Chardin, who theorized about a continual emergence toward an ultimate integration of spirit, humans, and nature. I use the term as theorized by Richard B. Norgaard and Erich Jantsch who conceive a theory of sustainability. I am thankful to Johannes M.M.H. Thijssen for clarifying the origins of the word.
10. Francesco Di Castri, David W. Goodall, and Raymond L. Spect, *Mediterranean-Types Shrublands* (Amsterdam: Elsevier Scientific Publishing Co., 1981), 3.
11. Peregrine Horden and Nicholas Purcell, *The Corrupting Sea: A Study of Mediterranean History* (Malden, Mass.: Blackwell Publishing Co., 2003), 410.
12. A. T. Grove and Oliver Rackham, *The Nature of Mediterranean Europe: An Ecological History* (New Haven, Conn.: Yale University Press, 2001), 10.
13. Ave Perevolotsky and No'am Seligman, "The Role of Grazing in Mediterranean Rangeland Ecosystems," *BioScience* 48 (December 1998): 1007.
14. Grove and Rackham, *The Nature of Mediterranean Europe*, 8.
15. Ibid., 364.
16. John Tomkinson, *Between Heaven & Earth: The Greek Church* (Athens: Anagnosis Publications, 2003), 109.
17. Andreas Troumbis, "Cattle, Sheep, and Goats, Ecological Role Of," *Encyclopedia of Biodiversity*, vol. 1 (San Diego, Calif.: Academic Press, 2001), 651.

18. Theodoros Iosifides and Theodoros Politidis, "Socio-Economic Dynamics, Local Development and Desertification in Western Lesvos, Greece," *Local Environment* 10 (2005): 496.
19. Ibid.
20. G. Giovannini, R. Vallejo, S. Lucchesi, S. Bautista, S. Ciompi, and J. Llovet, "Effects of Land Use and Eventual Fire on Soil Erodibility in Dry Mediterranean Conditions," *Forest Ecology and Management* 147 (2001): 15–23.
21. Horden and Purcell, *The Corrupting Sea,* 457.
22. Thomas Maxwell, "Considering Spirituality: Integral Spirituality, Deep Science, and Ecological Awareness," *Zygon* 38 (June 2003): 260.
23. Karl S. Zimmerer, "Human Geography and the 'New Ecology': The Prospect and Promise of Integration," *Annals of the Association of American Geographers* 48 (March 1994): 125.
24. Zev Naveh and Yohay Carmel, "The Evolution of the Cultural Mediterranean Landscape in Israel as Affected by Fire, Grazing, and Human Activities," *Bioscience* 50 (April 2000): 339.
25. Ibid., 340.
26. Kimberly Whitney, "Greening by Place: Sustaining Cultures, Ecologies, Communities," *Journal of Women and Religion* 19/20 (2003): 15.
27. The Very Reverend Stanley S. Harakas, "The Earth Is the Lord's: Orthodox Theology and the Environment," *The Greek Orthodox Theological Review* 44 (1999): 162.
28. Ibid.
29. Bartholomew I, *A Rich Heritage,* (Grand Rapids, Mich.: William B. Eerdmans Publishing Co., 2003), 219.

Bibliography

Braudel, Fernand. *Civilization and Capitalism.* New York: Harper & Row, 1981–1984.

Castree, Noel, and Bruce Craun, eds. *Social Nature: Theory, Practice, and Politics.* Malden, Mass.: Blackwell Publishers, 2001.

Castri, Francesco di, David W. Goodall, and Raymond L. Specht. *Mediterranean-Type Shrublands, Ecosystems of the World,* vol. 11. Amsterdam: Elsevier Scientific Publishing Co., 1981.

Chapple, Christopher Key, ed. *Ecological Perspectives: Scientific, Religious, and Aesthetic Perspectives.* Albany: State University of New York Press, 1994.

Chryssavgis, John. *Cosmic Grace and Humble Prayer: The Ecological Vision of the Green Patriarch Bartholomew I.* Grand Rapids, Mich.: William B. Eerdmans Publishing, 2003.

Constantelos, Demetrios J. *Understanding the Greek Orthodox Church: Its Faith, History, and Practice.* New York: Seabury Press, 1982.

Farina, Almo. "The Cultural Landscape as a Model for the Integration of Ecology and Economics." *BioScience* 50 (April 2000): 313–20.

Farina, A., A. R. Johnson, S. J. Tuner, and A. Belgrano. "'Full' World versus 'Empty' World Paradigm at the Time of Globalisation." *Ecological Economics* 45 (2003): 11–18.

Foxhall, Lin. "Cultures, Landscapes, and Identities in the Mediterranean World." *Mediterranean Historical Review* 18 (December 2003): 75–92.

Giourga, H., N. S. Margaris, and D. Vokou. "Effects of Grazing Pressure on Succession Process and Productivity in Old Fields on Mediterranean Islands." *Environmental Management* 22 (July/August 1998): 589–96.

Giovannini, G., R. Vallejo, S. Lucchesi, S. Bautista, S. Ciompi, and J. Llovet. "Effects of Land Use and Eventual Fire on Soil Erodibility in Dry Mediterranean Conditions." *Forest Ecology and Management* 147 (2001): 15–23.

Glacken, C. J. *Traces on the Rhodian Shore: Nature and Culture in Western Thought from Ancient Times to the End of the Eighteenth Century*. Berkeley: University of California Press, 1967.

Goudsblom, J. "The Human Monopoly on the Use of Fire: Its Origins and Conditions." *Human Evolution* 1 (1986): 517–23.

Grove, A. T., and Oliver Rackham. *The Nature of Mediterranean Europe: An Ecological History*. New Haven, Conn.: Yale University Press, 2001.

Harakas, Stanley. "The Earth Is the Lord's: Orthodox Theology and the Environment." *The Greek Orthodox Theological Review* 44 (1999): 149–62.

Haywood, John. *Historical Atlas of the Medieval World AD 600–1492*. New York: Barnes and Noble Books, 1998.

His All Holiness Ecumenical Patriarch Bartholomew. "The Abuse of Nature Is a Sin." *Earth Island Journal* 13 (Spring 1998): 43.

———. "Encyclical Letter, September 1, 1994." In Fr. John Chryssavgis, ed., *Cosmic Grace and Humble Prayer: The Ecological Vision of the Green Patriarch Bartholomew I*, 45–47. Grand Rapids, Mich.: William B. Eerdmans Publishing Co., 2003.

Hopwood, Keith. "Living on the Margin: Byzantine Farmers and Turkish Herders." *Journal of Mediterranean Studies* 10 (2000): 93–105.

Horden, Peregrine, and Nicholas Purcell. *The Corrupting Sea: A Study of Mediterranean History*. Malden, Mass.: Blackwell Publishing Co., 2003.

Hughes, J. Donald. *Pan's Travail: Environmental Problems of the Ancient Greeks and Romans*. Baltimore: Johns Hopkins University Press, 1994.

Huntington, Ellsworth. "The Burial of Olympia: A Study in Climate and History." *The Geographical Journal* 36 (December 1910): 657–86.

Iosifides, Theodoros, and Theodoros Politidis. "Socio-Economic Dynamics, Local Development and Desertification of Western Lesvos, Greece." *Local Environment* 10 (October 2005): 487–99.

Jameson, Michael H., Curtis N. Runnels, and Tjeerd H. van Andel. *A Greek Countryside: The Southern Argolid from Prehistory to the Present Day*. Stanford, Calif.: Stanford University Press, 1994.

Jantsch, Erich. *The Self-Organizing Universe: Scientific and Human Implications of the Emerging Paradigm of Evolution*. New York: Pergamon Press, 1980.

———, ed. *The Evolutionary Vision: Toward a Unifying Paradigm of Physical, Biological, and Sociocultural Evolution*. Boulder, Colo.: Westview Press, 1981.

Keller, Evelyn Fox. "Ecosystems, Organisms, and Machines." *Bioscience* 55 (December 2005): 1069–74.

Lavergne, Sebastien, Wilfried Thuiller, James Homina, and Max Debussche. "Environmental and Human Factors Influencing Rare Plant Local Occurrence, Extinction and Persistence: A 115-Year Study in the Mediterranean Region." *Journal of Biogeography* 32 (May 2005): 799–811.

Levin, Simon A. "Complex Adaptive Systems: Exploring the Known, the Unknown and the Unknowable." *The Bulletin of the American Mathematical Society* 40 (January 2002): 3–19.
Livingston, David N. *Nathaniel Southgate Shaler and the Culture of American Science.* Tuscaloosa: University of Alabama Press, 1987.
Louloudis, Leonidas, and Napolean Maraveyas. "Farmers and Agricultural Policy in Greece Since the Accession to the European Union." *Sociologia Ruralis* 37 (1997): 270–85.
Loumou, Angeliki, and Christina Giourga. "Olive Groves: The Life and Identity of the Mediterranean." *Agriculture and Human Values* 20 (Spring 2003): 87–95.
Malanson, George. "Considering Complexity." *The Annals of the Association of American Geographers* 89 (December 1999): 746–53.
Marsh, George Perkins. *Man and Nature.* Edited by David Lowenthal. Seattle: University of Washington Press, 2003.
Maxwell, Thomas. "Considering Spirituality: Integral Spirituality, Deep Science, and Ecological Awareness." *Zygon: Journal of Religion and Science* 38 (June 2003): 257–76.
McNeill, J. R. *Mountains of the Mediterranean World: An Environmental History.* Cambridge: Cambridge University Press, 1992.
Miller, Dean A. *The Byzantine Tradition.* New York: Harper & Row, 1966.
Morehouse, Barbara, and Martha L. Henderson, Kostas Kalabokidis, and Theodoros Iosifides, "The Social Context of Forest Fire Management on the Islands of Lesvos, Chioes, and Samos in Greece: A Report on Results of a Survey Conducted in Spring, 2004." Department of Geography, University of the Aegean, Lesvos, 2005.
Naveh, Zev. "The Total Human Ecosystems: Integrating Ecology and Economics." *Bioscience* 50 (April 2000): 357–61.
Naveh, Zev, and Yohay Carmel. "The Evolution of the Cultural Mediterranean Landscape in Israel as Affected by Fire, Grazing, and Human Activities." In S. Wasser, ed., *Evolutionary Theory and Processes: Modern Horizons, Papers in Honor of Eviator Nevo,* 337–80. Amsterdam: Kluwer Academic Publishers, 2003.
Norgaard, Richard B. *Development Betrayed: The End of Progress and a Co-Evolutionary Revisioning of the Future.* London: Rutledge Press, 1994.
Papanastasis, V., M. Arianoutsou, and G. Lyrintzis. "Management of Biotic Resources in Ancient Greece." In *Proceedings of the 10th MEDECOS Conference, April 25–May 1, 2004.* Rotterdam: Millpress, 2004.
Perevolotsky, Avi, and No'am Seligman. "The Role of Grazing in Mediterranean Rangeland Ecosystems." *BioScience* 48 (December 1998): 1007–17.
Phillips, Jonathan D. "Divergence, Convergence, and Self-Organization in Landscapes." *Annals of the Association of American Geographers* 89 (September 1999): 466–88.
Pyne, Stephan. *Vestal Fires: An Environmental History, Told Through Fire, of Europe and Europe's Encounter with the World.* Seattle: University of Washington Press, 1997.
Rackham, Oliver. "Fire in the European Mediterranean." *Aridlands Newsletter* 54 (November/December 2003): 1.
Robinson, Thomas M., and Laura Westra. *Thinking About the Environment.* Lanham, Md.: Lexington Books, 2002.
Sallares, Robert. *The Ecology of the Ancient Greek World.* Ithaca, N.Y.: Cornell University Press, 1991.
Semple, Ellen Churchill. *Geography of the Mediterranean Region.* New York: Henry Holt and Co., 1931.

Smith, Thomas S., and Gregory T. Stevens. "Emergence, Self-Organization, and Social Interaction: Arousal-Dependent Structure in Social Systems." *Sociological Theory* 14 (July 1996): 131–53.

Teilhard de Chardin, Pierre. *Man's Place in Nature: The Zoological Group*. New York: Harper & Row Publishing, 1966.

Thompson, John N. *The Coevolutionary Process*. Chicago: University of Chicago Press, 1994.

Tomkinson, John L. *Between Heaven & Earth: The Greek Church*. Athens: Anagnosis, 2003.

Troumbis, Andreas Y. "Battle, Sheep and Goats." In Simon A. Levin, ed., *Encyclopedia of Biodiversity*, vol. 1, 651–63. London: Academic Press, 2001.

Vernadsky, Valdimir. *The Biosphere*. Oracle, Ariz.: Synergetic Press, 1986; orig. 1926.

Westra, Laura, and Thomas M. Robinson, eds. *The Greeks and the Environment*. Lanham, Md.: Rowman & Littlefield Publishers, 1997.

Whitney, Kimberly. "Greening by Place: Sustaining Culture, Ecologies, Communities." *Journal of Women and Religion* 19 (2001): 11–26.

Yannitsaros, Artemois, and Ioannis Bazos. "A Brief Outline of the Flora and Vegetation of Lesvos Island." Report on Record in the Institute of Systematic Botany, Department of Biology, University of Athens.

Yeats, William B. "Sailing to Byzantium." In Richard J. Finneran, ed., *W. B. Yeats: The Poems*, 193. New York: McMillan Pub., 1983.

Zimmerer, Karl S. "Human Geography and the 'New Ecology': The Prospect and Promise of Integration." *Annals of the Association of American Geographers* 84 (March 1994): 108–25.

10

THE VISION OF MALLEABLE NATURE: A COMPLEX CONVERSATION

Andrew Lustig

Introduction

The New Visions project has elaborated five perspectives on nature that capture characteristic features of that concept as deployed by different disciplines. To its credit, the project has chosen to focus on the larger philosophical, scientific, and religious themes that frame such discussions. All five visions of nature raise fundamental questions of ontology, epistemology, ethics, and theology. The vision of malleable nature, however, highlights the nodes of intersection across those four subject areas in distinctive ways, because this perspective focuses on the complexity of normative judgments about the appropriate limits (if any) on human alterations of the natural world. Such judgments invoke different perspectives on moral theory (e.g., deontology, virtue theory, and utilitarianism) as well as different interpretations of the value of nature in these theoretical frameworks (e.g., intrinsic or instrumentalist). Moreover, as exemplified by recent advances in biotechnology and regenerative medicine, the task of normative line-drawing grows ever more problematic, because the ways that formerly "external" life processes are "internalized" in biotechnological applications pose fundamental challenges to the cogency of classical interpretations of nature.

The present chapter does not aim to resolve the substantive tensions among different metaphysical, epistemological, and normative perspectives on nature and the natural. Instead, it assumes the more modest task of conceptual cartography. Among the five visions, *malleable nature* is distinctive in the ways that it illustrates the network of relations across disciplines. In this essay, I will explore three fundamental aspects of that complex conversation. First, I will consider the polyvalence of nature as a basic term and the difficulties thereby generated in making ethical judgments about human interventions into the natural world. From the outset, that descriptive variety raises philosophical questions about our descriptions of nature as a given and our prescriptive conclusions about alterations of nature (the "is-ought" problem). Second, I will provide an overview of key emphases in recent critical studies that contribute to the complexity of our normative judgments about nature as malleable. These themes, especially as developed in feminist thought, challenge historical (and perhaps still dominant) perspectives on nature as appropriately open to human mastery in largely instrumental terms. Finally, and in somewhat greater detail, I will review various religious perspectives on nature that provide the general background to judgments about particular interventions or technologies. I will conclude that there are possibilities for modest convergence among seemingly disparate worldviews, as well as between religious and secular discussions about technology.

The Continuum of New Visions of Nature

It is helpful to plot the five visions of nature on something of a theoretical and practical continuum, with each having implications for the others. Viewing nature as "external" assumes that we can metaphysically distinguish aspects that are separate or apart from ourselves from those that are "constitutive" of us. This sense of nature as in some sense a given is a prominent feature of several of the New Visions perspectives on nature and, obviously, a core emphasis in environmental studies. Yet the concept of nature is always contextualized, as a number of other chapters in this volume forcefully demonstrate. For example, David Livingstone concludes that understandings of nature exhibit profound differences because of the distinctive ways that particular visions are appropriated in different settings.[1] Even more pointedly, John Hedley Brooke starkly poses the problem we face in our efforts to consistently or coherently invoke the term: "We are currently living through an era in which the traditional dualities [e.g., nature/man, nature/supernature, natural/unnatural] are failing and our grasp of what it means to speak of 'nature' at all is wavering."[2] And the range of interpretation is not limited to different historical understandings. Cultural readings also shape, and are shaped by, differing perspectives on space itself, as Martha Henderson's discussion of the geography of landscape makes clear.[3] Or again, as James Proctor points out, the term *environment,* like the term *nature,* is interpreted in different ways for a range of varied, and often conflicting, purposes.[4]

As postmodernists, we realize that all conceptualizations of nature inevitably influence the object being analyzed; hence, a naïve realism toward nature as a given is as misguided as claims for such realism more generally. Consider various examples. The vision of evolutionary nature is well confirmed by its specific explanatory power, and other visions of nature, to retain their currency, necessarily incorporate elements of neo-Darwinian thought. At the same time, the vision of evolutionary nature itself requires that we clearly distinguish mere alterations from changes that are meaningful in Darwinian terms. The latter judgments necessarily involve interpretive criteria for assessing the *relative* influences of natural selection and other contributors to niche-dependent adaptations. Similarly, the vision of nature as sacred includes differing judgments about divinity/spirituality vis-à-vis the world (with a range of religious and theological interpretations of immanence and transcendence as categories). Sacred nature also includes quasi-religious attitudes toward nature that, while shorn of the richer worldview that inspired them, remain deep influences on much current spirituality. Nature as culture may appear to be the exception here, because a commitment to thoroughgoing constructivism appears to call into question the character and cogency of the other four visions.[5] Even here, however, where human agency is deemed decisive, cultural constructs always require "materials" toward which human intentionality is directed and exercised. Thus, the vision of nature as culture, therefore, will draw on a range of theoretical commitments in its support. Such commitments may well include the stance of critical realism, which accepts a certain givenness to the world, even as it acknowledges the cultural embeddedness of all conceptual interpretations.

Malleable Nature as a Conceptual Midpoint

In plotting the five visions of nature on a continuum, malleable nature stands at something of a conceptual midpoint. It provides an intersection for attitudes and values that are primary emphases in other visions without thereby incorporating the (at least) prima facie descriptive and/or normative priorities evident in those accounts. To describe nature as malleable brings to the foreground the explicit normative dimensions of our discussion; moreover, it challenges the continuing relevance of earlier appeals to nature as in some sense a prescriptive norm. Contemporary judgments about the *appropriateness* of human alterations of nature seem to require that we move beyond earlier historical assumptions that nature, in itself, provides sufficient grounding for subsequent normative reflection. For reasons I will explore below, current accounts of nature, described naturalistically, seem incapable of providing a perspective on the good toward which we should strive or an answer to the question of what we ought to do.

The Is-Ought Distinction

In traditions that are understood as ways of life, description and evaluation defy easy separation. From the time of Stoic visions of nature as normative (if not before), "the origins of natural order [have been] invoked to justify" moral choices and actions.[6] However, the rise of modern science has called into question the conflation of the empirical and the moral, as captured by David Hume's famous criticism of arguments that move from description to prescription without clear warrant. It is important to note, despite popular misconceptions, that Hume was not prohibiting such arguments, but criticizing them for failing to provide the necessary "bridgework" between empirical facts and moral directives.[7] The more tentative thrust of Hume's criticism should be distinguished from the "naturalistic fallacy" as described by G. E. Moore. For Moore, moral terms *cannot* be defined in terms of nonmoral realities [emphasis mine].[8] A naturalistic ethic that finds its criterion of the good and its moral justification in what nature has established is incoherent. What *is* simply lacks the power to justify what we believe *ought to be*.

However, the proscription against inferring prescriptions from descriptions, and vice versa, cuts both ways. Read starkly, it makes science irrelevant to ethics, but it makes ethics equally irrelevant to science. For many in both camps, that conceptual segregation is problematic. On the one hand, efforts in the wake of Kant to formalize the logic of moral claims seem to undercut ascriptions of moral value to phenomenal nature. On the other hand, recent work in the history and philosophy of science underscores the vital role of values and value-laden presuppositions in prompting and shaping scientific inquiry. There are efforts, of course, to retain appeals to nature as the empirical basis of morality. However, those attempts have been seen by many as an illicit reduction of normative ethics to descriptive ethics, and therefore unpersuasive. Moreover, on any such account, the concept of nature remains ambiguous. The natural precedents for both competition and reciprocal altruism can be found in nature. Which should become our moral guide? In deciding, we will necessarily appeal to a criterion that transcends nature. "The choice between these attitudes is a moral choice," writes Mary Midgley, "so it cannot be determined by science."[9]

In all of this, it is important to acknowledge the variety of efforts, in science and philosophy of science, as well as in ethics and the humanities, to conceive of the import of the natural for the normative. Moreover, both the philosophical and religious literature may emphasize the differences between discussions of "human nature" and "nature in general." It is (or may be) possible to articulate some general concept of what it means to be "human" to guide or constrain ethical choices. However, it is far less clear, from the vantage point of current science or philosophy of science, that we are warranted in appealing to nature in ways that would provide equivalently clear ethical guidelines for interventions into it. Indeed, it may well be that what is "good" for nature is in conflict with what is good for humans.

Contributions from Critical Studies

Perspectives from critical studies raise a number of basic political and social questions concerning the status of nature as a descriptive and prescriptive concept. First, who has "constructed" the idea of nature—that is, which persons, professions, or social classes are its primary "authors"? Second, toward what audience is the concept directed and for what purposes? Third, as perhaps a more pointed restatement of the second question, who benefits from or is privileged by a particular account or interpretation? Fourth, in what ways does the concept function as a source and/or product of social relations? Finally, do those at the margins of the institution or tradition within which the concept is developed agree with its articulation and implications, or would they describe nature in different terms?

Recent feminist philosophers and historians have explored, often in richly nuanced studies of particular institutions and professions, how descriptions of nature have served to reinforce social inequalities in ways pernicious to the interests of women and other marginalized members of society. From that rapidly growing literature, five themes emerge as especially relevant to the discussion of malleable nature. First, there are implications for understandings of epistemology, with special attention to the ways that concepts and methods in philosophy of science will be expanded or recast in light of feminist concerns. Western culture has often equated rationality with masculinity and emotionality with femininity, with the result that traditional epistemology has tended to valorize "male" forms of cognition.[10]

Second, some recent feminist philosophers of science have challenged the status of scientific objectivity as defined by the Baconian paradigm that framed the mechanistic philosophy of the scientific revolution. On their reading, especially since the Enlightenment, the scientific method has been centrally linked to notions of prediction, control, and domination of nature. In contrast, these feminists seek to redescribe scientific success as the ability of scientists "to listen to nature's self-revelations." The shift of metaphor here—from power and domination to attentive and respectful listening—appears to suggest a different understanding of scientific objectivity itself, and of the methods that may be most appropriate for promoting scientific discovery.[11]

Third, social feminists have made identifying and rectifying persisting gender-based inequalities their central focus. Feminist historians have argued that the status of women has been largely determined by social structures and practices, especially practices related to reproduction and childrearing. Several feminists have explored the ways that some prominent liberal theorists have insisted on a biologism of innate differences between men and women in order to justify social inequalities in a manner amenable to their own political predilections. Indeed, a significant portion of anatomical research in nineteenth-century, institution-based science focused on the putatively natural bases of inequality between men and women (and between Caucasians and other races).[12]

Fourth, in the historical efforts of a male-dominated scientific establishment to identify natural differences in anatomy and biology as the basis of social inequality, the privileged place of white males in society and the professions was justified in light of those putative differences. In that division of labor, a larger working parallel also emerged: "[N]ature is to female as culture is to male."[13] The latter distinction has emerged as a basic construct in a number of disciplines, including literature, anthropology, and history, which have been subject to major critiques by feminists on that basis.

Finally, as a further elaboration of the above parallel, Carolyn Merchant has explored the various depictions of nature as female in the "gendering" of nonhuman nature.[14] Nature as "mother" emphasizes aspects of emotionality, nurture, and feeling, and was, according to Merchant, the metaphor that shaped earlier organic conceptions of nature. A second contrasting image, however, identified a female nature as the source of chaos and disorder; again, according to Merchant's reading, that image became increasingly important in the context of the scientific revolution, wherein science was viewed as the appropriate extension of human power over an unruly nature.[15]

The political implications of the above emphases have been explored by social feminists, especially in the "hermeneutic of suspicion" they direct against "patriarchal" claims that link social inequalities to differences in biology, and in their exploration of gender as a social construct or ideological projection. At the same time, ecofeminists have sought to critically retrieve an earlier vision of a more "organic" nature as a necessary corrective in articulating an ethic of environmental responsibility.[16]

The Complexity of Religious Conversations about Malleable Nature

The above perspectives all bear, explicitly or implicitly, on interpretations of nature as malleable. But especially influential in historical and current discussions of nature and the natural are religious interpretations, with three questions of special salience. First, precisely how are nature and the natural understood as religious concepts; that is, in what ways are they normative? Second, what interventions into nature are obligatory, permitted, or excluded by various interpretations of nature and the natural? In viewing nature as malleable, what are the nature and the scope of limits into human interventions? Third, what moral insights may be gained not only from the religious traditions themselves but also from the critical reappropriation of traditions in light of other perspectives?

I will review several traditional appeals to nature as a theological and moral category. Some traditions are more systematic in their moral methodology, while others are more broadly thematic than casuistical in their approaches. Some traditions focus primarily on general intentions rather than particular judgments about acts. Moreover, in many discussions of particular technologies, some apparently "secular" voices may borrow concepts and language about nature as isolated shards from much richer traditions of religious and ethical reflection.

What follows, then, is a brief summary of several religious traditions with two aims in mind. First, it is meant to provide certain conceptual, historical, and methodological touchstones by which to identify and situate particular religious traditions. Second, while it identifies distinctive characteristics among traditions, it also introduces themes that may suggest areas of common understanding.

Hinduism

The core emphasis in Hinduism is the identity of *Atman* (Self) and *Brahman* (Being), with a consequent lessening of focus on the physical body. Hindu dualism, however, is not analogous to the Cartesian divide between mind and body. Instead, Hindu thought emphasizes the difference between an individual's psychophysical consciousness and impersonal *Atman* that is one's true Self. Enlightenment (*moksha*) is the act of becoming fully aware of one's essential nature, which is identical with cosmic reality.

The implications of Hindu beliefs for human alterations of nature are therefore mixed. On the one hand, the transcendent nature of one's true self might imply that there is no principled reason to oppose interventions that change either phenomenal nature or the physical body, both of which are impermanent. On the other hand, the spiritual thrust of Hindu practice suggests caution in such alterations of nature, because an undue concern with either the natural world or the physical body may be at odds with the essentially spiritual nature of the quest for enlightenment. As a result, one cannot offer general affirmations or condemnations of human actions that alter nature from Hindu sources. Nor do these sources support negative judgments against "meddling with nature" or "playing God." Rather, Hindu sources will frame normative assessments within a holistic account of nature and human nature. All efforts to alter nature or human nature must be viewed in light of the core Hindu principle of *ahimsa*, which dictates nonviolence and the avoidance of harm.

Buddhism

Classical Buddhism is a nontheistic quest for liberation, based on teachings (*dharma*) that allow one to identify factors that hinder transcendental enlightenment. In contrast to Hinduism, which affirms a true and eternal self, central to Buddhism is the notion of no-self (*anatma*). Enlightenment (*nirvana*) involves identifying attachment and desire as the source of suffering and the overcoming of suffering by following the Eightfold Path. Buddhism is grounded in the emptiness (*sunyata*) of existence: All existence and existents are interdependent, impermanent, and nonsubstantial.

In his discourses, Sakyamuni Buddha (563–483 BCE) emphasized the so-called Four Noble Truths that are the touchstones of the quest for liberation and enlightenment. The First Noble Truth is the fact of the omnipresence of suffering (*dukkha*). The Second Noble Truth identifies the cause of suffering to be attachment and desire. The Third Noble Truth proclaims that there is cessation (*nirodha*) to suffering through the "vanishing of afflictions so that no passion remains."[17] The Fourth Noble Truth sets out the way to eliminate suffering and achieve enlightenment—through specific, practice-based strategies called the Eightfold Path.

The various schools of Buddhism are perhaps most clearly distinguished by their different interpretations of core Buddhist precepts. Theravada (found in Thailand, Cambodia, Vietnam, and Laos) takes a more literalistic stance. In the Mahayana and Vajrayana schools, the commitment to end suffering serves to contextualize the application of precepts in particular circumstances. With reference to the issues of altering nature through human interventions, the central Buddhist emphasis on the virtue of compassion would seem to have positive implications for interventions that are clearly therapeutic in their intent. At the same time, compassion has also been invoked to support the loving acceptance of disability, especially within the context of belief in the patient acceptance of particular karmic effects as an opportunity for growth toward enlightenment. All forms of intervention must be assessed in light of the central focus of Buddhist understanding, that is, self-transformation.

Judaism

Ethical values lie at the heart of Judaism and are central to all aspects of Jewish life. The core concepts of Judaism are "God; Torah, or 'Teaching'; and the community of Israel, the Jewish people."[18] The origin of Judaism is God's covenant (*brit*) with the Jewish people, and the giving of Torah as the pattern for community. Torah consists of Scripture (written Torah) and tradition (oral Torah). The written Torah comprises the first five books of the Hebrew Bible, while the oral Torah encompasses "all Jewish traditional teaching—in fact, all authentic Jewish thought and practice."[19]

The Jewish ethical method is characteristically legalistic, pluralistic, casuistic, and open to further interpretation and application.[20] When values come into conflict, Judaism adopts the method of casuistry: The scriptural norm is determined; rabbinic interpretations, especially in the responsa literature are studied; precedents are identified and compared; and principles and rules are derived and applied. In the context of God as Creator and author of the covenant, certain core ethical values shape a characteristically Jewish approach.

In light of our focus on malleable nature, two aspects of Jewish thought are especially salient. First, Jewish thought does not elaborate a doctrine of "natural law," as such. Thus, absent specific *halakhic* prohibition, humans are enjoined to alter and restore nature under the rubric of "repairing the world" (*tikkun olam*). Second, God is Creator of the world and author of the covenant established by giving Torah to Israel as the chosen people. God's creation, however, is imperfect; therefore, human beings are called to actively work in concert with God to achieve God's purposes on Earth. (Despite that recognition of creation's imperfection, there is no analogue to "original sin" as found in Christian theology.) The affirmation of life and healing are central to Jewish thought and practice as direct implications of repairing the world. Judaism, therefore, is positively disposed toward interventions into nature that offer therapeutic promise or human betterment.

Christianity

Perspectives on the concept of nature vary widely in the Christian tradition, with implications for arguments, both positive and negative, about the legitimacy of deliberate alterations of nature. Both nature and human nature are teleologically suffused in Roman Catholic thought. While original sin has distorted the human capacity to know and to do what is good, Catholic moral doctrine is distinguished from Protestantism by its greater confidence in our natural moral capacity to discern the good and our natural moral will to do the good (at least partially). In Catholic thought, God's Eternal Law has structured all existence, and human beings, through their rational capacity, participate in that law. Despite the disordering effects of Adam's original sin, people retain a capacity to discern the good and to act upon that discernment according to the dictates of natural law. All interventions into, or alterations of, nature are therefore to be viewed in light of the requirements of natural law, although significant discussion continues among theologians concerning the scope of natural law directives and how binding they are.

The implications of Catholic moral theology for specific interventions remain mixed. For example, official Roman Catholic teaching on new reproductive technologies is based on what has been called the "inseparability principle"; that is, procreative potential and conjugal intimacy should not be separated in any act of marital intercourse. Moreover, official Catholic teaching also rejects all manipulations (including embryonic stem cell research) that involve destruction of embryos as forms of early abortion. But in other areas of biotechnology (including genetics) that do not involve such destruction, particular developments will be assessed in light of the goods of human flourishing that are served or disserved in any given instance.

A distinctively Protestant form of ethical reflection proceeds from seeing Scripture as a foundational authority, although the ways that Scripture functions as an authority vary widely.[21] But in the process of moral discernment, several sources of tension have featured prominently in recent Protestant bioethics and are directly relevant to issues of altering nature. On the one hand, there has been, from the Reformation onward, a central Protestant emphasis on the freedom of the Christian to respond to God's call. However, that freedom is constrained by the fundamental distinction drawn between the realms of creation and redemption, and each variety of Protestant moral thought is based largely on different interpretations of the nature and scope of human responsibility in light of sin. Some thinkers tend to emphasize the continued goodness of God's creation, despite the Fall, and see in new technological possibilities the extension of traditional notions of responsible stewardship.[22] More recently, some Protestant thinkers have developed an account of positive Christian responsibility under the rubric of "cocreation" in accord with God's good purposes.[23] Other Protestant thinkers remain more literally conservative in their assessment of altering a malleable nature. They therefore tend to emphasize the pervasive effects of sin on human epistemological and moral capacities, and the likelihood of human hubris in usurping God's prerogative.[24] The linkage between their generally negative anthropology and specific judgments, however, often remains vague or unspecific. In addition, there are varying perspectives on human nature itself. The judgment about whether human nature is (relatively) fixed or open-ended will influence the interpretations of, warrants for, and limits on particular interventions.

Eastern Orthodox Christianity remains closer to Roman Catholic than to Protestant thought in its positive assessment of the continuing presence and exercise of natural moral capacities, despite the effects of sin. The Orthodox approach exhibits several characteristic features. First, the "living consciousness" or "mind" of the church provides the ethos for ethical reflection. Central to that experience, as noted above, is the Tradition writ large—the Bible, the Patristic writings, the decisions of the first seven church councils, and a rich legacy of mystical and ascetical writings. Second, Orthodoxy blends elements of both apophatic (negative) and kataphatic (positive) theology. Because God is always transcendent, he is "best described by what He is not."[25] At the same time, the Divine remains, through God's divine energies, "in touch" with the world,[26] and on that basis a positive (though limited) theology of God's attributes is possible. Moreover, from the "paradoxical relationship between God and the world, one that is both continuous and discontinuous," there emerges a "relative independence and autonomy" with respect to the world that allows for its use and development.[27] In this attribution of relative autonomy, Orthodoxy shares affinities with certain Calvinist strands of Protestant thought. Third, Orthodoxy emphasizes that humans are made in the image and likeness of God. Human "likeness" to God embodies a dynamism, a movement toward theosis, which leads Orthodoxy to interpret human nature in more open-ended than static terms. As a result, Orthodox assessments of altering nature shares affinities with Protestant interpretations of humans as "cocreators"[28] rather than with traditional, fairly static, Roman Catholic depictions of natural law and of human nature.

Islam

The historical debates within Islam are quite relevant to discussions of altering malleable nature. In the first half of the eighth century, a group of scholars called the Mu`tazilites emerged, and in their interpretations sought to show that there was "nothing repugnant to reason in the Islamic revelation."[29] By the ninth century, Hellenic philosophical writings had become available in Arabic translations, and the Mu`tazilites incorporated basic elements of Greek science and metaphysics into their discussions. In so doing, they introduced the idea of an inherent nature in entities that is discernible by reason and an account of natural causation according to physical principles and natural regularities. By contrast, the Ash`arite school of interpretation rejected such rationalism, stressing that apparent natural regularities are actually the result of Allah's direct actions, rather than secondary causes.

Different voices emerge in current Islamic discussions of malleable nature. One recent analysis recommends that scientific and technological developments be assessed according to several basic rules. One of them, in the form of a verse from the Qur'an about "changing God's creation," provides the context for assessing human efforts to alter nature. In an expression of his plans to lead humankind astray, the Qur'an quotes Satan addressing God about humanity: "Verily of thy servants I . . . shall lead them astray . . . [a]nd

I shall command them and they will change God's creation" (4:119). While this passage has been debated for centuries, according to the consensus of Islamic scholars, it should not be construed as a comprehensive ban on human interventions into nature, because a blanket proscription would preclude the development and disbursement of many of the benefits of modern science and medicine. Instead, such changes in nature should be assessed in light of a separate rule, whose text is the following: "Wherever the welfare exists, there stands the statute of God." Therefore, "interventions that foster human life, health, and welfare are not only possible but are deemed acts of commendable charity that will be rewarded by God." [30]

When considering recent technologies that alter nature, however, Abdulaziz Sachedina suggests that the application of traditional insights will be less straightforward for two reasons. First, while "*shari'a* reflect[s] Muslim endeavors to ensure that Islam pervade[s] the whole of life, . . . many areas of human existence, including the ethical problems connected with the medical treatment of ailments, received little systematic attention in the classical formulations of the legal thought."[31] Second, with modern technological advances, "Muslim jurists are faced with a crisis because, by its own standards, Islamic jurisprudence has ceased to progress toward some further stage of development. The method of inquiry and the forms of argument have [proved inadequate] to furnish solutions to concrete problems faced by the community."[32]

Summary

As our earlier analysis suggested, the concept of nature, both descriptively and prescriptively, is polyvalent. As the contributors to this volume have emphasized, nature will be contextualized quite differently in different historical, cultural, and geographical settings. In discussing the malleability of nature, I have sought to place that vision as the midpoint along a continuum. It has been said that ethics, as a systematic enterprise, is largely about line-drawing, that is, specifying the warrants for appropriate human choices and actions. Among the five visions of nature analyzed in the current project, the vision of malleable nature distinctively, and perhaps uniquely, foregrounds questions about the normative warrants for human alterations of nature. As we have seen, the postmodern discussion of nature necessarily moves beyond earlier assumptions that nature, of itself, provides sufficient grounding for subsequent normative reflection. Moreover, insights from critical studies tend to confirm skepticism about the possibility of a universally shared concept of nature.

In our review of religious views, I identified a wide range of perspectives about the licitness of human interventions into the natural world. For those traditions that emphasize creation as a theological category, nature will be evaluated positively or negatively as a source of moral insight, depending on how created nature is linked with other core aspects of theological understanding. Roman Catholic and Orthodox Christian perspectives, for example, generally share more positive assessments of created nature as a resource for ethical reflection, in contrast to Protestant and Jewish perspectives. At the same time, many religious arguments have focused less on the notion of natural limits to human alterations of nature than on other ethical issues, especially the likely maldistribution of benefits and burdens that conflicts with requirements of social justice.

In all such religious discussions, it is clear that the complexity of the conversation will not allow us to revert to unhelpful caricatures. We will not discover, upon careful review, new evidence for some perennial "warfare" between science and religion. Religious perspectives are no more uniform on questions of altering nature than are secular ones. Thus, we will find robustly voiced caution about human efforts to redesign nature as well as celebrations of human agency as an appropriate exercise of "cocreation." And in both the religious and secular conversations, we will need to attend to the variety of methodological and substantive commitments both within and across traditions of discourse. More practically, while popular discussions of altering nature have often been devoted to explicit public conflicts between religion and biotechnology, exploring the possibilities for cooperation is equally important. It is reasonable to suppose that the potential for cooperation emerges from the general agreement that intervention into nature for human benefit is justifiable in principle. How deep, then, is this agreement? Is it based on shared views of nature or on diverse but overlapping views? Are there as-yet-unrecognized grounds for cooperation? Often, only minimal connections are drawn between judgments about particular issues and the deeper traditions of religious reflection on the status of nature and the natural. What is still needed, as exemplified by the New Visions project, is a more systematic effort to link religious interpretations of created nature to other perspectives. Only by tracing the rich interdisciplinary aspects of the various visions of nature will their full potential as intellectual and moral heuristics be realized.

Notes

1. David Livingstone, "Locating New Visions," in this volume.
2. John Hedley Brooke, "Should the Word *Nature* Be Eliminated?" in this volume.
3. Martha Henderson, "Rereading a Landscape of Atonement on an Aegean Island," in this volume.
4. James Proctor, "Environment after Nature: Time for a New Vision," in this volume.
5. In each of the other four visions, the metaphysical and normative questions appear inevitably linked in sequential or dialectical fashion. How one defines nature will establish the ways that relations between nature and human nature are understood. Those relations, in turn, will offer, at least inchoately, a sense (intuitive or otherwise, depending on which version of moral theory one espouses) of where and on what basis one will draw normative lines between acceptable and unacceptable alterations of nature.
6. Robin Lovin and Frank Reynolds, *Cosmogony and Ethical Order* (Chicago: University of Chicago Press, 1985), 2.
7. David Hume, *A Treatise on Human Nature* (Oxford: Clarendon Press, 1973).
8. Lovin and Reynolds, *Cosmogony and Ethical Order*, 3.
9. Mary Midgley, "Criticizing the Cosmos," in William B. Drees, ed., *Is Nature Ever Evil?* (New York: Routledge, 2003), 24.
10. Raimo Tuomela, "Feminist Philosophy," in Robert Audi, ed., *The Cambridge Dictionary of Philosophy* (Cambridge: Cambridge University Press, 1995), 263.
11. "The question of women engaging in science is not just a question of equality—whether all people should have an equal opportunity to pursue careers of their choosing. Nor is it a question of 'manpower,' having enough scientists to sharpen a chosen nation's competitive advantage . . . It is a question of knowledge. Only recently have we begun to appreciate that who does science affects the kind of science that gets done. . . . Our knowledge of nature [has] been influenced by struggle determining who is included in science and who is excluded, which projects are pursued and which ignored, whose experiences are validated and whose are not. . . ." Londa Schiebinger, *Nature's Body: Gender in the Making of Modern Science* (Boston: Beacon Press, 1993), 3.
12. E.g., "The laws of divine and natural order reveal the female sex to be incapable of cultivating knowledge, and this is especially true in the fields of natural sciences and medicine." Theodor L. W. von Bischoff, *Das Studium und due Ausubung der Medicin durch Frauen*, quoted in Londa Schiebinger, ed., *Feminism and the Body* (New York: Oxford University Press, 2000), 45.
13. Sherry Ortner, "Is Female to Male as Nature Is to Culture?" *Feminist Studies* 1 (1972): 5–31.
14. Carolyn Merchant, *The Death of Nature* (San Francisco: Harper and Row, 1980), 1–41.
15. "Two new ideas, those of mechanism and of the domination and mastery of nature,

became core concepts of the modern world. An organically oriented mentality in which female principles played an important role was undermined and replaced by a mechanically oriented mentality that either eliminated or used female principles in an exploitative manner. As Western culture became increasingly mechanized in the 1600s, the female earth and the virgin spirit were subdued by the machine." Ibid., 2.

16. See, e.g., Greta Gaard, ed., *Ecofeminism: Women, Animals, and Nature* (Philadelphia: Temple University Press, 1993); Carolyn Merchant, *Earthcare: Women and the Environment* (New York: Routledge, 1995); Rosemary Radford Ruether, *Gaia and God: An Ecofeminist Theology of Earth Healing* (San Francisco: Harper & Row, 1992); and Karen Warren, ed., *Ecological Feminist Philosophies* (Bloomington: Indiana University Press, 1996).
17. Hajime Nakamura, "Buddhism," in Warren Reich, ed., *Encyclopedia of Bioethics*, 4 vols. (New York: Free Press, 1978), 1:135
18. Aaron L. Mackler, *Introduction to Jewish and Catholic Bioethics* (Washington, D.C.: Georgetown University Press, 2003), 44.
19. Ibid.
20. Baruch Brody, "Themes and Methods in Jewish Ethics," unpublished lecture as part of the "Recovering the Traditions" conference (Houston: The Institute of Religion, 1998).
21. Allen Verhey, "The Bible in Christian Ethics," in James Childress and John MacQuarrie, eds., *The Westminster Dictionary of Christian Ethics* (Philadelphia: Westminster Press, 1986), 57.
22. E.g., Ronald Cole-Turner, *The New Genesis: Theology and the Genetic Revolution* (Louisville: Westminster, 1993).
23. E.g., Ted Peters, *Playing God: Genetic Determinism and Human Freedom* (New York: Routledge, 1997).
24. Paul Ramsey, *Fabricated Man* (New Haven, Conn.: Yale University Press, 1970).
25. Stanley Harakas, *Wholeness of Faith and Life: Orthodox Christian Ethics* (Brookline, Mass.: Holy Cross Orthodox Press, 1999), 90.
26. Ibid.
27. Ibid.
28. Peters, *Playing God.*
29. Abdulaziz Sachedina, "Islam," in Warren Reich, ed., *Encyclopedia of Bioethics*, 2nd ed., vol. 3 (New York: Macmillan, 1995), 1294.
30. Hassan Hathout and B. Andrew Lustig, "Bioethical Developments in Islam," in B. Andrew Lustig, ed., *Bioethics Yearbook*, vol. 3 (Dordrecht, The Netherlands: Kluwer, 1993), 133.
31. Sachedina, "Islam," 1293.
32. Ibid., 1295–96.

Bibliography

Brody, Baruch. "Themes and Methods in Jewish Ethics." Unpublished lecture as part of the "Recovering the Traditions" conference. Houston: The Institute of Religion, 1998.

Cole-Turner, Ronald. *The New Genesis: Theology and the Genetic Revolution.* Louisville: Westminster, 1993.

Gaard, Greta, ed. *Ecofeminism: Women, Animals, and Nature.* Philadelphia: Temple University Press, 1993.

Harakas, Stanley. *Wholeness of Faith and Life: Orthodox Christian Ethics*. Brookline, Mass.: Holy Cross Orthodox Press, 1999.
Hathout, Hassan, and Andrew Lustig. "Bioethical Developments in Islam." In Lustig, ed., *Bioethics Yearbook*, vol. 3, 133–48. Dordrecht, The Netherlands: Kluwer, 1993.
Hume, David. *A Treatise on Human Nature*. Oxford: Clarendon Press, 1973.
Lovin, Robin, and Frank Reynolds. *Cosmogony and Ethical Order*. Chicago: University of Chicago Press, 1985.
Mackler, Aaron. *Introduction to Jewish and Catholic Bioethics*. Washington, D.C.: Georgetown University Press, 2003.
Merchant, Carolyn. *The Death of Nature*. San Francisco: Harper & Row, 1980.
———. *Earthcare: Women and the Environment*. New York: Routledge, 1995.
Midgley, Mary. "Criticizing the Cosmos." In William B. Drees, ed., *Is Nature Ever Evil?* 11–26. New York: Routledge, 2003.
Nakamura, Hajime. "Buddhism." In Warren Reich, ed., *Encyclopedia of Bioethics*, vol. 1, 135. New York: Free Press, 1978.
Ortner, Sherry. "Is Female to Male as Nature Is to Culture?" *Feminist Studies* 1 (1972): 5–31.
Peters, Ted. *God: Genetic Determinism and Human Freedom*. New York: Routledge, 1997.
Ramsey, Paul. *Fabricated Man*. New Haven, Conn.: Yale University Press, 1970.
Ruether, Rosemary Radford. *Gaia and God: An Ecofeminist Theology of Earth Healing*. San Francisco: Harper and Row, 1992.
Sachedina, Abdulaziz. "Islam." In Warren Reich, ed., *Encyclopedia of Bioethics*, 2nd ed., vol. 3, 1289–97. New York: Macmillan, 1995.
Schiebinger, Londa, ed. *Nature's Body: Gender in the Making of Modern Science*. Boston: Beacon Press, 1993.
———. *Feminism and the Body*. New York: Oxford University Press, 2000.
Tuomela, Raimo. "Feminist Philosophy." In Robert Audi, ed., *The Cambridge Dictionary of Philosophy*, 263. Cambridge: Cambridge University Press, 1995.
Verhey, Allen. "The Bible in Christian Ethics." In James Childress and John MacQuarrie, eds., *The Westminster Dictionary of Christian Ethics*. Philadelphia: Westminster Press, 1986.
Warren, Karen, ed. *Ecological Feminist Philosophies*. Bloomington: Indiana University Press, 1996.

11

VISIONS OF A SOURCE OF WONDER

Fred D. Ledley

Any vision of nature, science, and religion is inevitably colored by a prism of scholarly, scientific, or spiritual perspectives. John Hedley Brooke's chapter in this book, for example,[1] titled "Should the Word *Nature* Be Eliminated?" challenges the use of the word *nature* as one that is inextricably related to specific cultural, scientific, religious, and historical contexts. Andrew Lustig goes further,[2] arguing that the reality of nature itself is malleable and that human choices and interventions, which are driven by particular conceptions of nature and even different religious perspectives, may alter the essence of nature itself.

My vision of nature is certainly no less conditional than the views of my colleagues. Perhaps my vision is even more constrained because, as a scientist and a physician, I am trained to look at nature through the lens of a specific, scientific method for the utilitarian purpose of healing. I am also indoctrinated in a liberal Judaic tradition that associates creation and nature, as well as its workings and wonders, with a transcendent deity.

I offer this caveat not only because I am aware of being informed, and thus potentially limited, by these perspectives or prejudices. I offer it because I am conscious of trying to achieve a vision of nature that is not colored by the prisms of these developed ways of knowing, a vision that exists prior to the effects that these structured disciplines may exert on my perspectives on nature or even the essence of nature itself. I am also drawn to ask questions about nature that may teach me something about science and religion themselves, and their place in nature and its evolution.

Specifically, I am drawn to explore the sense of wonder that is often associated with nature. There is a long tradition in Western thought that ascribes the origin of science, religion, and philosophy to a sense of wonder. For example, Aristotle wrote, "For it is owing to their wonder that men both now begin and at first began to philosophize; they wondered originally at the obvious difficulties, then advanced little by little and stated difficulties about the greater matters, e.g., about the phenomena of the moon and those of the sun and the stars, and about the genesis of the universe."[3] Adam Smith weighed in as follows: "Wonder, therefore, and not any expectation of advantage from its discoveries, is the first principle which prompts mankind to the study of Philosophy, of that science which pretends to lay open the concealed connections that unite the various appearances of nature."[4] According to Thomas Carlyle, "Wonder is the basis of worship."[5] In Ralph Waldo Emerson's view, "Men love to wonder, and that is the seed of science."[6] Abraham Joshua Heschel noted that "Awareness of the divine begins with wonder."[7]

Is there a source of wonder extant in nature that inspires this sense of wonder, our science and our wisdom, our spirituality and our worship? Is this part of our vision of nature, or is it added on to what is natural? Can it be explained through scientific and religious ways of knowing that are inspired by our wonder, or must it be explored without reference to these derivative disciplines? If, as Barbara King suggests in her chapter in this volume, chimpanzees or great apes have the ability to perceive wonder and conceive spirituality,[8] do we share homologous neurological pathways for sensing wonder, and, if so, why have such pathways evolved?

The Sublime as a Source of Wonder

There are many suggestions in Western thought that sources of wonder exist in nature separate from the subjects of science and religion. Wolfgang von Goethe, writing in *The Theory of Color,* notes that "Nature also speaks to other senses which lie even deeper, to known, misunderstood, and unknown senses."[9] Henry David Thoreau retreated to Walden Pond "to anticipate, not the sunrise and the dawn merely, but, if possible Nature herself!"[10] Albert Einstein pointed out that "To me it suffices to wonder at these secrets and to attempt humbly to grasp with my mind a mere image of the lofty structure of all that there is."[11] The psalmist sang, "The heavens are telling the glory of God; and the firmament proclaims his handiwork. Day to day pours forth speech, and night to night declares knowledge. There is no speech, nor are there words; their voice is not heard; yet their voice goes out through all the earth, and their words to the end of the world."[12] William Wordsworth expressed a variant of this thought in verse:

> And I have felt
> A presence that disturbs me with the joy
> Of elevated thoughts; a sense sublime
> Of something far more deeply interfused,
> Whose dwelling is the light of setting suns,
> And the round ocean and the living air,
> And the blue sky, and in the mind of man;
> A motion and a spirit, that impels
> All thinking things, all objects of all thought,
> And rolls through all things.[13]

It is here, between poetry and prose, at the boundary of observation and the limits of language, that we may begin to envision the source of our sense of wonder. That nature provides a multitude of sights, sounds, smells, tastes, and tactile sensations is accepted—but we nonetheless wonder at their existence. That nature can separate light into the colors of the rainbow, produce thunder and lightning from vapors and heat, and support life and intelligence is amazing—and we wonder why these things can happen. That nature comprises seemingly endless numbers of stars and galaxies and geological formations and forms of life is ineffable—and we wonder at the complexity. That nature can present us with cosmic impacts, eruptions, and earthquakes that can level mountains and extinguish species is awesome—and we wonder at nature's incomprehensible power.

Our sense of wonder is evident in questions about nature that have no formal subject. It exists in questions that inspire human imagination and inquiry, contemplation and curiosity. It arises from attributes of nature that transcend substance—a nature that Romantic writers like Wordsworth described as "sublime."

Wordsworth described the sublime[14] as causing a sense of exaltation and awe, a sense of duration in which "individuality is lost in the general sense of duration belonging to the Earth itself," "impressions of power," feelings of apprehension, dread, fear, or wonder, and a sensation that "calls upon the mind to grasp at something towards which it can make approaches but which is incapable of attaining." He writes:

> I enumerated the qualities which must be perceived in a mountain before a sense of sublimity can be received from it. Individuality of form is the primary requisite; and the form must be of that character that deeply impresses the sense of power. And power produces the sublime either as it is thought of as a thing to be dreaded, to be resisted, or that can be participated.[15]

Immanuel Kant defined the sublime as "the name given to what is absolutely great," and as exhibiting the characteristics of being both "mathematically sublime"[16] and "dynamically sublime."[17] To Kant, that which is mathematically sublime cannot be bounded by the material attributes of nature or measurable quantities or forms, but is rather evident in its effect in the mind. "The sublime is that, the mere capacity of thinking which evidences a faculty of mind transcending every standard of sense. The mind feels itself set in motion in the representation of the sublime in nature."[18] That which is dynamically sublime incorporates also a reaction to the awesome power of nature, a sense of fear, and a sense of awe. "It is an object (of nature) the representation of which determines the mind to regard the elevation of nature beyond our reach as equivalent to a presentation of ideas."[19]

Arthur Schopenhauer also recognized that the sublime embodies an interplay between objects in nature and their effect on the mind.

> He may comprehend only their Idea that is foreign to all relation, gladly linger over its contemplation and be elevated precisely in this way above himself, his person, his willing, and all willing. In that case, he is then filled

> with the feeling of the sublime; he is in a state of exaltation and therefore the object that causes such a state is called sublime.[20]

For Schopenhauer, the sublime in nature is evident in infinite greatness, objects that are threatening and terrible, and the feelings of isolation, insignificance, fear, and foreboding that arise from these objects.[21] He goes further, recognizing degrees of feeling the sublime. Faint traces of the sublime can be recognized in "scenes of mere solitude and profound peace" that induce feelings of alarm and isolation. Stronger feelings of the sublime are related to "nature in turbulent and tempestuous motion" and "ideas in those very objects that are threatening and terrible to the will."[22]

Other writers have recognized the sublime in nature by negation, namely from the sense that something can be subtracted from nature and that the sensations associated with the sublime can be disrupted or lost. Theodore Roszak describes the loss of this "other, greater reality" when commercial aviation took the mystery out of flight.

> I watch . . . and ponder the wealth of artistic and religious symbolism that lies hidden beneath this popular, potted history of aviation. Classic images of flight and ascendance fill my thoughts . . . the skylark of the Romantic poets . . . Geruda, the divine vulture, the vehicle of the Yogis . . . St. Bonaventure's ascent of the mind to God . . . the Taoist holy men who rode the wind in carriages drawn by flying dragons . . . the prophet Elijah in his chariot of fire . . . the Vedic priest scaling his tall pillar, crying from its summit with arms outspread, "We have come to the heaven, to the gods, we have become immortal." I think of the angels pointed with flam, the furies, genies, and wing-footed gods . . . and behind all these, shining through from the dawn of human consciousness, reflected in a thousand mythical images, the shamanistic vision-flight, the ecstasy with winds. All of this passes through my mind, the real but subterranean meaning of flight, buried now beneath a glamorous technology of jet planes and moon rockets, lost, totally lost from the common awareness; . . .
>
> We have made It real—materially—historically real but at the sense of some other, greater reality . . . something that eludes us for all our power and cunning.[23]

What is this sublime aspect of nature that can set our mind in motion, stimulate our fears as well as our sense of awe and wonder, and leave us with a sense of incompleteness when it is lost? Is this the source of our wonder that inspires our science and spirituality? How can we make this source of wonder part of our vision of nature?

Here I must again pause to offer explanation and apology, because I am limited to looking for answers to these questions within my knowledge of science and religion. That is not to say that the insights of science and religion, inspired by a sense of wonder, do not provide insights into the source of this wonder. In fact, both science and religion recognize boundaries to the natural world that can be described and measured. The spiritual world, by contrast, is subject to canonical revelation and faith. Both also retain vestiges or atavisms of a primal appeal to wonder, which may inform a vision of nature.

Why Does the Bible Begin with the Letter *Bet* (ב)?

Rabbinic commentaries on creation provide an illustrative example of how sources of wonder exist at the boundaries of canonical religion. In Judaism, there is a tradition of rabbinical exegesis that delves beyond the biblical accounts of creation into aspects that are not explicitly revealed. Such exegesis often involves searching for nonliteral meaning in the Hebrew words themselves. In this search, every word, every combination of words, every syllable, every letter, every line comprising the shapes of the letter, and even the decorative crowns that are traditionally placed on certain letters in the hand-transcribed text of the Torah are considered to be of divine origin and are presumed to have both literal and abstract meanings. So, too, the numerical values associated with the Hebrew letters, which also serve as numerals, are studied to discern sums and sequences that reveal hidden associations and ideas.

The first words of the book of Genesis (in Hebrew: *Breishit* or *Breishis*) are a particularly rich subject for such exegesis.[24] The biblical account of creation begins with the words *Breishit bara Elohim,* a phrase traditionally translated as "In the beginning, God created . . ." This is not, however, a literal translation of the Hebrew text, which is difficult to translate because the construction is inherently ungrammatical and the etymology of the words enigmatic.

One of the challenges to understanding this phrase is that the etymology and meaning of the word *breishit* is unclear. The word *breishit* incorporates the root *reish*, which can refer to the "head" or, metaphorically, something at the top or the beginning. Hence, the common translation of the word *breishit*, referring to the "beginning."

Grammatically, however, the word *breishit* is an adjective, whose meaning may be approximated by "before."[25] The phrase *Breishit bara Elohim*, however, does not contain an associated noun. Instead, the word *breishit* is followed by the verb *bara*, meaning "created," and *Elohim*, referring to God—thus, "God created." The translation "Before God created" may be more literally accurate, but it is inconsistent with the subsequent passages, which present a chronology of events subsequent to the initiation of creation.

The medieval scholar Rashi (Solomon Bar Isaac), whose commentary on the Torah is traditionally studied alongside the Hebrew text, suggested that if the first words were intended to introduce a chronological account of creation, the text should have read *Bi-rashonah bara Elohim*—"At first God created." Since this is not the language of the Torah, the rabbis concluded that the word *breishit* must have a different, nonliteral meaning.[26]

The philosopher Maimonides (Moshe Ben Maimon) interpreted the root *reishit* as referring to a principle from which other things in creation will follow.[27] This is similar to interpretations in the Midrash Rabbah, a sixth-century CE text that codified the early rabbinic commentaries on the book of Genesis.[28] One midrash suggests that the word *breishit* could be read *b'reishit* with the word *reishit* interpreted as an allusion to the Torah and thus read as "With Torah, God created . . ."[29] Since the Torah is also understood as the Law, this midrash suggests that the opening paragraph of Genesis could be translated as "With Law, God created . . ." Another midrash suggests that the word *breishit* could be read as *bara shit* and the phrase translated as "With six [things], God created . . ."[30] From this the rabbis inferred that as many as six things preceded creation that are not explicitly described in the text.

Other midrashim on the word *breishit* ask why the Torah begins with the letter *bet*. It would seem logical, the rabbis noted, for God to begin the Torah and, by implication, creation, not with the second letter of the alphabet, *bet*, which has the numerical value of two, but rather with the first letter of the alphabet, *aleph*, which has the numerical value of one. The Torah could have begun *Elohim bara breishit* (or "God created at the beginning"), which would have started logically with an *aleph* and conveyed a similar meaning.[31] Why, they ask, didn't creation begin at the beginning?

One midrash suggests that the Law given to Moses, the Decalogue, begins with the letter *aleph,* and that this law is, in fact, the principle of creation, which came before *Breishit.*[32] Another possible explanation is that creation may indeed have begun with *aleph,* which is a silent letter in Hebrew. The silence of this letter and its absence suggest the presence of primordial mysteries before creation, and hence before nature, which are silent, unknowable, unrevealed, and ineffable.

Each of these midrashim suggests that while the creation of heaven and Earth may have begun *ex nihilo* with *Breishit,* there are predicates to this creation that are not explicitly described or revealed. Traditional, rabbinic Judaism, however, forbids consideration of what may have come before creation. The Midrash Rabbah derives this prohibition from the fact that the Torah begins with the letter *bet* (ב), which is closed on the right side and open on the left. Since Hebrew is read from right to left, the shape of this letter was interpreted as revealing that the biblical account of creation was open to study, but that which came before was closed to understanding. Just as the *bet* is closed on all sides and open only in front (at which point the Torah begins), so, too, you are not permitted to speculate on what is above and what is below, what is before and what is behind, but only on the day of creation and onwards.[33] In formally excluding that which may have come before creation, the boundaries of religion are explicitly delimited to the space and time of the created world and to objects of creation that are described or revealed in Torah. That which may have come before *Breishit* is expressly and explicitly deemed unknown and unknowable.

Centuries of mystics, however, have ignored such constraints and explored elements of creation that are not evident in the literal text. The kabbalistic text Sefer Yetzirah (Book of Creation), for example, derives the sequence of creation from the intrinsic shapes and properties of the Hebrew letters that comprise the text of the Torah, not from the words or their literal meaning. "Twenty-two letters: He carved them, hewed them, refined them, weighed them, and combined them, and He made of them the entire creation and everything to be created in the future."[34]

In the Zohar, the classic text of Kaballah, mystics would provide a description of the creation that was the source and subject of their mysticism.

> Time had begun. Its great pendulum, whose beats are the ages, commenced to vibrate. The era of creation or manifestation had at last arrived. The *nekuda reshima* [primal point or nucleus] appeared. From it emanated and expanded the primary substance, the illimitable phosphorescent ether, of the nature of light, formless, colorless, being neither black nor green nor red. In it, latent yet potentially as in a mighty womb, lay the myriad prototypes and numberless forms of all created things as yet indiscernible, indistinguishable.[35]

I am not arguing for the veracity of exegesis of ancient texts, nor suggesting that such exegesis can provide any valid insights or new visions of nature. Rather, I have recounted this tradition as an example of a perspective on nature that lies explicitly outside the canons and concerns of religion—one that arises from the mystical experience. It is a view of a nature that is sublime in the sense defined by Wordsworth, Kant, or Schopenhauer; a nature that is a source of exaltation, awe, and wonder.

The mystical vision of the Kabbalists and the Zohar is not unique, but contains elements that recur in many forms of mystical experience and in myth. James Frazier, Joseph Campbell, Claude Levi-Strauss, and others have identified common elements of mythology and mysticism in diverse human cultures.[36] Many of these visions contain elements that fit the description of the sublime.[37]

Mystical Images of the Sublime

To the modern mind, the notion that insights into creation or visions of nature can be gleaned from the shapes of sacred letters or their numerical values seems ridiculous. Such mysticism is commonly dismissed as simply a primitive form of vision, one that is uninformed by the more modern insights of science and religion.

When mystical accounts of creation are considered at all, it is common to simply identify elements in the syntax of the mystical vision that are shared with those of more orthodox science or Scripture and conclude that the mystical vision of nature is congruent to that of science and religion.[38] Thus, similarities are emphasized between the biblical account of creation *ex nihilo*, the graphic images of the Zohar, and the modern cosmology of the Big Bang. In this view, the "primary substance," or *tohu v'vohu*, is compared to plasma, the separation of light and dark is compared to light escaping from the gravitational pull of primordial plasma, and the "prototypes and numberless forms" compared to atoms and molecules that comprise the matter of the universe.

Such reasoning, however, misses the point. Mysticism is not intended to achieve an objective description of a material and measurable nature. It is explicitly unconstrained by either the constructions of science or the canons of religion and is indifferent to either the discipline of the scientific method or the doctrines of faith.

Mysticism is, in its essence, an exploration designed to transcend the limits of physical sensation and observation. It is embodied in a "faculty of mind," rather than facts and measurements. It encompasses ideas and ideations that may be set in motion not by perception, but by abstraction, imagination, and a sense of wonder.

Inherent in the mystical vision of nature is the suggestion that there is something in nature that stimulates the senses while bypassing sensation, something transcendent and universal, something that induces awe and wonder while bypassing reason and religion, and something that defies metes or measures. It is a response to something ineffable, "a motion and a spirit, that impels all thinking things, all objects of thought," to something that may have no material point of reference outside the mind of the mystic—something that may be best described as sublime.

Mysticism, Religion, and Science

The mystic's appreciation of sublime elements in nature is closely related to the sense of wonder that has been identified as an inspiration for science and religion. Moreover, mysticism itself may be an important historical and ontological bridge between the sense of sublime aspects of nature and the formative role ascribed to wonder in the development of both religion and science.

Commenting on this, William James wrote, "One may say truly, I think, that personal religious experience has its root and centre in mystical states of consciousness."[39] With regard to Judaism, Gershom Scholem has written:

> Mysticism is a definite stage in the historical development of religion and makes its appearance under certain well defined conditions. It is connected

with, and separable from, a certain stage of the religious consciousness. It is also incompatible with certain other stages which leave no room for mysticism.[40]

Vestiges of mysticism still exist in many religions. Drug-induced rituals and ascetic practices designed to enhance mystical experiences remain central to some non-Western religions. Even in Western monotheism, which tends to eschew the phantasmagorical stories and traditions of mysticism in favor of appeals to tradition and the structured exegesis of sacred text, principles and practices reminiscent of mysticism coexist comfortably with canon that is consistent with the sensibilities of a modern world. In contemporary, liberal Judaism, for example, the oral tradition explicitly incorporates not only *Halakha*, text focused on the formal elaboration of the Law, but also *Aggadah*, comprising a rich collection of often fantastic legend, imagery, and myth.

Science and mysticism are seemingly less compatible. Contemporary science defines itself in self-referential terms as being based on a "scientific method" that limits the domain of science to that which can be objectively described in measurement or mathematics, thus excluding anything that may have come before. Mysticism is seemingly antithetical to this discipline, and atavistic elements of mysticism and myth are actively purged from accepted practice whenever possible.

The history of science, however, provides many examples of mysticism and mystic intuition having been a source of inspiration and ideas for science. Euclidean geometry did not arise within the context of a refined system of Cartesian coordinates or symbolic algebra. Rather, this essential foundation of modern mathematics and science arose from an essentially mystical search for harmonies, symmetries, and aesthetics among the Pythagorean philosophers and their successors, much of which was based on Delphic mysticism and Olympian mythology. Johannes Kepler's laws are famously related, in his own writings, to an almost deified image of the sun and a neo-Platonic search for harmonies in the motion of the planets and geometry. Similarly, Leibniz's discovery of calculus arose in the context of a metaphysical philosophy of monads, which incorporated concepts of infinitesimals developed originally by scholars to describe the nature of angels.

Nor has science been able to fully eliminate mysticism, insight, and intuition from the scientific method. Bertrand Russell describes a relationship between instinct (mysticism) and reason, which remains dynamic.

> But in fact the opposition of instinct and reason is mainly illusory. Instinct, intuition or insight is what first leads to the beliefs which subsequent reason confirms or confutes; but the confirmation, where it is possible, consists, in the last analysis, of agreement with other beliefs not less instinctive. Reason is a harmonizing, controlling force rather than a creative one. Even in the most purely logical realm, it is insight that first arrives at what is new.[41]

He also wrote:

> Theories which turn out to be true and important are first suggested to the minds of their discoverers by considerations which are utterly wild and absurd. The fact is that it is difficult to think of the right hypothesis, and no technique exists to facilitate this most essential step in scientific progress.[42]

Those scientists who have had the good fortune to experience a moment of scientific discovery, an "Aha!" or "Eureka!" moment, in which columns of numbers and scribbled notations suddenly morph into understanding, recognize that discovery cannot be completely explained by reason alone. Those who sense elegance in calculus and computation, energy and evolution, symmetries and strings respond to an essence that is ineffable. Those who contemplate Gödel's Theorem of Incompleteness or Heisenberg's Principle of Indeterminancy recognize that there are limits to what can be established by experimentation or explained by mathematics.[43] Those who dedicate their careers to research implicitly accept that there is always something more, something that remains unknown and unfathomable, something greater, something sublime to pursue and explain. As Einstein observed:

> The finest emotion of which we are capable is the mystic emotion. Herein lies the germ of all art and all true science. Anyone to whom this feeling is alien, who is no longer capable of wonderment and lives in a state of fear is a dead man. To know that what is impenetrable for us really exists and manifests itself as the highest wisdom and the most radiant beauty, whose gross forms alone are intelligible to our poor faculties—this knowledge, this feeling . . . that is the core of the true religious sentiment.[44]

Edna St. Vincent Millay captured this sense in poetry:

> Euclid alone has looked on Beauty bare.
> Let all who prate of Beauty hold their peace,
> And lay them prone upon the earth and cease
> To ponder on themselves, the while they stare

At nothing, intricately drawn nowhere
In shapes of shifting lineage.[45]

The sense of wonder that has been identified as an inspiration for science and religion may be congruent with the mystical experience, which, in both the historical record and in practice, is a wellspring for both religious and scientific thought. So, too, the sublime elements in nature that are the subject of the mystics' vision of nature may be the source of this sense of wonder.

There is ample evidence for an essence of nature that exists outside the realm of formal scientific studies or religious reflection, which can stimulate our sense of wonder. This essence of nature is evident in the stillness of the desert and the thunder of the clouds, the silence of the *aleph* and the Big Bang. It is something that Wordsworth described above Tintern Abbey and Thoreau sensed at Walden Pond; something that Goethe, Einstein, and Russell recognized beyond the boundaries of scientific experimentation, something that the rabbis who wrote Midrash Rabbah, the Psalmist, and the Kabbalists saw before creation, and something that Roszak feared was at risk from technological progress. It is an essence of nature that may be immaterial; one that may exist only in the questions that inspire human imagination and inquiry, explicable through mysticism but beyond the boundaries of science and religion. It is an aspect of nature that can be described best as sublime—something that is "absolutely great," inherently beyond measure, something that may be manifest only in stimulating our sense of wonder, our science and our wisdom, our spirituality and our worship.

An Evolved Sense of Wonder

What is most notable about this vision of a sublime aspect of nature is that it straddles the boundary of the material world and perception. It may arise from experiencing deserts or mountains, but is not embodied in the geology of the rock. It is related to the immense scope of the universe, but is not bounded by mass or space or time. In localizing this vision, the question is whether the sublime essence of nature exists as an object of nature itself or whether it is something that is objectified by the intrinsic properties of the mind. Are there material aspects of nature that excite the synapses of the brain, causing us to experience a sense of wonder, as there are properties of nature that excite our sensations of sight or sound or smell, enabling us to experience vision, hearing, and taste? Or is the experience of wonder a cognitive phenomenon, constructed without reference to the material world?

Some psychologists and anthropologists have suggested that the common patterns of mysticism and myth observed in different societies reflect intrinsic properties of the human mind. Carl Jung[46] proposed that the human brain preserved archetypes and primordial images from human evolution. Claude Levi-Strauss[47] believed that myth arose from the structures of the human brain.

Contemporary psychometric instruments determine factors of personality and innate patterns of human behavior that correlate with mystical experiences. For example, one commonly used psychometric instrument, the Temperament and Character Inventory (TCI), delineates dimensions of temperament and character that are hypothesized to reflect the function of specific neurobiological pathways in the brain.[48] These dimensions include four measures of temperament, including "novelty seeking," "harm avoidance," "reward dependence," and "persistence," and three measures of character, including "self-directedness," "cooperativeness," and "self-transcendence." Self-transcendence, as measured by the TCI, comprises attributes such as self-forgetfulness, transpersonal identification, and spiritual acceptance or mysticism. Other psychometric tests have also been used to assess measures associated with mysticism, spirituality, and religion.[49]

Several lines of evidence suggest that measures of spirituality and transcendence are associated with discrete biological functions in the brain. One study, for example, used positron emission tomography (PET scanning) of the brain to associate self-transcendence, as measured by TCI, with the function of the neurotransmitter serotonin and the density of the serotonin 5-HT_{1A} receptor.[50] Genetic studies suggest that these personality characteristics may be partly heritable.

Heritability is a statistical measure of the relative contribution of inherited and environmental factors contributing to a particular trait—specifically, the percentage of the interpersonal variation in a trait that can be accounted for by inheritance. The heritability of spirituality has been calculated by performing psychometric tests in identical twins and in fraternal twins. Identical twins share all their genetic information; thus, those traits that differentiate one twin from another are presumed to arise from environmental, rather than inherited factors. Fraternal twins share only half their genetic information, and can be differentiated by both environmental and inherited factors. Factors that are more highly inherited will be more likely to be shared by identical twins, while those that are less highly inherited are less likely to be shared.

Twin studies demonstrate that there are clear genetic contributions to psychometric parameters of behavior and personality, even after taking into account factors of common upbringing and environment. In particular, studies using a variety of psychometric tests suggest that there is a tendency for identical twins to share spiritual characteristics more closely than nonidentical twins. The calculated heritability of spirituality is between 39 and 43 percent.[51] This magnitude is similar to the heritability of major depression, 35–40 percent,[52] but lower than the heritability of Alzheimer's disease (dementia),[53] 58–79 percent; schizophrenia, 41–87 percent;[54] and bipolar disorders (manic depression), 73–93 percent.[55]

Recently, several studies have attempted to go beyond this statistical measure of heritability to identify specific genes that correlate with psychometric measures of spirituality. Doug Comings has identified an association between spiritual acceptance (measured by TCI) and the DRD-4 gene that encodes a brain receptor for the neurotransmitter dopamine.[56] Dean Hamer has identified a separate association between genetic variations in the VMAT-2 (Vesicular Monoamine Transporter-2) gene, which encodes for a protein that regulates the release of dopamine from dopaminergic synapses.[57]

These genes were of particular interest to researchers since they encode elements of the pathways for the neurotransmitter dopamine. These pathways are also involved in psychiatric disorders, such as schizophrenia, which are characterized by hallucinations that are often considered analogous to mystical experiences. Moreover, dopaminergic pathways, as well as the serotonergic pathways that are also associated with spirituality in some studies, are targets for psychomimetic drugs, such as psilocybin, mescaline (peyote), and LSD, that can induce hallucinogenic and psychedelic states resembling mystical experiences. Several papers have recently analyzed the apparent similarities between drug-induced hallucinogenic states and non–drug-induced mystical experiences and concluded that there are, in fact, significant parallels between these two psychosomatic experiences as well as the neurobiological mechanisms by which these experiences arise.[58]

Hallucinogenic substances have a long association with both mysticism and religious practices. The prophecies of the Oracle of Delphi may have been induced by breathing ethylene gas, which naturally permeated the cave where the priestess retreated to receive her prophecies.[59] Many non-Western religions incorporate hallucinogens in religious rituals. Moreover, the religious traditions of many monastic and ascetic orders incorporate controlled breathing, dietary restrictions, and experiences at altitudes and environments that could induce neurobiological effects similar to those induced by hallucinogenic drugs.[60] In the 1960s, psilocybin was used in the infamous "Good Friday Experiment" in which Christian divinity students who received the drug prior to a religious service reported more profound religious experiences than those who received a chemically inert placebo.[61] The association between hallucinogenic drugs and religious experiences has led some to propose that drug-induced mystical experiences may have been involved in the origin of religion itself.[62]

Hamer's recent book, *The God Gene: How Faith Is Hard Wired into Our Genes,* follows in the tradition of those who have discerned a relationship between mystical experiences and religious expression. His argument is that genetic variations in genes that encode the pathways for dopamine or serotonin metabolism—the pathways that may mediate mystical and spiritual experiences—may make individuals more sensitive to such experiences. Certain individuals may thus be more strongly predisposed for religious experiences and "softwired for God."[63]

The fact that there are pathways in the brain that may mediate a mystical recognition of a sublime essence in nature does not mean that there is no material source to this perception. Neil Goodman writes, in his review of the relationship between drug-induced psychomimetic symptoms and the mystical experience, "The mystics themselves would surely be loathe to hear that their experiences can be explained in purely materialistic ways, but this need not be the case. It is like understanding the way the eye receives and the brain processes visual information, which does not take away from the experience and the importance of seeing."[64]

Goodman could go further. The fact that images can be "seen" during dreams when the eyes are closed, that pictures can be recovered from molecular-based memory, that dementia can bring visions of things that are not real, that genetic variation makes some people see images that appear to be blurred and others images that are bereft of color, and that migraines, seizures, and schizophrenia can cause the perception of light when there is none, does not imply that vision is not a material phenomenon of nature associated with a particular electromagnetic spectrum and the impact on photons on the photoreceptors of the eye. Similarly, the fact that images may be processed into syntactical elements in the brain or fractals in a computer, conveyed by action potentials involving the flux of ions across the membranes of neurons in the brain, and converted into binary code to be transmitted as pulses of monochromatic and coherent light through fiber-optic cables, does not deny that there was a material source for such images.

The presence of neurobiological pathways that mediate the perception of a sublime aspect to nature would seem to suggest that this aspect of nature is not illusory. The observations that these pathways may be disturbed by drugs or disease, that genetic variation may alter the sensitivity and acuity of perception, and that cultural and environmental factors may influence perception and expression does not militate against the possibility that these pathways may be processing a unique spectrum of information. This afferent information is certainly not electromagnetic or even chemical in form. Nevertheless, there is no reason to assume that the pathways in the brain that are generating mystical imagery and mystical experiences are not receiving and processing information in a way that is analogous to the way sight, sound, and smell are received and processed into vision, hearing, and taste.

From a biological perspective, the presence of pathways that can sense sublime elements of nature suggests that such elements, in fact, exist in nature. Sensory organs and the pathways that process sensory stimuli in the brain often require sensory stimuli in their development to stimulate the necessary gene expression, cell growth, and synaptic formation. Conversely, such pathways often atrophy when such stimuli are absent.

The same pattern is evident over evolutionary time. Animals that evolve in darkness tend to gradually lose genetic elements that are required for vision when they have no selective function. Pathways that perceive and process sensations become more advanced when these stimuli are present and provide a competitive advantage.

There may be significant evolutionary implications to the ability to sense the sublime elements of nature and experience a sense of wonder. Among evolutionary biologists, it is widely believed that the selective advantage of *Homo sapiens* does not lie primarily in physical attributes, such as our large brain, upright posture, opposing thumb, and vocal cords, but rather in the aptitudes of our brain. Richard Dawkins has reviewed the advantages that accrue from our ability to construct and vocalize language, to recognize symbols and syntax, to draw representations of aspects of nature, to unconsciously compute and project objects on long trajectories, to impart our accumulated knowledge and culture to our progeny through memes or culturagens, and to see analogies and express symbolic meanings.[65] I would suggest that the evolution of pathways that sense the sublime essence of nature and generate sensations of wonder may be even more important.

Barbara King's chapter in this volume ("God from Nature: Evolution or Emergence?") builds on the observation that great apes exhibit a capacity for wonder in the same way that they exhibit the capacity for language, symbolism, and the use of tools in a way that is qualitatively, if not quantitatively, similar to the capabilities of humans. Given the close genetic similarity of humans and great apes, it is very likely that the serotonergic and dopaminergic pathways that are involved in mysticism, spirituality, and the sense of wonder in humans are present in homologous form in apes and carry out similar functions. The great apes whom Jane Goodall observed staring at waterfalls with an apparent sense of wonder[66] may be engaged in a behavior that is truly homologous to Wordsworth sensing the sublime as he observed the Wye River winding past Tintern Abbey.

The human response to wonder, however, does not stop with perception. The sense of the sublime that Wordsworth sensed above Tintern Abbey not only inspired his poetry, but may have also inspired medieval clergy to construct the first Cistercian abbey on the site in 1131 CE, the great Gothic abbey on the same site two centuries later, and J. M. W. Turner's painting of the abbey centuries after that.[67] So, too, the sense of wonder inspires the pursuit of science and religion.

No less than the gravitational fields that are responsible for the motion of the planets around the sun, the composition of our atmosphere, and the morphing of our geography, the sublime is responsible for the human-made moons that are in orbit around the planets, the changing chemistry of the atmosphere and climate, and the changing frontiers of forests, deserts, rivers, and coastlines in response to human development. The forces put in motion by the sublime are also evident in the lights that radiate into space from our planet, the construction of artificial habitats, and the genetic engineering of new life forms. Just as gravitational fields generate forces in nature only through their impact on bodies with mass,[68] the sublime generates forces in nature through its impact on the mind. These forces are, nevertheless, vital and potent forces in nature, and important elements of a vision of nature that is evolving and malleable.

Conclusion

This vision of nature is one in which the sublime is an essential part of nature and is not superadded or supernatural. The sublime essence of nature exists. It stimulates an evolved sense of wonder. It is "absolutely great," infinite and ineffable, wondrous and awesome, "a motion and a spirit, that impels all thinking things, all objects of all thought, and rolls through all things." It sets the mind in motion, inspires science and religion, and is an intrinsic, material force in an evolutionary and malleable nature.

The vision of a sublime aspect within nature raises two questions that are relevant to this thesis, but beyond the scope of this chapter. First, the ineffable and infinite attributes of the sublime, as well as the mystical way of knowing, are often associated with descriptions of God. In fact, many of the writings on the sublime and the mystical are intended in this context. Nothing in the present analysis, however, presupposes that the sublime is holy or separate from nature. In fact, the argument that I am making is that the sublime can be envisioned as an essence within nature.

Second, many of those who have written about the sublime have argued that this ineffable essence of nature is diminished when it is subsumed within scientific or religious understanding. Inasmuch as the sublime may exist only as objectified in the mind, then removing it from mysticism to understanding may fundamentally alter its sublime essence. If, however, the sublime is an integral and infinite object of nature itself, as argued here, then it is not delimited by understanding, and its essence is not altered by science or religion.

Notes

1. John Hedley Brooke, "Should the Word *Nature* Be Eliminated?" in this volume.
2. Andrew Lustig, "The Vision of Malleable Nature: A Complex Conversation," in this volume.
3. Aristotle, *Metaphysics*, Book 1, Section 2.
4. Adam Smith, *The History of Astronomy*, Section III.
5. Thomas Carlyle, *Sartor Resartus*, Book 1.
6. Ralph Waldo Emerson, *The Complete Works of Ralph Waldo Emerson*, vol. VII, *Society and Solitude*, chapter VII, Works and Days.
7. Abraham Heschel, *Between God and Man*, 41.
8. Barbara King, "God from Nature: Evolution or Emergence?" in this volume.
9. Wolfgang von Goethe, *Preface to Theory of Color*, 158.
10. Henry David Thoreau, *Walden*, 31.
11. Albert Einstein, *My Credo*.
12. Psalm 19, according to Masoretic Text, *The Holy Scriptures* (Philadelphia: Jewish Publication Society of America, 1955), 877.
13. William Wordsworth, "Lines Composed Above Tintern Abbey," *The Complete Poetical Works*.
14. William Wordsworth, "The Sublime and the Beautiful," *Selected Prose*, 263–74.
15. Ibid., 271.
16. Immanuel Kant, "The Mathematically Sublime," *The Critique of Judgment*, A. SS 25–27. Kant describes the mathematically sublime as follows: "The above definition may also be expressed in this way: that is sublime in comparison with which all else is small. Here we readily see that nothing can be given in nature, no matter how great we may judge it to be, which, regarded in some other relation, may not be degraded to the level of the infinitely little, and nothing so small which in comparison with some still smaller standard may not for our imagination be enlarged to the greatness of a world. Telescopes have put within our reach an abundance of material to go upon in making the first observation, and microscopes the same in making the second. Nothing, therefore, which can be an object of the senses is to be termed sublime when treated on this footing. But precisely because there is a striving in our imagination towards progress ad infinitum, while reason demands absolute totality, as a real idea, that same inability on the part of our faculty for the estimation of the magnitude of things of the world of sense to attain to this idea, is the awakening of a feeling of a supersensible faculty within us; and it is the use to which judgment naturally puts particular objects on behalf of this latter feeling, and not the object of sense, that is absolutely great, and every other contrasted employment small. Consequently it is the disposition of soul evoked by a particular representation engaging the attention of the reflective judgment, and not the object, that is to be called sublime. . . . The foregoing formulae defining the sublime may, therefore, be supplemented by yet another: The sublime is that, the mere capacity of thinking which evidences a faculty of mind transcending every standard of sense."
17. Ibid., SS 28–29. Kant describes the dynamically sublime as follows: "Bold, overhanging, and, as it were, threatening rocks, thunderclouds piled up the vault of heaven, borne along with flashes and peals, volcanoes in all their violence of destruction, hurricanes leaving desolation in their track, the boundless ocean rising with rebellious force, the high waterfall of some mighty river, and the like, make our power of resistance of trifling moment in

comparison with their might. But, provided our own position is secure, their aspect is all the more attractive for its fearfulness; and we readily call these objects sublime, because they raise the forces of the soul above the height of vulgar commonplace, and discover within us a power of resistance of quite another kind, which gives us courage to be able to measure ourselves against the seeming omnipotence of nature."

18. Ibid., SS 25.
19. Ibid., SS 28.
20. Arthur Schopenhauer, *The World as Will and Representation,* vol. 1, 201.
21. Ibid., 205: "If we lose ourselves in contemplation of the infinite greatness of the universe in space and time, meditate on the past millennia and on those to come; or if the heavens at night actually bring innumerable immensity of the universe, we feel ourselves reduced to nothing; we feel ourselves as individuals, as living bodies, as transient phenomena of will, like drops in the ocean, dwindling and dissolving into nothing. But against such a ghost of our own nothingness, against such a lying impossibility, there arises the immediate consciousness that all these worlds exist in our representation, only as modifications of the eternal subject of pure knowing. This we find ourselves to be, as soon as we forget individuality; it is the necessary, conditional supporter of all worlds and of all periods of time. The vastness of the world, which previously disturbed our peace of mind, now rests within us; our dependence on it is now annulled by its dependence on us. All this, however, does not come into reflection at once, but shows itself as a consciousness, merely felt, that in some sense or other (made clear only in philosophy) we are one with the world, and are therefore not oppressed but exalted by its immensity."
22. Ibid., 202–4.
23. Theodore Roszak, *Where the Wasteland Ends,* 319.
24. *English translation of the Chumash, Rashi Commentary, and Text Footnotes from Metsudah Chumash/Rashi,* translated by Avrohom Davis, Nachum Y. Kornfeld, and Abraham B. Walzer (New York: Ktav Pub., 1997); http://www.tachash.org/texis/vtx/chumash (accessed July 2006).
25. Wilfred Suchat, *The Creation According to the Midrash Rabbah,* 3.
26. Rashi, *Breshis,* online at http://www.tachash.org/metsudah/b01r.html.
27. Moses Maimonides, *The Guide for the Perplexed,* translated by M. Friedlander, Part II, Chapter XXX. Maimonides writes: "For this reason, Scripture employs the term *breshit* (in a principle), in which the *bet* is a preposition denoting 'in.' The true explanation of the first verse of Genesis is as follows: In [creating] a principle God created the beings above and the things below."
28. Wilfred Shuchat, *The Creation According to the Midrash Rabbah,* 3.
29. Ibid., 3–5.
30. Ibid., 15–22.
31. Ibid., 46–53.
32. This explanation occurs in different forms in historical texts, including the Midrash (*The Midrash,* translated by Wynn Eestcott, S. K. L. Mathers, Herman Adler, Adolf Neubauer, Samuel Rappaport, and Michael Friedlander), and is expanded on by Shuchat, *The Creation According to the Midrash Rabbah,* 49–50.
33. Cited in Suchat, *The Creation According to the Midrash Rabbah,* 46. Another version from Midrash: "It is forbidden to inquire what existed before creation, as Moses distinctly tells us (Deut. 4:32): 'Ask now of the days that are past which were before thee, since the day

God created man upon earth.' Thus the scope of inquiry is limited to the time since the Creation." *The Midrash*, translated by Wynn Eestcott, S. K. L. Mathers, Herman Adler, Adolf Neubauer, Samuel Rappaport, and Michael Friedlander.

34. *Sepher Yetzirah* (Book of Creation), chapter IV, 4.
35. Simeon Ben Jochai, *The Book of Light* (The Zohar), 85.
36. Among the now-classic analyses of the myths and mysticisms of different cultures were James Frazier, *The Golden Bough*; Joseph Campbell, *The Hero with a Thousand Faces*; and Claude-Levi Strauss, *The Raw and the Cooked*.
37. Various authors have tried to categorize aspects of the mystical experience. William James identified mysticism as being "ineffable"—transcending description, "noetic"—apprehended by the mind, "transcient"—not a continuous state of mind or vision, and "passive"—with the mystic feeling subject to a superior power. William James, "The Varieties of Religious Experience: A Study in Human Nature," Gifford Lectures on Natural Religion, Edinburgh in 1901–1902. Walter Pahnke developed a typology to quantify mystical experiences in the "Good Friday Experiment," which attempted to compare natural and drug-induced mystical experiences. His typology includes "Unity"—a loss of self and sense of underlying oneness; "Transcendence of Time and Space"; "Deeply Felt Positive Mood"—a sense of joy, blessedness, and peace; "Sense of Sacredness"—a nonrational response of awe, wonder, and reverence; "Objectivity and Reality"—an intuitive, insightful knowledge of an ultimate reality and certainty that such knowledge is real; "Paradoxicality"—logically contradictory descriptions and interpretations; "Alleged Ineffability"—an experience beyond description and words; "Transiency"—a temporary experience in contrast to the permanence of experience; "Persisting Positive Changes in Attitude and Behavior." Walter Pahnke, "Drugs and Mysticism."
38. A good example of this approach to rationalizing mysticism with science and religion is found in Nathan Aviezer, *In the Beginning: Biblical Creation and Science*.
39. James, "The Varieties of Religious Experience: A Study in Human Nature," lecture XVI.
40. Gershom Sholem, *Major Trends in Jewish Mysticism*.
41. Bertrand Russell, *Mysticism and Logic*, 17.
42. Bertrand Russell, *Religion and Science*, 27.
43. This has been explored in Douglas Hofstadter, *Gödel, Escher, Bach: An Eternal Golden Braid*.
44. Helen Dukas and Banesh Hoffman, eds., *Albert Einstein the Human Side* 66.
45. Edna St. Vincent Millay, "Euclid alone has looked at Beauty bare"; online at http://www.sonnets.org/millay.htm#217.
46. Carl Jung, *Man and His Symbols*.
47. Claude Levi-Strauss, *The Raw and the Cooked*.
48. Joel Paris, "Neurobiological Dimensional Models of Personality: A Review of the Models of Cloniger, Depue, and Siever."
49. Another important test frequently used in studies of spirituality is the Minnesota Multiphasic Personality Inventory-2. D. A. MacDonald and D. Holland, *Spirituality and the MMPI-2*.
50. Jacqueline Borg, Andree Begnt, Henrik Sodersrtom, and Lars Farde, *The Serotonin System and Spiritual Experiences*.
51. Thomas Bouchard, David Lykken, Matthew McGuire, Nancy Segal, and Auke Tellegen, "Sources of Human Psychological Differences: The Minnesota Study of Twins Reared

Apart"; Thomas Bouchard, Matt McGue, David Lykken, and Aukie Tellegen, "Intrinsic and Extrinsic Religiousness: Genetic and Environmental Influences and Personality Correlates"; Ming Stung, Wesley Williams, John Simpson, and Michael Lyons, "Pilot Study of Spirituality and Mental Health in Twins."

52. Kenneth Kendler, Margaret Gatz, Charles Gardner, and Nancy Pedersen, "A Swedish National Twin Study of Lifetime Major Depression."
53. Margaret Gatz, Chandra Reynolds, Laura Fratiglioni, Boo Johansson, James Mortimer, Stig Berg, Amy Fiske, and Nancy Pedersen, "Role of Genes and Environments for Explaining Alzheimer Disease."
54. Tyrone Cannon, Jakko Kaprio, Muttunen Lonnqvist, and Markku Koshenvuo. "The Genetic Epidemiology of Schizophrenia in a Finnish Twin Cohort."
55. Peter McGuffin, Fruhling Rijdijk, Martin Andrew, Rand Katz, and Alastair Cardno, "The Heritability of Bipolar Affective Disorder and the Genetic Relationship to Unipolar Depression."
56. D. E. Comings, N. Gonzales, G. Saucier, J. Johnson, and J. MacMurray. "The DRD4 Gene and the Spiritual Transcendence Scale of the Character Temperament Index."
57. Dean Hamer, *The God Gene: How Faith Is Hardwired into Our Genes.*
58. These arguments are classically associated with the drug culture of the 1960s and writers such as Terence McKenna, Gordon Wasson, and Timothy Leary. Recently, advances in neurobiology have spawned new scholarly interest in the parallels between drug-induced states and mystical experiences. For example, Neil Goodman, "The Serotenergic System and Mysticism: Could LSD and the Nondrug-Induced Mystical Experience Share Common Neural Mechanisms"; R. R. Griffiths, W. A. Richards, U. McCann, and R. Jess, "Psilocybin Can Occasion Mystical-Type Experiences Having Substantial and Sustained Personal Meaning and Spiritual Significance."
59. J. Z. deBoer, J. R. Jale, and J. Canton, "New Evidence for the Geological Origins of the Ancient Delphic Oracle (Greece)."
60. S. Arzy, M. Idel, T. Landis, and O. Blanke, "Why Revelations Have Occurred on Mountains? Linking Mystical Experiences and Cognitive Neuroscience."
61. This infamous experiment, conducted in association with Timothy Leary, involved administering psilocybin or a placebo to divinity students immediately before a Good Friday service. The results of this controlled, double-blind experiment were described in Walter Pahnke, *Drugs and Mysticism.* A follow-up study of this population was published in 1991. Rick Doblin, "Pahnke's 'Good Friday Experiment': A Long-Term Follow-Up and Methodological Critique."
62. Claude Levi-Strauss, *Structural Anthropology*, vol. 2, 222–37.
63. Hamer, *The God Gene*, 211.
64. Neil Goodman, "The Serotonergic System and Mysticism: Could LSD and the Nondrug-Induced Mystical Experience Share Common Neural Mechanisms?"
65. Richard Dawkins, *Unweaving the Rainbow, Science Delusion and the Appetite for Wonder.*
66. Jane Goodall, presentation at conference, "The Quest for Knowledge, Truth, and Values in Science & Religion: Science and the Spiritual Quest II," Harvard Divinity School's Center for World Religions, October 2001; online at http://www.hno.harvard.edu/gazette/2001/10.25/16-science.html. "We find that chimps do have a sense of wonder, of awe," Goodall said. "I think we can see the roots of some kind of religion from chimp behavior—[inhibited] by their lack of language."

67. History from http://www.cadw.wales.gov.uk/upload/resourcepool/Tintern710.pdf.
68. This is true in a Newtonian model of gravitational force.

Bibliography

Aristotle. *Metaphysics*, translated by W. D. Ross. Adelaide, Australia: University of Adelaide. Online at http://etext.library.adelaide.edu.au/a/aristotle/metaphysics/ (accessed July 2006).

Arzy, Shahar, Moshe Idel, Theodor Landis, and Olaf Blanke. "Why Revelations Have Occurred on Mountains? Linking Mystical Experiences and Cognitive Neuroscience." *Medical Hypothesis* 65 (2005): 841–45.

Aviezer, Nathan. *In the Beginning: Biblical Creation and Science*. Hoboken, N.J.: KTAV Publishing, 1990.

Borg, Jacqueline, Andree Bengt, Henrik Soderstom, and Lars Farde. "The Serotonin System and Spiritual Experiences." *American Journal of Psychiatry* 160 (November 2003): 1965–1969.

Bouchard, Thomas, David Lykken, Matthew McGuire, Nancy Segal, and Auke Tellegen. "Sources of Human Psychological Differences: The Minnesota Study of Twins Reared Apart." *Science* 250 (October 1990): 223–28.

Bouchard, Thomas, Matt McGue, David Lykken, and Auke Tellegen. "Intrinsic and Extrinsic Religiousness: Genetic and Environmental Influences and Personality Correlates." *Twin Research* 2 (1999): 88–98.

Campbell, Joseph. *The Hero with a Thousand Faces*. Princeton, N.J.: Bollingen, 1973.

Cannon, Tyrone, Jakko Kaprio, Muttenen Lonnqvist, and Markku Koshenvuo. "The Genetic Epidemiology of Schizophrenia in a Finnish Twin Cohort." *Archives of General Psychiatry* 55 (July 1998): 67–74.

Carlyle, Thomas. "*Sartor Resartus*." Great Literature Online, http://carlyle.classicauthors.net/SartorResartus/SartorResartus10.html (accessed July 2006).

Comings, D. E., N. Gonzales, G. Saucier, J. Johnson, and J. MacMurray. "The DRD4 Gene and the Spiritual Transcendence Scale of the Character Temperament Index." *Psychiatric Genetics* 10 (December 2000): 185–89.

Davis, Avrohom, Nachum Kornfeld, and Abraham Walzer. *Chumash, Rashi Commentary, and Text Footnotes from Metsudah Chumash/Rashi*. New York: Ktav Pub. Inc., 1997. Online at http://www.tachash.org/texis/vtx/chumash (accessed July 2006).

Dawkins, Richard. *Unweaving the Rainbow, Science Delusion and the Appetite for Wonder*. New York: Houghton Mifflin, 1998.

deBoer, J. Z., J. R. Jale, and J. Chanton. "New Evidence for the Geological Origins of the Ancient Delphic Oracle (Greece)." *Geology* 29 (August 2001): 707–10.

Doblin, Rick. "Pahnke's 'Good Friday Experiment': A Long-Term Follow-Up and Methodological Critique." *Journal of Transpersonal Psychology* 23 (1991). Online at http://www.druglibrary.org/schaffer/LSD/doblin.htm (accessed December 8, 2008).

Einstein, Albert. *My Credo*. Online at http://www.einstein-website.de/z_biography/credo.html (accessed July 2006).

Emerson, Ralph Waldo. *The Complete Works of Ralph Waldo Emerson*, vol. VII—Society and Solitude (1870). Online at http://www.rwe.org/comm/index.php?option=com_

content&task=view&id=37&Itemid=213 (accessed July 2006).
Frazier, James. *The Golden Bough.* New York: Macmillan, 1963.
Gatz, Margaret, Chandra Reynolds, Laura Fratiglioni, Boo Johansson, James Mortimer, Stig Berg, Amy Fiske, and Nancy Pedersen. "Role of Genes and Environments for Explaining Alzheimer Disease." *Archives of General Psychiatry* 63 (February 2006): 168–74.
Goethe, Wolfgang von. Preface to *Theory of Color,* in Johann Wolfgang von Goethe, *Scientific Studies,* edited and translated by Douglas Miller. New York: Suhrkamp Publishers, 1988.
Goodman, Neil. "The Serotonergic System and Mysticism: Could LSD and the Nondrug-Induced Mystical Experience Share Common Neural Mechanisms?" *Journal of Psychoactive Drugs* 34 (July 2002): 263–72.
Griffiths, R. R., W. A. Richards, U. McCann, and R. Jesse. "Psilocybin Can Occasion Mystical-Type Experiences Having Substantial and Sustained Personal Meaning and Spiritual Significance." *Psychopharmacology,* reprint ed. Online at http://www.springerlink.com/media/hhwx2b1mrh6wym5ewvtp/contributions/v/2/1/7/v2175688r1w4862x.pdf (accessed July 2006).
Hamer, Dean. *The God Gene: How Faith Is Hardwired into Our Genes.* New York: Doubleday, 2004.
Heschel, Abraham. *Between God and Man.* New York: Free Press, 1997.
Hofstadter, Douglas. *Godel, Escher, Bach: An Eternal Golden Braid.* New York: Vintage Books, 1980.
Holy Scriptures. Philadelphia: Jewish Publication Society of America, 1955.
James, William. "The Varieties of Religious Experience. A Study in Human Nature," Gifford Lectures on Natural Religion, Delivered at Edinburgh in 1901–1902, lecture XVI. Online at http://www.csp.org/experience/james-varieties/james-varieties.html (accessed December 8, 2008).
Jung, Carl. *Man and His Symbols.* New York: Dell, 1964.
Kant, Immanuel. *The Critique of Judgment,* translated by James Creed Meredith. Adelaide, Australia: eBooks@Adelaide, 2004. Online at http://etext.library.adelaide.edu.au/k/kant/immanuel/k16j/ (accessed July 2006).
Kendler, Kenneth, Margaret Gatz, Charles Gardner, and Nancy Pedersen. "A Swedish National Twin Study of Lifetime Major Depression." *American Journal of Psychiatry* 163 (January 2006): 109–14.
Levi-Strauss, Claude. *Structural Anthropology,* vol. 2. Translated by Monique Layton. Chicago: University of Chicago Press, 1976.
———. *The Raw and the Cooked.* Translated by John and Doreen Weightman. Chicago: University of Chicago Press, 1969.
MacDonald, D. A., and D. Holland. "Spirituality and the MMPI-2." *Journal of Clinical Psychology* 59 (2003): 399–410.
Maimonides, Moses. *The Guide for the Perplexed,* 2nd ed. (1904). Translated by M. Friedlander. Online at http://www.sacred-texts.com/jud/gfp/gfp.htm (accessed July 14, 2006).
McGuffin, Peter, Fruhling Rijdijk, Martin Andrew, Rand Katz, and Alaistair Cardno. "The Heritability of Bipolar Affective Disorder and the Genetic Relationship to Unipolar Depression." *Archives of General Psychiatry* 60 (May 2003): 497–502.

Midrash. Translated by Wynn Eestcott, S.K.L. Mathers, Herman Adler, Adolf Neubauer, Samuel Rappaport, and Michael Friedlander. New York: Parke, Austin, and Lipscomb, 1917. Online at http://www.sacred-texts.com/jud/mhl/mhl00.htm (accessed December 8, 2008).

Pahnke, Walter. "Drugs and Mysticism." *International Journal of Parapsychology* VIII (Spring 1966): 295–313. Online at http://leda.lycaeum.org/?ID=16461 (accessed December 8, 2008).

Paris, Joel. "Neurobiological Dimensional Models of Personality: A Review of the Models of Cloniger, Depue, and Siever." *Journal of Personality Disorders* 192 (February 2005): 156–70.

Rashi (Solomon Bar Isaac). *Bereishis*. In Avrohom Davis, Nachum Kornfeld, and Abraham Walzer, eds., *Chumash, Rashi Commentary, and Text Footnotes from Metsudah Chumash/Rashi*. New York: Ktav Pub., Inc., 1997. Online at http://www.tachash.org/metsudah/b01r.html (accessed July 2006).

Roszak, Theodore. *Where the Wasteland Ends: Politics and Transcendence in Postindustrial Society*. Garden City, N.Y.: Anchor Books, 1973.

Russell, Bertrand. *Mysticism and Logic*. Totowa, N.J.: Barnes & Noble, 1981.

———. *Religion and Science*. New York: Oxford University Press, 1997.

Schopenhauer, Arthur. *The World as Will and Representation*, vol. 1. Translated by E.F.J. Payne. New York: Dover, 1969.

Sepher Yetzirah (Book of Creation), translated by W. W. Wescott. Online at http://www.sacred-texts.com/jud/yetzirah.htm (accessed July 14, 2006).

Sholem, Gershom. *Major Trends in Jewish Mysticism*. New York: Schocken Books, 1973.

Simeon (Ben Jochai). *The Book of Light* (The Zohar). Translated by Nurho de Manhar. New York: Theosophical Publishing Company, 1900. Online at http://www.sacred-texts.com/jud/zdm/zdm010.htm (accessed July 2006).

Smith, Adam. "The History of Astronomy." In W.P.D. Wightman and J. C. Bryce, eds., *Essays on Philosophical Subjects*, vol. III, Glasgow Edition of the Works and Correspondence of Adam Smith. Indianapolis: Liberty Fund (1982). Online at http://oll.libertyfund.org/Texts/LFBooks/Smith0232/GlasgowEdition/PhilosophicalSubjects/0141-04_Bk.html (accessed July 2006).

Suchat, Wilfred. *The Creation According to the Midrash Rabbah*. Jerusalem: Devora Publishing, 2002.

Thoreau, Henry David. *Walden*. New York: Brimhall House, 1951.

Tsuang, Ming. Wesley Williams, John Simpson, and Michael Lyons. "Pilot Study of Spirituality and Mental Health in Twins." *American Journal of Psychiatry* 159 (March 2002): 486–88.

Wordsworth, William. *The Complete Poetical Works*. London: Macmillan and Co., 1888; reprinted 1999 at Bartleby.com. Online at http://www.bartleby.com/145/ww138.html (accessed July 2006).

———. *Selected Prose*. New York: Penguin, 1988.

12

NATURE AS CULTURE: THE EXAMPLE OF ANIMAL BEHAVIOR AND HUMAN MORALITY

Nicolaas A. Rupke

Introduction

Nature is a word that has been given many meanings throughout the history of modern scholarship (see, e.g., John Hedley Brooke's piece in this volume, "Should the Word *Nature* Be Eliminated?"). Today, the concept of nature continues to be invested with connotations that range widely, "from wonder to ecological crisis."[1] The distinction of *nature* from *culture* has been problematized, especially in recent environmental history scholarship (James Proctor's essay in this volume, "Environment after Nature: Time for a New Vision"). The point has repeatedly been made that whichever way we may wish to conceive of nature, our views of it invariably and inevitably partake of culture.

But what of nature in the sense of Humboldt's *cosmos*, defined as "physical reality"? Does not "the description of the physical universe" present us with an objective view of nature? Or is the contrary true—that the scientific study of observable reality presents us with an enculturated form of the physical world? If each of the alternatives has some validity, to what extent?

The issue is important, if only because it relates to that of the so-called moral authority of nature[2] or, to be precise, to the moral authority of the scientific study of nature. If science gives us an objective view of nature,[3] including human existence, then the view of "culture as nature" that acknowledges nature's moral authority is strengthened. By contrast, if it can be shown that scientists in their explorations of the tangible world significantly invest this with the cultural predilections of their place and time, and thus that we are dealing with "nature as culture," the presumed moral authority of nature would appear to be weakened, indirectly adding legitimacy to other claims, such as those of religion.

In this chapter, I join those who previously have made a case for nature as culture.[4] At a more general level, I further side with colleagues such as David Livingstone[5] (also in this volume), who have put forward the view that location, taken in a broad sense and including cultural location, has a constitutive significance in the production, distribution, and reception of views of nature. Scientific knowledge, he argues, is not "the view from Nowhere," but always "a view from Somewhere." In support of this contention, I offer an example taken from the scientific study of animal behavior and human morality, which I am exploring in a project that covers the relevant Anglo-American, French, and Batavo-German literature from the past two centuries. My starting point is the time of the early professionalization of animal behavior studies, during the early part of the nineteenth century, when various entomological societies were established and specialized journals began to appear. In this chapter I present a partial and preliminary account of the project by examining in some detail the work of one early author, to show how his science was inseparable from his philosophico-political preoccupations, thereby indicating the ambiguity and instability of the perceived significance for human morality of animal behavior.

The example is an effective one for making my point of nature as culture, but one might well object by asking this: Does the subject of animal behavior and human morality really belong to the natural rather than the social sciences? Ever since Herbert Spencer (1820–1903) and the beginnings of social Darwinism, the biologizing of culture has been common in these fields,[6] and a case study taken from them would be less convincing. It is true that, from a disciplinary point of view, my example occupies an intermediary area between the biological and the social sciences, and that in the history of animal behavior studies a tug-of-war over who controlled the field has taken place between psychologists, sociologists, and zoologists. Yet the phenomenon of animal behavior has, from the very beginning of the professionalization of entomology, attracted leading biologists, a trend that later was indicated by the change of animal psychology to ethology, the introduction of quantitative analytical techniques, and the award in 1973 of the Nobel Prize in Physiology or Medicine to the trio of ethologists Karl von Frisch (1886–1982), Konrad Lorenz (1903–1989), and Nikolaas Tinbergen (1907–1988).

Many of the professionals who study animal behavior themselves see the story of their discipline as a development driven by the dynamics of an increase in methodological rigor, from anthropomorphic animal psychology to disengaged-objective ethology.[7] From the point of view they take, this is a perfectly valid perception. Yet their account of how today's ethology came to be can equally well be captured by a wholly different narrative. Issues of cognitive-methodological progress can be subsumed under a nonprogressive story of successive appropriations of animal behavior on behalf of contemporaneous interests. The story in this case is not one of scientific progress but of sociopolitical change, of successive presentist indications of societal interests that dictated the central message spoken by animal behavior to questions of human conduct.

For the connection between animal behavior and human morality to be compelling, zoologists first had to accept a fundamental animal-human commonality, in particular with respect to mind. A much-debated issue in the animal behavior literature of the eighteenth century and the early nineteenth century was whether or not animals have "souls" (a "mind," "intelligence," etc.) or are just instinct-machines.[8] In France, Charles Georges Leroy (1723–1789), for example, in his classic *Lettres sur l'intelligence et la perfectibilité des animaux: avec quelques lettres sur l'homme* (1802) (*The Intelligence and Perfectibility of Animals from a Philosophic Point of View*, 1870), addressed this question, as did the entomologist Pierre André Latreille

(1762–1833), who wrote on ants and other insects. Yet at that time, inferences from animal behavior for human affairs remained by and large allegorical and incidental.[9]

In Great Britain, around 1800, where the story of my project begins, the still-vibrant tradition of natural theology went hand in hand with a persistent belief in a fundamental difference between animals and humans. The Anglican clergyman and entomologist William Kirby (1759–1850), in his Bridgewater Treatise *On the Power Wisdom and Goodness of God as Manifested in the Creation of Animals and in their History Habits and Instincts* (1835), maintained that humans stand apart from the animal kingdom by having a share not only of nature but also of the Divine: "He is the only member of the Animal Kingdom that partakes both of a heavenly and of an earthly nature,—that belongs both to a material and an immaterial world: and on this account it was that God, when he had created man, constituted him king over the whole sphere of animals with which he had peopled this globe that we inhabit."[10] In this exalted position, humans were to rule animals and sit in judgment of their behavior (as had been done quite literally during medieval and early-modern times in the case of "animal trials").[11] If, at times, animals appear to behave intelligently, Kirby asserted, this did not indicate the existence of a mental commonality with humans but was the result of direct divine intervention, the creature being instrumentalized by God to fulfill a particular purpose. These occasions were "interpositions of the Deity by which the law of instinct is suspended, to answer a particular purpose of his Providence."[12] Given that nature was divinely designed, humans might profitably trace patterns of proper behavior in animals, but a bearing of animal behavior on human morality continued to be kept by and large at the level of allegory, as in Jean de la Fontaine's famous "The Cicada and the Ant" and other fables.

Carl Vogt's Untersuchungen über Thierstaaten *(1851)*

One of the first treatises—if not *the* first treatise—in which the animal-human barrier was broken down and human society was discussed in terms of animal behavior was written by the earth and life scientist Carl Vogt (1817–1895). A student of Justus von Liebig (1803–1873) at Giessen, Vogt was appointed professor of zoology at that university in 1846. He was at the forefront of the European revolutionary struggle for *Freiheit und Einheit* (freedom and national unification),[13] a leading member of the German Ante-Parliament and the German National Assembly in Frankfurt and subsequently in Stuttgart. His political radicalism during and in the wake of the 1848–1849 Revolution, which took the form of anarchism, forced him to flee to Switzerland, where, in 1852, he became a professor of geology at the Geneva Academy. There he significantly contributed during his later years to that institution's transformation in 1873 into the city's university. Both politically and philosophically, he was a radical, known, together with Ludwig Büchner (1824–1899) and Jakob Moleschott (1822–1893), for his so-called "vulgar" materialism.[14]

Vogt's contributions to the earth and life sciences ranged widely, and included influential popular books, textbooks, as well as specialized studies in marine biology. Not long after his appointment to the Giessen chair of zoology, Vogt started a translation into German of the anonymous and notorious *Vestiges of the Natural History of Creation*, written by the Edinburgh publisher Robert Chambers (1802–1871), a translation that was intended as part of Vogt's antimonarchist and materialist campaign.[15] In the wake of the appearance of Darwin's *On the Origin of Species*, Vogt became a fervent supporter of Darwin and took an active part in the Darwinian debates. At about the same time that Thomas Henry Huxley published his *Evidence as to Man's Place in Nature* (1863), Vogt produced *Vorlesungen über den Menschen, seine Stellung in der Schöpfung und in der Geschichte der Erde* (1863) (*Lectures on Man, His Place in Creation, and in the History of the Earth*, 1864).

Vogt ridiculed the belief that humans possess an immaterial soul, and during the 1850s he took a leading role in the so-called *Materialismusstreit* (conflict over materialism), famously exchanging blows with the Göttingen professor of anatomy Rudolph Wagner, who took a traditional Christian stance. He expanded his attack on Wagner's human soul metaphysics in an essay on "animal souls," arguing that there existed no fundamental difference between animals and humans and that, in both, mental activities—"mind" or "soul"—are nothing but a collective expression of neural activities that come to an end when the organism—human as well as animal—ceases to exist. The human mind, like that of animals, is a material product, an excretion of the organ of the brain.[16] Thus, humans and animals were the same in the sense that neither possessed a mental entity that could be called a "soul." The question is not whether animals have intelligence, Vogt thundered during the tumultuous uproar of the 1848–1849 Revolution, but whether humans do; and he illustrated his point with a book on *Untersuchungen über Thierstaaten* (*Investigations of Animal States*) (1851).

The book consisted of three essays, the first and most substantial of these dealing with the *Bienenstaat* (colonies of bees), the second with *Heuschrecken und verwandtes Gesindel* (grasshoppers and related riffraff), and the third with *Blasenträger*, or Siphonophora, colonial marine hydrozoa that resemble jellyfish and of which a well-known species is the so-called Portuguese Man O' War. His main example of "animal states" were colonies of honeybees, however, which—Vogt maintained—are monarchies or, more precisely, constitutional monarchies. He described the working of bee colonies in terms of this model: "I will show you this constitutional monarchy in the animal kingdom, with the autocratic ruler at the top, who in order to remain on the throne kills even her own children; with the hereditary peers, upheld by a non-obligation to do any work; with the poor oppressed people who have to direct their kind care towards the nursing of the children and the feeding of the offspring, and who now and then struggle free from their slavery, only to fall back into it!"[17]

The three castes—workers, drones, and queen bee—did not represent three kinds of bees that cooperate in a collective effort for survival, but rather a divided society with oppressed victims on the one hand and privileged oppressors on the other. The workers constitute a proletariat (from 15,000 to a rare maximum of 30,000), the drones are landed gentry/nobility/hereditary peers (600 to 1,000), and the single queen bee is the monarch. The workers are female bees who become workers as a result of an inadequate and limited supply of food. This lack of attention to their dietary needs is actually put into practice by the workers themselves, which shows the extent to which oppression of the members of the lower classes produces compliance. It also shows that the proletariat is not born but raised, kept in its lowly state by the "poison of neglect."[18] In addition to caring for the young, the workers serve the drones, licking them, brushing them upon their return home, and keeping the queen warm, when needed, by surrounding her in thick heaps of devoted subjects.

The cycle of a bee colony, Vogt continued, starts at the moment of swarming, is then followed by a period of stable existence and finally ends with the partition of the group. A proletarian bee emerges from its birth cell after only twenty days; a noble drone takes longer—twenty-four days. The drones immediately lay claim to their privileges and take no part in the work, not in the building of the combs nor in the care of the young, but fly out to feed themselves at leisure, returning satiated. The queen bee is the sum total of the colony, laying eggs, first of workers (many), then of drones (a middling number), and lastly of young queen bees (just a few). As soon as the latter mature, considerable agitation develops. Large numbers of workers, especially young ones, cover the few queen-bee cells, until at last a new, young queen bee emerges. The "wing-telegraph" of the bees spreads the exciting news, and the queen mother, surrounded by old workers and drones, approaches the birthplace, not to welcome her offspring into this world but to try and kill her own daughters with her sting. The young workers, however, rally to protect the pretender. The old queen bee, if unsuccessful, leaves the hive, accompanied by loyal workers and drones, swarming and settling elsewhere to start a new colony.

In the meantime, while the older generation is leaving the hive, a young queen emerges. Her first act, however, is not one of regal liberality, but of fratricide. She scrambles to the adjacent incubation cells and plunges her poisonous sting into the as-yet-unborn queen pupae, or the nearly developed queen imagos. Their corpses are removed from the cells, thrown out of the hive, and the workers demolish the empty royal cells. On occasion, two young queen bees emerge simultaneously. Would they fight each other with the help of large armies? No, the bees, being more clever than humans, let the pretenders fight it out among themselves in a queen-on-queen battle. The victorious one is courted by the drones. The company moves to the entrance of the hive and, leaving the workers behind, flies up into the air. What happens then had not yet been discovered, Vogt reported; but, he added, "a Victoria always finds her Prince Albert" ("*Aber eine Viktoria findet immer ihren Prinzen Albert*").[19] After a few hours the company returns, disheveled, exhausted, covered in pollen, and is received by the workers who treat their oppressors with caring devotion. The "wing telegraph," again, appears to transmit the message: "Her Majesty has deigned to marry" ("*Ihre Majestät haben geruht sich zu vermählen*").[20]

"Oh my German fatherland, how much do you resemble bee societies!"

("*O mein deutsches Vaterland, wie sehr gleichst du den Bienenstaaten!*")[21]—Vogt exclaimed. Gradually, he turned his narrative account of the beehive into a causal account of the failed Revolution of 1848–1849 and into a road map for future activism. Force would never suffice to bring about a permanent change, especially not a change of mind. Ultimately, violence was not the solution to the political crisis, because the monarchy is founded on the loyalty and the way of thinking of the working class in the first place. Those who suffer under the yoke of servitude, when freed, seek a new yoke. Moreover—Vogt warned the constitutional reformers—a constitutional monarchy, when it comes about by way of the rule of law, will inevitably turn into a Russian kind of autocratic rule. What is the difference between Russia under Catherine II and a bee "state," he asked? The conundrum's solution was to be found in the behavior of workers when, at rare occasions, they come to the realization of their true condition. Bees prefer the inside of their hive to be dark, and if one places a colony in a glass hive, the bees cover the glass with wax. If one continues to remove this, a revolutionary agitation develops among the workers. "The proletarians gain insight into their situation" ("*Die Proletarier erhalten Einsicht in ihre Lage*").[22] They stop feeding the young and caring for the queen bee and, after a while, swarm into the open, living the rest of their lives in free and happy anarchy.

Vogt was familiar with the contemporary professional literature on entomology, including the relevant writings by Hermann Burmeister (1807–1892) and Lorenz Oken (1779–1851).[23] Moreover, the third part of his *Thierstaaten* was based on his own internationally acclaimed research into marine hydrozoa.[24] There existed, in other words, a definite scientific foundation to his book. Yet the places where he wrote and researched its three essays, Bern (the first two) and Nice (the last), were the initial stations of his post-Revolution exile, and the politics of the time, in which Vogt was deeply immersed, were as much, if not more, of a factor in his choice of animals and of his highlighting of particular behavioral features than any unencumbered scientific curiosity. Vogt's interest in animal "psychology" by and large coincided with—and appears to have been driven by—the turmoil of 1848–1849, when he took an active part in the politics of the time, convinced that the results of the scientific study of nature would help direct attempts to overthrow the old monarchical order, whether autocratic or constitutional, to establish a true democracy of *Volkssouveränität* (people's sovereignty), and bring about political change—freedom and the antimonarchist unification of Germany. His *Thierstaaten* denoted the disillusionment with the reactionary politics that followed in the wake of the brief revolutionary successes. Animal behavior speaks to us about good and bad social organization and should inform our political conduct in the case of revolution and reaction, he insisted.

In Vogt's later, less revolutionary and more bourgeois-establishmentarian period in Geneva, little can be found that takes up the thread of his innovative *Thierstaaten* of 1851. Not even in his equally innovative applied entomological *Vorlesungen über nützliche und schädliche, verkannte und verläumdete Thiere* (*Lectures on Useful and Harmful, Unappreciated and Maligned Animals*) (1864) did he return to the political topicality of his 1851 treatise.[25] Only at the very end of these lectures did Vogt add a characteristically sarcastic, Vogtian allusion to animal behavior in the context of human affairs, connecting slavery among ants, known since Pierre Huber's (1777–1840) *Recherches sur les moeurs des fourmis indigènes* (*Researches into the Customs of Indigenous Ants*) (1810), to the slavery issue in the United States. Why do Christian slave owners not add the phenomenon of animal slavery to the ammunition of their defense, he queried? If God has created the institution of slavery among insects, surely humans should be allowed to imitate this, they could argue.[26]

Vogt will have known the answer to his own question. Christians did not accept that nature held moral authority over human actions. On the contrary, humans were the executive arm of God's moral governance of the world and, as Kirby had maintained in his Bridgewater Treatise, should sit in judgment of animal behavior. In Vogt's view, the opposite was true. Animals—insects—had attained greater perfection than humans in matters of social organization. Given the length of geological time that certain animal groups had existed—much longer than humans—they had found solutions to problems that were superior to ours. They possess all forms of sociopolitical systems: socialist, republican, aristocratic, monarchical, and the like. The least advanced of these dispenses with the individual, whereas the most advanced respects individuality. The debasement of individuals can go as far as to make them seem like no more than organs of a larger organic totality, without free movement or independence of any kind, as in the case of Siphonophora. By contrast, the individual becomes increasingly more perfect when a development toward anarchy takes place, in a process of emancipation from the state. "Anarchy steels the organs, sharpens the senses, increases mental strength—in a state of anarchy the single individual exercises all his organs and abilities, fighting the elements as well as his enemies, to achieve in combat the independence he needs."[27] A fine example of an anarchist is the clever fox, who anarchically "does his thing" in a self-built hole and only during a short childhood lives under parental, patriarchal discipline. Anarchy was freedom, and human progress had to be a striving toward a state of anarchy. That was the message of animal behavior studies. Vogt's researches, ranging from honeybees to Siphonophora, made the animal world speak a loud and single message: "Anarchy! Anarchy!"[28]

Bird's-Eye View of Later Stages

We would not capture the meaning of *Thierstaaten* if we were to depict this book as, in essence, zoology-driven. What shaped Vogt's account was his location in the cultural landscape of the time. By reading animal behavior in terms of the contemporaneous political discourse, Vogt appropriated nature on behalf of societal ideals. His was a particularly blatant instance of appropriation, but—I shall argue in a full account of my project—similar instances have dominated the entire subsequent history of animal behavior and human morality. Let us here, in preliminary substantiation of this claim, continue with a brief mention of a few of these instances, restricted to historical, no longer living, representatives of the subject.

Among Vogt's contemporaries was Ludwig Büchner, (in)famous for his *Kraft und Stoff* (*Force and Matter*) (1854), in which he, too, affirmed the fundamental materiality and sameness of the animal and human minds. A fellow materialist, he engaged in writing treatises similar to Vogt's *Thierstaaten*. His *Aus dem Geistesleben der Thiere oder Staaten und Thaten der Kleinen* (*The Mental Life of Animals or Societies and Feats of the Little Ones*) (1876) added to Vogt's example of the bees that of the ants, interpreting ant society as a republic and, in reference to the most famous republic of all, the United States, discussing *in extenso* the phenomenon of slavery among ants.[29] However, with the founding of the German Reich in 1871, the preoccupation with issues of German national politics disappeared, to be replaced by others, and Büchner broadened his reading of animal behavior in his *Liebe und Liebes-Leben in der Thierwelt* (*Love and Love-Life in the Animal World*) (1879) to include the institutions and variants of love, such as courtship, marriage, family, parental love, altruism, friendship, jealousy, sociability, and patriotism.

Much of what Büchner wrote appeared in the wake of Darwin's *On the Origin of Species*, and the basis for animal-human comparisons was no longer just the popular, raw materialism of Büchner-Moleschott-Vogt but the belief in human descent from apelike ancestors. During the early decades of Darwinism, by far the larger part of animal behavior and human morality became subsumed under new concerns, such as colonies, empire, civilization, and the various institutions of human society, in addition to those of politics—in particular, that of marriage. In Darwin's immediate vicinity was the Liberal politician, anthropologist, and entomologist John Lubbock, first Lord Avebury (1834–1913). In his highly popular *Ants, Bees and Wasps* (1882), he only incidentally, not systematically, connected insect society to human society; yet this was not due to a lack of interest in tracing the animal roots of human institutions. On the contrary: In line with evolutionary racism, and in the context of British Empire politics, Lubbock shifted the boundary between animals and humans to draw it between "savages" and "civilized men." Thus, the animal antecedents of civilized conduct were to be discovered by studying the behavior of "modern savages"—that is, indigenous peoples in colonized territories—as Lubbock did in his *Prehistoric Times* (1865; plus many later editions) and *The Origin of Civilization* (1870).[30]

By and large, sexuality remained a taboo subject, in many instances touched upon only indirectly by means of a discussion of more "proper" subjects, such as matrimony. Büchner's *Liebe und Liebes-Leben in der Thierwelt,* for example, only tangentially discussed love in terms of sexual love. The person to break this taboo was the redoubtable popularizer of Darwinism in Germany, Wilhelm Bölsche (1861–1939). An admiring follower of Darwin's "Rottweiler," Ernst Haeckel (1834–1919), a committed member of the Monist League and a central figure of the Friedrichshagen colony of sexually libertarian and liberated artists,[31] Bölsche adopted the role of science popularizer, promoting the Darwinian worldview as a secular alternative to Christianity. One of his strategies was to relocate human morality, removing it from its ecclesiastical foundations to base it on science. In his most successful book, *Das Liebesleben in der Natur* (*Love Life in Nature*) (1898), he discussed human sexual mores as an integral part of the totality of sexual behavior in nature. In the process, Bölsche relativized a range of social institutions, from marriage to monastic celibacy, opening up avenues of progress for liberal causes, such as the suffragist movement, and at length discussed prostitution, which he condemned, stressing at the same time that puritanical conventions had driven young men to get their first sexual experiences in the "degenerate" context of sex for money.[32]

This assault on the Christian foundations of society was parried by contemporaries of Bölsche. The oeuvre of the entomologist and termite expert Erich Wasmann (1859–1931), for example, can be seen as the Catholic answer to the Darwinism of Bölsche and the Monist League. A Jesuit priest, Wasmann, too, addressed issues of comparative behavior in animals and humans but, in his *Vergleichende Studien über das Seelenleben der Ameisen und der höhern Thiere* (*Comparative Studies of the Psychology of Ants and Higher Animals*) (1897) and later studies, he reaffirmed the church's position on the fundamental separateness of animals from humans, even though he had become a convinced evolutionist. Animals are not mere reflex machines, Wasmann admitted, yet they do not have a spiritual nature as associated with immortal souls, the possession of which sets humans apart. He sustained this line of argument by demonstrating that animals have no intelligence. His very interest in ants may well go back to a critical review he wrote of Lubbock's insect book in German translation, *Ameisen, Bienen und Wespen Ants, (Bees and Wasps)* (1883), a book in which Lubbock had made great claims for animal intelligence.[33] Yet Wasmann saw ant society as a product of instincts. Not entirely consistent with this, Wasmann discussed the phenomenon of nonsexual castes in ant and termite colonies in terms that seemed like an apologia of the Catholic institution of monastic celibacy with its vows of poverty, chastity, and obedience.[34]

The trauma of World War I significantly determined which connections were made between animal behavior and human society. One author to be influenced that way was the Swiss Calvinist myrmecologist and psychiatrist Auguste Henri Forel (1848–1931). A professor of psychiatry at the University of Zürich (1879–1898), he was prominently involved in social reform, and concerned about sexually transmitted diseases and alcoholism. Like Bölsche, he was a member of the Monist League, and in his *Die sexuelle Frage* (1905) (*The Sexual Question*, 1922) broke contemporaneous taboos. As a convert to the Baha'i faith, he was a committed pacifist who advocated a "United States of the World." Forel nurtured a lifelong interest in ants, seeing their social behavior as superior to that of humans, in particular their attempts to solve the problem of war. In his *Le monde social des fourmis du globe comparé à celui de l'homme* (1923) (*The Social World of Ants Compared with that of Man* [1928]), he pointed out that colonial ants live in organized anarchy, without leaders, rulers or governments. Chauvinism, Forel believed, was the root of all atrocity, and the human world, to rid itself of this evil, should become like a giant, worldwide ant colony. He prominently promoted the League of Nations, which he conceived as a global parallel to the republics of Switzerland and the United States. Ants provided us with a successful experiment in peaceful living.[35]

As we know, Forel's pacifist ideas did not carry the day. What did live on following his death was his eugenics work, which during the Third Reich became incorporated in National Socialist (Nazi) programs of euthanasia.[36] One person to contribute to this narrowing of Forel's legacy was the applied entomologist and termite expert Karl Escherich (1871–1951). A fervent Nazi and early supporter of Adolf Hitler (1889–1945), he delivered a rousing speech, on his inauguration as president of Munich University, in which he denounced the "*Termitenwahn*" (termite madness) of the Bolsheviks, by which he meant their totalitarian annihilation of individuality, contrasting this with the "voluntary subordination" of citizens under the Nazi regime.[37]

The death and destruction of World War I and World War II formed the backdrop to questions asked of animal behavior by the new ethology in its preoccupation with aggression and territoriality. The position of innate aggressiveness was argued by a range of well-known names, from the popular writer Robert Ardrey (1908–1980) to the Nobel laureate Nikolaas Tinbergen. Already during World War II, the early German representative of ethology, Werner Fischel (1900–1977), addressed *Die kämpferische Auseinandersetzung in der Tierwelt* (*Combat in the Animal World*) (written 1943; published 1947). One of the most famous and respected representatives of the ethological approach, who significantly contributed to relocating the study of animal behavior from psychology to zoology, was Konrad Lorenz, whose book on the natural history of aggression, *Das sogenannte Böse* (1963)

(*On Aggression*, 1966), gave expression to the contemporaneous need for facing up to and dealing with the disturbing fact of wars between "civilized" nations. Humans were innately aggressive, he believed, but a study of how aggressive behavior is regulated in higher animals might help in averting future calamities of warfare between nation-states.[38]

Conclusion

This bird's-eye view of animal behavior and human morality can do no justice to the complexity and richness of the full story, nor does it include the more recent preoccupation with altruism and, as Barbara King discusses in this volume, with emotionality ("God from Nature: Evolution or Emergence?"). Moreover, the broad questions posed at the beginning of this chapter about whether or not, and the extent to which, science presents us with an objective view of nature cannot be comprehensively answered on the basis of just this case study. Yet, however narrow the compass of my account, it underpins, I believe, the argument for nature as culture—for the view that the scientific study of nature is embedded in a cultural matrix that permeates it.

The crucial question to ask is this: Has the scientific study of animal behavior given us a set of essential norms of moral conduct? Or—softening the demand made on science—have the connections between animal behavior and human morality, although made by scientists, proved independent from them, not bound by their personality or culture, to use Charles Gillispie's criterion for what characterizes objective science?[39] The answer would appear to be "no". The particular link, described by Vogt and his comrades-in-arms, served and was made in the cause of the ideals of the Revolution of 1848–1849. Subsequent generations latched on to aspects of animal behavior in ways that betrayed the societal preoccupations of their places and times. Over and again, animal behavior has been made to speak a message that mediated presentist group concerns. To put it theatrically: The scientist/zoologist appears to have acted like a puppetmaster, making nature speak his or her own ventriloquist message.

One could conjecture that the connection between animal behavior and human conduct was constructed at a less profound and more trivial level than here presented, and that Vogt did not read animal behavior through the sociopolitical lens of his place and time, but merely tried to explain it in terms of an anthropomorphic model that just readily presented itself. Instances of such anthropomorphism can probably be found, such as when Vogt compared the nuptials of a queen bee to Victoria finding her Albert.[40] In this case his intent may well have been less to relate insect life in a critical way to the British monarchy than to select from contemporaneous royalty two conspicuous names in commenting on a delicate aspect of apiology. Yet, Vogt possibly intended some antimonarchist mocking here as well. Be that as it may, the main thrust of his approach was the reading of insect behavior in terms of Europe's revolutionary politics of 1848–1849. To Vogt, the study of insect behavior offered a prescription for solving the political problems of human society. Nature was made a source of imperatives for reform. Moreover, if the political turmoil of the time merely suggested to him an obvious model for his insect studies, my general point remains unchanged, namely, that from one space-and-time location to another, apiologists, myrmecologists, and other animal behavior experts read in nature different meanings and that these invariably reflected transitory preoccupations, often related to current affairs during troubled times.

In trying to construe culture in terms of nature, Vogt's cultural location proved of constitutive significance, and the converse of what he attempted came about: Nature was construed in terms of culture—by him and his contemporaries, as well as by later animal behavior scientists. As I have attempted to show, the scientific advocates of culture as nature in this story, of whom Vogt was an early exemplar, were entrapped in an "egocentric predicament," and proved to be "servitors to their own prepossessions," not unlike historians when they write about the past.[41]

Notes

1. See, for example, Hans Bak and Walter W. Hölbling, eds., *"Nature's Nation" Revisited: American Concepts of Nature from Wonder to Ecological Crisis* (Amsterdam: VU University Press, 2003).
2. See, for example, Lorraine Daston and Fernando Vidal, eds., *The Moral Authority of Nature* (Chicago: University of Chicago Press, 2004).
3. The claim to scientific objectivity has been made by, among others, the doyen of the history of science, Princeton University's Charles Gillispie, as long ago as his *The Edge of Objectivity* (1960), but also recently in the epilogue to a collection of his classic papers (Charles C. Gillispie, "l'Envoi," *Essays and Reviews in History and History of Science* [Phil-

adelphia: American Philosophical Society, 2007]). Defining science as "the developing body of knowledge concerning the structure, forces and conditions of nature," Gillispie contrasts it with the fine arts: A work of art pertains to the artist or writer in a way that a work of science does not pertain to the scientist. Scientific knowledge, he maintains, although created by people, is independent of its creators, in spite of personal or national styles. Science may well be produced "out of personality and in culture," but is not then bound by personality or culture. It is impersonal and general and, together with the technology that attends it, science is, moreover, the only aspect of Western civilization that other civilizations have fully assimilated—unlike religion, philosophy, Western arts and letters—and can in this sense be described as objective—a body of knowledge made by people, but made about the world and not about themselves.

4. For example, Gregg Mitman, *The State of Nature: Ecology, Community, and American Social Thought, 1900–1950* (Chicago: University of Chicago Press, 1992), 9. See also Charlotte Sleigh's excellent *Six Legs Better: A Cultural History of Myrmecology* (Baltimore: Johns Hopkins University Press, 2007), *passim*.
5. See David Livingstone, *Putting Science in Its Place: Geographies of Scientific Knowledge* (Chicago: University of Chicago Press, 2003).
6. For a discussion of earlier scientific influences on the social sciences, see Johan Heilbron, "Social Thought and Natural Science," in Theodore M. Porter and Dorothy Ross, eds., *The Cambridge History of Science*, vol. 7, *The Modern Social Sciences* (Cambridge: Cambridge University Press), 40–56.
7. William Homan Thorpe, *The Origins and Rise of Ethology: The Science of the Natural Behaviour of Animals* (London: Heinemann, and New York: Praeger, 1979).
8. Hermann Samuel Reimarus, *Allgemeine Betrachtungen über die Triebe der Thiere, hauptsächlich über ihre Kunst-Triebe: Zum Erkenntniss des Zusammenhanges der Welt, des Schöpfers und unser selbst* (Hamburg: Bohn, 1760).
9. For a discussion of the role of bees in eighteenth-century political discourse and Bernard Mandeville's (1670–1733) *The Fable of the Bees* (1714), see Danielle Allen, "Burning *The Fable of the Bees*: The Incendiary Authority of Nature," in Daston and Vidal, *The Moral Authority of Nature*, 74–99. An early author to draw support for republicanism from insect societies was the Genevan Pierre Huber (1777–1840), author of *Recherches sur les moeurs des fourmis indigènes* (1810).
10. William Kirby, *On the Power, Wisdom and Goodness of God as Manifested in the Creation of Animals and in their History, Habits and Instincts*, 2 vols. (London: Pickering, 1835), 2:519.
11. Esther Cohen, *The Crossroads of Justice: Law and Culture in Late Medieval France* (Leiden, The Netherlands: Brill, 1993).
12. Kirby, *History, Habits and Instincts of Animals*, 2:280. Kirby's views on the nature of instinct were not shared, however, by William Spence, with whom he coauthored the four-volume work, *An Introduction to Entomology* (London: Longman, 1815–26).
13. Heinrich Best, "'*Que faire avec un tel peuple?*' Carl Vogt et la révolution allemande de 1848–1849," in Jean-Claude Pont et al., eds., *Carl Vogt; science, philosophie et politique (1817–1895)* (Chêne-Bourg, Switzerland: Georg, 1998), 13–29.
14. Frederick Gregory, *Scientific Materialism in Nineteenth Century Germany* (Dordrecht, The Netherlands: Reidel, 1977); Pont et al., *Carl Vogt*.
15. Nicolaas A. Rupke, "Translation Studies in the History of Science: The Example of *Ves-*

tiges," British Journal for the History of Science 33 (2000): 209–22. Later, Vogt bid, although in vain, to translate Darwin's *Descent of Man* and *Variation of Plants and Animals;* see Adrian Desmond and James Moore, *Darwin* (London: Penguin, 1992), 543, 573.

16. Nicolaas A. Rupke, *Richard Owen* (New Haven, Conn.: Yale University Press, 1994), 307–8.
17. "*Ich will sie Euch zeigen, diese konstitutionelle Monarchie im Thierreiche, mit dem Alleinherrscher an der Spitze, der sogar seine eigenen Kinder tödtet, um sich auf dem Throne zu erhalten, mit der erblichen Pairie* [hereditary peers], *gestützt auf die Nichtverpflichtung zur Arbeit, mit dem armen, gedrückten Volke, das seine rührende Sorge auf Pflegung der Kinder und Ernährung der Nachkommenschaft richten muß, und das nur zuweilen aus der Sklaverei sich aufrafft, um auf's neue wieder darin zu versinken!*" Carl Vogt, *Untersuchungen über Thierstaaten* (Frankfurt am Main: Literarische Anstalt, 1851), 37.
18. Vogt, *Thierstaaten*, 49.
19. Ibid., 97.
20. Ibid., 98.
21. Ibid., 86.
22. Ibid., 57.
23. Hermann Burmeister, *Handbuch der Entomologie*, 5 vols. (Berlin: Reimer, 1832–1855); Lorenz Oken, *Allgemeine Naturgeschichte für alle Stände*, 7 vols. (Stuttgart: Hofmann, 1833–1845), 5:2.
24. Carl Vogt, "Note sur les Siphonophores," *Annales des sciences naturelles. Zoologie* 3rd ser. 18 (1852): 273–78; "Über Siphonophoren," *Zeitschrift für wissenschaftliche Zoologie* 3 (1853): 522–25; "Mémoire sur les Siphonophores de la mer de Nice," *Mémoires de l'Institut National Genevois* 1 (1853) (published 1854): 1–164.
25. On the politics of the scientific notion of "insect pests," see Sarah Jansen, *"Schädlinge": Geschichte eines wissenschaftlichen und politischen Konstrukts 1840–1920* (Frankfurt am Main: Campus, 2003).
26. "*Zum Oeftern schon habe ich mich gewundert, diese von Natur wegen bei gewissen Ameisen eingeführte Sclaverei nicht unter den Argumenten zu finden, welche die Sclavenhalter Nordamerikas zu ihren Gunsten anzuführen gewohnt sind. Sie haben als fromme Christen und rechtgläubige Menschen die Bibel bis auf den letzten Boden ausgeschöpft, um die Sclaverei als eine göttliche Institution, vom Heiland gebilligt und von den Aposteln gepredigt, hinzustellen;—sie haben sich eigens Naturforscher selbst aus Europa kommen lassen, die des Ehrgefühls so bar und ledig waren, daß sie die Berechtigung der höher stehenden Menschenspecies, des Kaukasiers, zur Knechtung der niederen Race, des Negers, aus zoologischen Grundsätzen und Unterschieden zu deduciren suchten;—warum nicht auch noch die Natur als Dritte in den Bund rufen, wenn Glaube und Wissenschaft schon ihnen beispringen? Da hätte man ja, bei den Ameisen, Alles im schönsten Spiegelbilde—eine rothblonde, stärkere Race, die nur genießt, höchstens zum Zeitvertreibe einmal Krieg führt und Raubzüge unternimmt, und eine schwarzgraue, schwächere, dienende Race, die für ihre Herren arbeitet, sie füttert, umherschleppt und ihre Nachkommenschaft auferzieht und pflegt, wie wenn es Ihresgleichen wäre! Was will man mehr tun, als den Schöpfer nachahmen?*" Carl Vogt, *Vorlesungen über nützliche und schädliche, verkannte und verläumdete Thiere* (Leipzig: Keil, 1864), 179–80.
27. "*Die Anarchie stählt die Organe, schärft die Sinne, vermehrt die geistige Kraft—als Einzelnes den Elementen wie seinen Feinden gegenüber übt das Individuum in der Anarchie alle seine*

Organe und Fähigkeiten, um im Kampfe die Selbständigkeit zu erringen, deren es bedarf." Vogt, *Thierstaaten*, 29.

28. Ibid., 32.
29. Ludwig Büchner, *Aus dem Geistesleben der Thiere oder Staaten und Thaten der Kleinen* (Berlin: Hofmann, 1876), 131–62. Slavery among ants had been known ever since Pierre Huber's *Recherches sur les moeurs des fourmis indigènes* (Paris: Paschoud, 1810).
30. Representative of the period, too, was *Des sociétés animales* (Paris: Baillière, 1877) by the Bordeaux sociologist Alfred Victor Espinas (1844–1922), a classic of its kind, highly thought of by Espinas' colleague Émile Durkheim (1858–1917). A. S. Byatt, in her novella "Morpho Eugenia," in *Angels and Insects* (London: Chatto & Windus, 1992), offers a fictional account of Victorian fascination with ant behavior.
31. See, for details, Rolf Kauffeldt and Gertrude Cepl-Kaufmann, *Berlin-Friedrichshagen: Literaturhauptstadt um die Jahrhundertwende* (Munich: Boer, 1994).
32. Wilhelm Bölsche, *Das Liebesleben in der Natur. Eine Entwickelungsgeschichte der Liebe*, 3 vols. (Leipzig: Diederichs, 1898–1903).
33. For more on Forel, see A. J. Lustig, "Ants and the Nature of Nature in Auguste Forel, Erich Wasmann, and William Morton Wheeler," in Daston and Vidal, *The Moral Authority of Nature*, 282–307, on 291–98.
34. Erich Wasmann, *Die Ameisen, die Termiten und ihre Gäste: Vergleichende Bilder aus dem Seelenleben von Mensch und Tier* (Regensburg: Manz, 1934), 6–7.
35. See further Lustig, "Ants and the Nature of Nature," 285–90.
36. Auguste Forel, *Le monde social des fourmis du globe comparé à celui de l'homme*, 5 vols. (Geneva: Kundig, 1921–23). For a more detailed discussion of Forel, see Torsten Rüting, "Von der Erbsünde zur Genetik, von Ameisen zu Menschen—Auguste Henri Forel (1848–1931)," *Acta Historica Leopoldina* 39 (2004): 227–65.
37. Karl Escherich, *Termitenwahn: Rede gehalten beim Antritt des Rektorats der Ludwig-Maximilians-Universität am 25 Nov. 1933* (Munich: Langen & Müller, 1934). Much on Escherich is to be found in Jansen, "*Schädlinge*," note 25.
38. Konrad Lorenz, *On Aggression* (London: Methuen, 1966). In the secondary literature, the use of biological determinism, of seeing culture as nature, is more often than not associated with the political right. Lorenz, whose Nazi past has in recent years come back to haunt his memory, is a good example of this association (see Theodora J. Kalikow, "Die ethologische Theorie von Konrad Lorenz: Erklärung und Ideologie, 1938 bis 1943," in Herbert Mehrtens and Steffen Richter, eds., *Naturwissenschaft, Technik und NS-Ideologie. Beiträge zur Wissenschaftsgeschichte der Dritten Reichs* (Frankfurt am Main: Suhrkamp, 1980), 189–214; Benedikt Föger and Klaus Taschwer, *Die andere Seite des Spiegels: Konrad Lorenz und der Nationalsozialismus* (Vienna: Czernin, 2001); Eugeniusz Nowak, *Wissenschaftler in turbulenten Zeiten. Erinnerungen an Ornithologen, Naturschützer und andere Naturkundler* (Schwerin: Stock & Stein, 2005), 203–13. As Gregg Mitman reminds us, however, the use of biological determinism has not been the exclusive province of the political right, but also occurred on the left of center. The case of Vogt supports his point. At the beginnings of the "biologizing of society," we find Vogt and his radical left-wing, socialist politics. Bölsche, too, was a pronounced left-winger. Biological determinism, it would appear, has not been the preserve of any one part of the political spectrum but found wide-ranging advocacy. See further *Aggression: The Myth of the Beast Within* (New York: Wiley, 1988), by the composite author John Klama, which explores the issue with historical insights.

39. See note 3 above.
40. On anthropomorphism, see Lorraine Daston and Gregg Mitman, eds., *Thinking with Animals: New Perspectives on Anthropomorphism* (New York: Columbia University Press, 2005).
41. The words in quotation marks are by Arthur Schlesinger Jr., "History and National Stupidity," *The New York Review of Books* (27 April 2006): 14, discussing the nature of historical scholarship. Science is not fundamentally different from the history of science—that is, from scholarship in the humanities. In a comparison of the disciplines of science and history, Otto Gerhard Oexle of Göttingen's Max Planck Institute of History has described how, by around 1900, in opposition to the dominant tradition of nineteenth-century empiricist thought, a new historical *Kulturwissenschaft* emerged, represented by Heidelberg University's Max Weber (1864–1920) and others. History, they contended, is not the Rankean "*Wie es eigentlich gewesen*"; it is not a *re*construction of factual reality, but a *con*struction of mental connections. This does not mean that there are no historical facts or that history is fictional, but it does imply that its truth is relational, that it is determined by the question. Thus, the scholar is an intrinsic part of the scholarship, just as in the Copenhagen interpretation of quantum mechanics and Werner Heisenberg's Uncertainty Principle the physicist in observing subatomic properties is part of the observed result (Otto Gerhard Oexle, "Begriff und Experiment. Überlegungen zum Verhältnis von Natur- und Geschichtswissenschaft," in Vittoria Borsò and Christoph Kann, eds., *Geschichtsdarstellung. Medien—Methoden—Strategien* [Cologne: Böhlau, 2004], 19–56). In recent decades, following the challenge of postmodernism, these issues have been debated with renewed vigor. Mainstream historians, by reconnecting with the work on collective memory by the philosopher-sociologist Maurice Halbwachs (1877–1945), have defined historiography in terms of memory cultures and, in the process, have "deconstructed" many a previously canonical narrative. Historians of science, in the meantime, deflected the brunt of postmodernism from their own work and made the subjects of their study, the scientists, bear it, by arguing that science is socially "constructed." Yet the narrative representation of the scientific past by the historians of science are constructions no less, something I recently illustrated in the form of a metabiographical study of Alexander von Humboldt, the different biographical representations of whom can be understood as sociopolitical appropriations of him (Nicolaas A. Rupke, *Alexander von Humboldt: A Metabiography* [Frankfurt am Main: Peter Lang, 2005]). As Harvard University's Stevin Shapin puts it: "The past is never wholly a foreign country, and we make and remake past lives to suit our present purposes" (Steven Shapin, "Lives after Death," *Nature* 441 [2006]: 286).

Bibliography

Allen, Danielle. "Burning *The Fable of the Bees*: The Incendiary Authority of Nature." In Lorraine Daston and Fernando Vidal, eds., *The Moral Authority of Nature*, 74–99. Chicago: University of Chicago Press, 2004.

Bak, Hans, and Walter W. Hölbling, eds. *"Nature's Nation" Revisited: American Concepts of Nature from Wonder to Ecological Crisis*. Amsterdam: VU University Press, 2003.

Best, Heinrich. " '*Que faire avec un tel peuple?' Carl Vogt et la révolution allemande de*

1848–1849." In Jean-Claude Pont et al, eds., *Carl Vogt; science, philosophie et politique (1817–1895)*, 13–29. Chêne-Bourg, Switzerland: Georg, 1998.
Bölsche, Wilhelm. *Das Liebesleben in der Natur. Eine Entwickelungsgeschichte der Liebe*, 3 vols. Leipzig: Diederichs, 1898–1903.
———. *Love-life in Nature. The Story of the Evolution of Love*. New York: Boni, 1926; London: Cape, 1931.
Borsò, Vittoria, and Christoph Kann, eds., *Geschichtsdarstellung. Medien—Methoden—Strategien*. Cologne: Böhlau, 2004.
Bröker, Werner. *Politische Motive naturwissenschaftlicher Argumentation gegen Religion und Kirche im 19. Jahrhundert: dargestellt am "Materialisten" Karl Vogt (1817–1895)*. Münster: Aschendorff, 1973.
Büchner, Ludwig. *Aus dem Geistesleben der Thiere oder Staaten und Thaten der Kleinen*. Berlin: Hofmann, 1876.
———. *Liebe und Liebes-Leben in der Thierwelt*. Berlin: Hofmann, 1879.
Burmeister, Hermann. *Handbuch der Entomologie*, 5 vols. Berlin: Reimer, 1832–1855.
Byatt, A. S. *Angels and Insects*. London: Chatto & Windus, 1992.
[Chambers, Robert]. *Natürliche Geschichte der Schöpfung des Weltalls, der Erde und der auf ihr befindlichen Organismen: begründet auf die durch die Wissenschaft errungenen Thatsachen*. Translated by Carl Vogt. Braunschweig, Germany: Vieweg, 1851.
Cohen, Esther. *The Crossroads of Justice: Law and Culture in Late Medieval France*. Leiden, The Netherlands: Brill, 1993.
Daston, Lorraine, and Gregg Mitman, eds. *Thinking with Animals: New Perspectives on Anthropomorphism*. New York: Columbia University Press, 2005.
Daston, Lorraine, and Fernando Vidal, eds. *The Moral Authority of Nature*. Chicago: University of Chicago Press, 2004.
Degen, Heinz. "Vor hundert Jahren: Die Naturforscherversammlung zu Göttingen und der Materialismusstreit." *Naturwissenschaftliche Rundschau* (1954): 271–77.
Escherich, Karl. *Termitenwahn: Rede gehalten beim Antritt des Rektorats der Ludwig-Maximilians-Universität am 25 Nov. 1933*. Munich: Langen & Müller, 1934.
Fischel, Werner. *Die kämpferische Auseinandersetzung in der Tierwelt*. Leipzig: Barth, 1947.
Forel, Auguste. *Le monde social des fourmis du globe comparé à celui de l'homme*. 5 vols. Geneva: Kundig, 1921–23.
———. *Die sexuelle Frage: Eine wissenschaftliche Studie*. Munich: Reinhardt, 17th ed, 1942.
Gillispie, Charles C. *Essays and Reviews in History and History of Science*. Philadelphia: American Philosophical Society, 2007.
Gregory, Frederick. *Scientific Materialism in Nineteenth-Century Germany*. Dordrecht, The Netherlands: Reidel, 1977.
Hirschmann, Elise. "*Karl Vogt als Politiker*." PhD diss., Frankfurt am Main University, 1925.
Huber, Pierre. *Recherches sur les moeurs des fourmis indigènes*. Paris: Paschoud, 1810.
Jansen, Sarah. "*Schädlinge*." *Geschichte eines wissenschaftlichen und politischen Konstrukts 1840–1920*. Frankfurt am Main: Campus, 2003.
Kalikow, Theodora J. "Die ethologische Theorie von Konrad Lorenz: Erklärung und Ideologie, 1938 bis 1943." In Herbert Mehrtens and Steffen Richter, eds., *Naturwissenschaft, Technik und NS-Ideologie. Beiträge zur Wissenschaftsgeschichte der Dritten Reichs*, 189–214. Frankfurt am Main: Suhrkamp, 1980.

Kauffeldt, Rolf, and Gertrude Cepl-Kaufmann. *Berlin-Friedrichshagen: Literaturhauptstadt um die Jahrhundertwende. Der Friedrichshagener Dichterkreis*. Munich: Boer, 1994.

Kirby, William. *On the Power, Wisdom and Goodness of God as Manifested in the Creation of Animals and in Their History Habits and Instincts*, 2 vols. London: Pickering, 1835.

Klama, John. *Aggression: The Myth of the Beast Within*. New York: Wiley, 1988.

Latreille, Pierre André. *Histoire naturelle des fourmis, et recueil de mémoires et d'observations sur les abeilles, les araignées, les faucheurs, et autres insectes*. Paris: Barrois, 1802.

Leroy, Charles Georges. *Lettres philosophiques sur l'intelligence et la perfectibilité des animaux: avec quelques lettres sur l'homme*. Paris: Bossange, Masson, Besson, 1802.

Livingstone, David N. *Putting Science in Its Place: Geographies of Scientific Knowledge*. Chicago: University of Chicago Press, 2003.

Lorenz, Konrad. *On Aggression*. London: Methuen, 1966.

Lubbock, John. *Prehistoric Times, as Illustrated by Ancient Remains and the Manners and Customs of Modern Savages*. London: Williams and Norgate, 1913 (orig., 1865).

———. *The Origin of Civilisation and the Primitive Condition of Man*. Edited and with an Introduction by Peter Rivière. Chicago: University of Chicago Press, 1978 (orig., 1870).

———. *Ameisen, Bienen und Wespen: Beobachtungen über die Lebensweise der geselligen Hymenopteren*. Leipzig: Brockhaus, 1883.

Lustig, A. J. "Ants and the Nature of Nature in Auguste Forel, Erich Wasmann, and William Morton Wheeler." In Daston and Vidal, eds., *The Moral Authority of Nature*, 282–307.

Mehrtens, Herbert, and Steffen Richter, eds. *Naturwissenschaft, Technik und NS-Ideologie. Beiträge zur Wissenschaftsgeschichte der Dritten Reichs*. Frankfurt am Main: Suhrkamp, 1980.

Misteli, Hermann. *Carl Vogt. Seine Entwicklung vom angehenden naturwissenschaftlichen Materialisten zum idealen Politiker der Paulskirche (1817–1849)*. Zürich: Leemann, 1938.

Mitman, Gregg. *The State of Nature: Ecology, Community, and American Social Thought, 1900–1950*. Chicago: University of Chicago Press, 1992.

Näf, Werner, ed. *Deutschland und die Schweiz in ihren kulturellen und politischen Beziehungen während der ersten Hälfte des 19. Jahrhunderts*. Bern: Lang, 1936.

———. "Abrechnung mit der deutschen Revolution von 1848/49. Aufzeichnungen Carl Vogts." In Näf, ed., *Deutschland und die Schweiz in ihren kulturellen und politischen Beziehungen während der ersten Hälfte des 19. Jahrhunderts*, 193–217.

Nowak, Eugeniusz. *Wissenschaftler in turbulenten Zeiten. Erinnerungen an Ornithologen, Naturschützer und andere Naturkundler*. Schwerin, Germany: Stock & Stein, 2005.

Oexle, Otto Gerhard. "Begriff und Experiment. Überlegungen zum Verhältnis von Natur- und Geschichtswissenschaft." In Vittoria Borsò and Christoph Kann, eds., *Geschichtsdarstellung. Medien—Methoden—Strategien*, 19–56. Cologne: Böhlau, 2004.

Oken, Lorenz. *Allgemeine Naturgeschichte für alle Stände*, 7 vols. Stuttgart: Hofmann, 1833–1845.

Pont, Jean-Claude et al., eds. *Carl Vogt: Science, philosophie et politique (1817–1895)*. Chêne-Bourg, Switzerland: Georg, 1998.

Reimarus, Hermann Samuel. *Allgemeine Betrachtungen über die Triebe der Thiere, hauptsächlich über ihre Kunst-Triebe: zum Erkenntniss des Zusammenhanges der Welt, des Schöpfers und unser selbst*. Hamburg: Bohn, 1760.

Rupke, Nicolaas A. *Richard Owen*. New Haven, Conn., and London: Yale University Press, 1994.

———. "Translation Studies in the History of Science: The Example of *Vestiges*." *British Journal for the History of Science* 33 (2000): 209–22.

———. *Alexander von Humboldt: A Metabiography*. Frankfurt am Main: Lang, 2005.

Rüting, Torsten. "*Von der Erbsünde zur Genetik, von Ameisen zu Menschen—Auguste Henri Forel (1848–1931).*" *Acta Historica Leopoldina* 39 (2004): 227–65.

Schmidt, Rolf. *Die Auffassung der Sinnlichkeit und die Einstellung zur Sexualität bei Wilhelm Bölsche*. Munich: Uni-Druck, 1964.

Shapin, Steven. "Lives after Death." *Nature* 441 (2006): 286.

Sleigh, Charlotte. *Six Legs Better: A Cultural History of Myrmecology*. Baltimore: Johns Hopkins University Press, 2007.

Thorpe, William Homan. *The Origins and Rise of Ethology: The Science of the Natural Behaviour of Animals*. London: Heinemann; and New York: Praeger, 1979.

Vogt, Carl. *Untersuchungen über Thierstaaten*. Frankfurt am Main: Literarische Anstalt, 1851.

———. "Note sur les Siphonophores." *Annales des sciences naturelles. Zoologie*, 3rd ser., no. 18 (1852): 273–78.

———. "Über Siphonophoren." *Zeitschrift für wissenschaftliche Zoologie* 3 (1853): 522–25.

———. "Mémoire sur les Siphonophores de la mer de Nice." *Mémoires de l'Institut National Genevois* 1 (1853): 1–164 (published 1854).

———. *Köhlerglaube und Wissenschaft. Eine Streitschrift gegen Hofrath Rudolph Wagner in Göttingen*. Giessen, Germany: Ricker, 1855.

———. *Altes und Neues aus Thier- und Menschenleben*, 2 vols. Frankfurt am Main: Literarische Anstalt, 1859.

———. *Vorlesungen über den Menschen: seine Stellung in der Schöpfung und in der Geschichte der Erde*. 2 vols. Giessen, Germany: Ricker, 1863.

———. *Vorlesungen über nützliche und schädliche, verkannte und verläumdete Thiere*. Leipzig: Keil, 1864.

Wasmann, Erich. *Vergleichende Studien über das Seelenleben der Ameisen und der höhern Thiere*. Freiburg im Breisgau: Herder, 1897.

———. *Die Ameisen, die Termiten und ihre Gäste: vergleichende Bilder aus dem Seelenleben von Mensch und Tier*. Regensburg: Manz, 1934.

13

ENVIRONMENT AFTER NATURE: TIME FOR A NEW VISION

James D. Proctor

Environmentalism is today more about protecting a supposed "thing"—the "environment"—than advancing the worldview articulated by Sierra Club founder John Muir, who nearly a century ago observed, "When we try to pick out anything by itself, we find it hitched to everything else in the Universe."[1]

Introduction

Recently I left an enviable faculty position of thirteen years, sold my house on the ocean, and became director of an environmental studies program at a small liberal arts school in the U.S. Pacific Northwest. I say this, so that when I say what I will say next you will not ignore me as some rabid anti-environmentalist:

I am anti-environment.

At least in the sense that *environment* is generally understood today, a taken-for-granted notion underlying everything from environmentalism to "environmentology."[2] Somehow our notion of environment got wrapped up in our notion of nature, and with it came a whole host of conceptual binaries that effectively drive a wedge through any lasting resolution of environmental problems.

This is not a new argument; in fact, everything I cite to support my claim is someone else's idea, not mine.[3] What is surprising is that so little of it has found its way into environmentalism. Thirtysomething years ago, around the time of the birth of the modern environmental movement, there was a great deal of low-hanging fruit to be picked, lots of obvious environmental problems that had pretty much been ignored up to then. Maybe this is why environmentalists didn't want to spend too much time forging new conceptual tools: Nature was imperiled, in some cases people were imperiled as a result, and the imperative was action, not talk. Well, thirtysomething years later, it's no revelation that there are environmental problems; we've all tasted the low-hanging varietals. And, sadly, it's no secret that many have proven rather indigestible, while the higher-up fruit has been virtually impossible to reach, let alone digest. Maybe it's time to rethink—to paraphrase Neil Evernden—what exactly *is* this environment we struggle so hard to save.[4]

I offer no magic here: Environmental problems will not go away once we forge a new vision of environment. Indeed, when you start walking down this path you may feel more and more uncertain as to where we are going. I've learned that many environmentalists are impatient folks: climate change demands an immediate and lasting response, biodiversity is being lost at an alarming rate, more toxins are finding their nefarious way into pregnant women, more families in marginal countries are working more degraded lands, harvesting less and less each year. The imperative of decisive and timely action is inarguable. And this, perhaps, is the very problem: We have been so busy talking about strategy, so deeply committed to proclaiming facts and prescribing action, that we've not taken the risk to think deeply. When we do, as I will suggest below, we may find that not only is our notion of environment in need of repair, but also our notions of the sources of authority upon which we often justify environmental concern, science and religion being the most prominent.

If there is one thing I want to reclaim about environment, it is the vision of connectedness articulated—perhaps a bit expansively—in John Muir's famous quote cited above. Connections matter empirically, morally, politically. The best of our knowledge of nature, of scientific inquiry, and of religious wisdom is the sense of connection they offer. The worst of environment in the contemporary sense is the binary disconnect it presumes by its very utterance. As such, environment echoes a recent usage of *nature* as a biophysical reality separate from culture—whether above or below us in beauty, intelligence, worth, or moral considerability.[5] Even to say that we are connected to nature/environment itself presumes a disconnect. Would you ever need to argue that there is a connection between mother and child, lover and lover, predator and prey? No, because these are relational terms: A mother is a mother by virtue of mothering a child, a predator seeks prey, which avoids the predator.

Recently, there has been a great deal of talk about whether or not there are connections between science and religion, but here, too, the very discussion presumes a disconnect—one enforced by the disconnection police, those who, for many good reasons, wish science to remain distinct from, say, creationism, or those who wish religion to be more than, say, an empirical experiment. And the disconnect between science and religion is perfectly analogous to the disconnect between nature and culture, a conceptual gap that leaves us resorting to the unimaginative language of seeking some "balance" between environment and society.

Connection is emphasized in other essays in this volume, though it may take different forms. The geographer David Livingstone's attention to locating our visions of nature, science, and religion in space, and the historian John Hedley Brooke's focus on doing the same in time, suggest that concepts arise from the connections we engage in via our situated practice. Whether or not we emphasize connectedness, it is our located connections that help explain what we emphasize in the first place. Connection is also emphasized by a wide array of fellow contributors struggling to define the wonderful yet convoluted relationship between science, religion, and nature, whether this involves delving into the inner lives of primates (Barbara King), the complex hybrids of genetic technology (Andrew Lustig), the imagination and practice of landscape in the Mediterranean (Martha Henderson), a new metaphysics of complexity (Robert

Ulanowicz), or the binding of facts and values into visions of nature as theologies (Willem Drees). What I wish to discuss here is how one prominent popular sense of nature as the biophysical world has fostered a disconnect in contemporary environmentalist thought and practice; thus, in restoring a sense of connection, environmentalists may well have to leave behind this treasured vision of nature.

To restore relationality, connection, to environment requires a new vision only insofar as it would entail listening to all the existing visions that entail relationality—this is not an entirely new vision! When we do this in a wholehearted way, we find that our other guiding landmarks, such as science and religion, change, too. This is threatening to the very best of us. It sounds abstract, of little practical value. There are lots of good reasons to turn away. Now would be a good time. But if you can be patient for a little while, I can at least sketch a broad-brushstroke picture for you.

The Environment, Defined

You would think that one easy way to resolve this semantic issue would simply be to go to a dictionary. What we find is that *environment* is a relatively recent word, by no means as old as *nature*, so perhaps it is understandable that environment has increasingly been understood in terms of nature. But this was not always so. The *Oxford English Dictionary*[6] traces environment back to the early seventeenth century, a mere four hundred years ago: At that incipient moment, and for the next two centuries, *environment* was used in its etymologically direct sense as that which surrounds. Yet by the mid-nineteenth century a more specific sense had arisen, of environment as consisting of those surroundings necessary to the development of moral character or biological life. This suggests a binary, stated explicitly in the *Webster's Dictionary*[7] second definition (the first being "that which surrounds") as environment consisting of either (a) the biophysical factors that determine the life of an organism, or (b) the sociocultural conditions that influence human life. What we see already, going back to the mid-nineteenth century, is a parsing out of *environment* into natural or cultural surroundings, both of which are significant but dissimilar enough to warrant distinction.

The third step in the evolution of our notion of environment is much more recent, possibly dating back only several decades. To get at this most recent step, consider Rachel Carson's *Silent Spring*,[8] which to many marks the birth of the modern environmental movement. Carson begins her famous book with a brief fable, then immediately presents her notion of environment at the start of the second chapter: "The history of life on earth has been a history of interaction between living things and their surroundings. To a large extent, the physical form and the habits of the earth's vegetation and its animal life have been molded by the environment" (p. 5). To Carson, the environment is *our* environment, our biophysical surroundings, to which we are connected. *Silent Spring* in many ways echoes Carson's 1951 best-selling *The Sea Around Us* in stressing connection.[9]

Fast-forward to the major environmental agencies today. As noted on its Website, the U.S. Environmental Protection Agency aims to "protect human health and the environment," working for "a cleaner, healthier environment for the American people." Is this not Carson's very wish, born of the damage wrought by DDT to chicks and now spread to a thousand and one connections necessary to life and well-being? Or what of the UK Environment Agency, whose job it is to "look after your environment and make it a better place—for you, and for future generations"? Or what of the United Nations Environment Programme, whose motto is "Environment for Development"? These agencies have by no means turned their backs on the received notion of environment as a connection with our biophysical surroundings.

But a closer look behind the rhetoric reveals a common set of compartments into which the environment is divided: The European Environment Agency website features content on acidification, air quality, biodiversity change, chemicals, climate change, and so forth. Each of these compartments is given due scientific and policy scrutiny. This attention to the environment would have been lauded by Carson, but there is something that has been lost, too, in its bureaucratized management: the environment as connection. Now it is an object among other objects to be managed. Environment is little different from roads, or the economy, or a disease outbreak. All of these are things, important to human well-being to be sure, but things nonetheless.

What has happened? Environment started as a relation, a sense of connection, then turned into a thing. First a generic expression of one's surroundings, then either one's connection to biophysical or sociocultural surroundings (the two tacitly understood as distinct), then just the biophysical stuff constituting our surroundings: air, water, and so forth. Each of these three steps is related: from surroundings to biophysical (or, alternately, human) surroundings to simply biophysical reality—what many people would call *nature*. The result is a double disconnect: first, moving from environment as surroundings to environment as a thing, and second, splitting environment into nature and culture sets of things along the way.

This observation that the term *environment* has not simply dropped out of the sky in some immutable form has not been entirely lost on its commentators. In his comprehensive history of the environmental sciences, for instance, Peter Bowler admits that this very category is a recent reconstruction of those physical and biological sciences that are relevant to understanding environmental problems. Yet he, too, conflates environment with Nature (his capitalization) in tracing its cultural history, thus failing to problematize the very process by which environment became understood as biophysical nature.[10]

Yet I am not sure you will be convinced by the above, especially in how I have characterized the final chapter of this drama. After all, don't we hear environmentalists often speak of our connection with nature in a Carsonesque way? What is fascinating about contemporary connection-talk, however, is that it can lead in such differing directions: The environmentalist asserts our connection with nature in order to bring the environment back as a dominant feature of the human equation, whereas many of those derided as antienvironmentalists assert our connection with nature in order to bring people back as a dominant feature of the environmental equation![11] Both of these efforts invoke a rhetoric of connection to assert not connection but reduction, championing biophysical or human reality in environmental controversies, and thus adopting one of the two poles implied in the evolution of the concept of environment.

The Dual Authorities: Science and Religion

The problem with our concept of environment is not restricted to environmental issues; indeed, it did not arise there. To get a broader sense of how environment came to be understood as (threatened) biophysical stuff, we need to reconsider our ways of understanding science and religion as well. Why science and religion? One important reason is that they play a major role as domains of epistemic, moral, cultural, indeed even political authority. How do we decide what is true? What (or who) is right? The contribution of science and religion to these questions is immense. How many times have you read some pronouncement on the global environment, or for that matter war, or stem cell research, or consumerism, or sexual behavior that cited a major scientist or religious organization? Though not all pronouncements of true and right are grounded immediately in science and religion, their authoritative role in many sectors of society is inarguable, and the environment is certainly no exception.

There has been considerable recent interest in the relationship between science and religion. Though many of us wish to maintain a respect and openness to both science and religion, we often suspect that they are supporting rather different pronouncements on current issues, and none of us enjoys cognitive dissonance, so the inevitable question arises: Can science and religion somehow be harmonized? Are they inevitably in conflict? How could we possibly live our lives in accordance with some version of both?

Recently, I had the pleasure of organizing a research and lecture series on the topic, culminating in the volume *Science, Religion, and the Human Experience*.[12] What I learned in examining popular and scholarly beliefs about science and religion is that they are generally assumed to fall into either one or two domains. If science and religion fall in one domain, they can be understood as either in conflict or in harmony; if two domains, they are understood as essentially independent, thus without conflict (nor much harmony). It is common to hear of conflict accounts: Think, for instance, of struggles in the United States over incorporating evolutionary theory in school science curricula. Here, science and religion are understood as vying for the same turf (the truth about the origin of human life); hence, conflict.

One common way to avoid these conflicts between science and religion is to separate them; hence the popularity of independence accounts, pronouncements that science and religion are two entirely different things mapping onto two entirely separate domains. A typical approach is that championed by the late scientist Stephen Jay Gould in his book *Rocks of Ages*.[13] Gould's NOMA (nonoverlapping magisteria) account relegates science to the realm of facts and religion to that of values: Both are essential in understanding the human condition, but science does a bad job when it gets mixed with values, and religion has little business making pronouncements of fact.[14]

Yet many people desire more than this separatist account of science and religion. Surely facts and values, that is science and religion, are not as distinct as the independence account would have it. Following this inclination, we find many current attempts to harmonize science and religion, to bring them into one domain. What is especially significant here is that this shared domain is often the environment. Consider this opening paragraph to a statement signed in 1992 by nearly ninety major American scientists and religious leaders:[15]

> We are people of faith and of science who, for centuries, often have traveled different roads. In a time of environmental crisis, we find these roads converging. . . . Our two ancient, sometimes antagonistic, traditions now reach out to one another in a common endeavor to preserve the home we share.

The statement continues:

> We believe that science and religion, working together, have an essential contribution to make toward any significant mitigation and resolution of the world environmental crisis. What good are the most fervent moral imperatives if we do not understand the dangers and how to avoid them? What good is all the data in the world without a steadfast moral compass?

We hear in this statement shades of Gould's magisteria: Science provides understanding and data, religion provides a moral compass. But, though they have their distinct identities, they have now come together in a common agenda of environmental protection.

This statement is not unique: It echoes a broad sentiment to harmonize science and religion in building coalitions to save the environment. A more recent version, titled "Earth's Climate Embraces Us All,"[16] was signed by a large number of prominent religious leaders and scientists to support the Climate Stewardship Act then under consideration by the U.S. Senate. This statement similarly acknowledges the differences between science and religion, for instance, in stating that "We do not have to agree on how and why the world was created in order to work together to preserve it for posterity." Yet it posits the global environment, specifically climate, to be a unifying domain of concern.

What sort of notion of environment is implied in these accounts of the relationship between science and religion? Do we detect a sense of environment as connection with that which surrounds? as our physical surroundings? as the biophysical realm itself, without necessary connection to us? Here the distinctions become more subtle. It would seem that the recent regathering of science and religion over environmental concern suggests the very sense of connection I stressed above. But science and religion themselves continue to be relegated to Gould's magisteria, retaining their dual-frequency heartbeat even in the harmony accounts. And I have not told the full story: For every publication bringing science and religion together over the environment, there is another expressing alarm that bringing religion in will compromise scientific rationality in environmental decision making.[17]

No, we have not yet dug deeply enough to understand how environment became a thing, became stuff. Science and religion have been subjected to the same forge that cast environment in its new shape, a forge in which the binary of nature and culture has served as a two-compartment hammering mold. Evidence comes from the mind of Gould himself, the multisyllabic NOMA echoing a simple dichotomy between fact and value that pervades even the harmony accounts of science and religion. What is a fact? What is a value? How are they different? The simplest way to understand their difference is to say that facts cling to nature, and values to culture: Facts are true by virtue of their correspondence to reality, that is, biophysical nature; values are meaningful by virtue of their connection to the valuer, that is, culture.

To the extent that environment is understood as nature, and to the extent that nature is understood as revealed by science, environment inevitably carries an objectivist tinge, a sense of environment as stuff, and a separation is assumed between us and environment that must presumably be overcome. We may be connected to the environment, we are concerned about it, we and it are perhaps even one, but we and it started as two, and this assumed binary point of origin inevitably weakens any sense of connection implied in environment. *Environment* becomes a schizophrenic term when hard, cold scientific rationality is paired up with soft, warm, spiritual impulse, not because science and religion are necessarily so distinct, but because they, too, have been progressively defined over time in relation to this sacred binary of nature and culture. In many ways, environment as connection was doomed, because relation is a fragile thing in an age ruled by the dichotomous key of nature versus culture. Environment was faced with two choices: Become the whim, the desire, the imagination of people, or become a hard reality, as separate from people as quarks. Environment, under the guiding authority of science and religion, has in some contexts become both, and though this looks like connection, it retains the very seed of alienation it attempts to overcome.

Latour's New Vision: Counting beyond Two

So, where did this binary of nature and culture come from, this bimonocular vision of science, religion, and environment? Many environmental accounts lament modern society's rejection of its dependence on nature, often implying some oneness of old, perhaps citing organismic metaphors such as Gaia[18] as reinforcement. These accounts may be true but they are certainly also trite: Yes, the social organization of modernity is more complex than some earlier societies, and their connections with biophysical reality have become more distanciated in space and time.[19] But these accounts have, in invoking modernist notions of nature, retained the very conceptual binary they wish to dispel. It is not a trivial point that virtually all premodern societies have no word for nature: Of course, they have plenty of words for birds, and soil, and climate, and medicinal herbs, but no overarching category of nature in which nonhuman items are lumped. Theirs was/is a natured culture, a cultured nature, a set of polluted categories, in modernist terms. We thus need to dig more deeply than these accounts to understand how environment has become stuff, how science and religion cannot in their current form help us save the environment.

I wish to summarize some of the work of Bruno Latour in digging more deeply. Latour has discovered a curious paradox about modernity: The more we mix nature and culture, the more we speak of purifying the two. As but one example, he cites in *We Have Never Been Modern*:[20]

> The smallest AIDS virus takes you from sex to the unconscious, then to Africa, tissue cultures, DNA and San Francisco, but the analysts, thinkers, journalists and decision-makers will slice the delicate network traced by the virus for you into tidy compartments where you will find only science, only economy, only social phenomena, only local news, only sentiment, only sex.

And all of these compartments ultimately parse into one of two boxes, as if two very powerful magnets draw any particle attempting the treacherous middle path to one or the other pole. These magnets are recharged bit by bit, each time we defend evolutionary science as objective (i.e., untainted by cultural bias), each time we approach American evangelical religious belief as a matter of personal faith (i.e., unconstrained by the laws of nature). Unwittingly, we join the disconnection police in conceptually disentangling an increasingly entangled nature-culture hybrid. And when it comes time to make sense of environmental problems, what options do we have but to turn environment either into natural reality, or cultural construct? If the former, environment awaits the testimony of science as to its objectively verifiable plight; if the latter, environment serves at best as an important source of personal inspiration, or at worst a white, middle-class special interest. Scientific realism and cultural constructivism are to Latour the only tongues an impoverished modernity can speak.

Latour has been known as a cofounder of actor-network theory, or ANT. It is a term he has both repudiated and embraced.[21] But it implies a sense of things as a result of connection. Actors could be anything: the AIDS virus, African farmers, the San Francisco mayor, simian demographics, sexual desire. Each is what it is only in relation to the others, thus the network. Latour's work stresses process: Things could be different. It also stresses hybridity: The categories of nature and culture (and other binaries, principally local and global) are discarded as useless. Time to count beyond two.

Contrast a network with a system—a concept beloved of many environmentalists.[22] A system has stocks (something being stored) and flows (something being moved from one stock to another). The system only makes sense where there is a relative purity to the thing being stored and moved around. It could be water, or carbon dioxide, or genetic information. This also gives it a consistent language to use, such as the language of fluid mechanics. The environment as a giant biophysical system is composed of many natural systems, such as hydrology, climate, and geomorphology, affected by cultural systems like politics, economics, and demography. Each has a relative purity, though all connect to the extent that one can be translated into the other. Ultimately, any one thing (a water droplet, a political uprising) is relegated to its particular system, then ultimately to nature or culture as a result of the system to which it belongs. Systems make sense as analytical constructs, but their implied purity actually takes us away from the sense of connection environmentalists may intend by invoking a "systems approach" or "whole-systems view." Systems are, in short, highly refined networks, sort of a refined-sugar way of looking at the hybrid reality environmentalists confront daily.

Science and religion are often understood as two major, contrasting ways of knowing, and this interests Latour because of what it reveals about modernity. Latour speaks of the modernist notion of knowledge versus belief in terms of the duality of "fact" and "fetish."[23] The modernist is to Latour a hammer-wielding iconoclast, a critic ready and able to smash idols to bits. The fact must be shown to be unfabricated; the fetish must be shown to be arising from some autonomous god or gods (an even more difficult challenge!). In both cases, illusion is understood as fabrication masking as reality.

Given modernity's article of faith prohibiting fabrication, science alone stands as our guide to reality, and religion becomes consigned to the cultural curio shop of inspirational moments. Yet Latour is unmoved by the impulse that led to this tidy duality:

> Once theory has made its analytical cut, once the noise of the breaking bones has been heard, it is no longer possible to account for how we know, how we construct, how to live the Good Life. We are left to try and patch back together subjects and objects, words and world, society and nature, mind and matter—those shards that were made to render any reconciliation impossible. (267)

What, then, is Latour's solution? To be sure, he recognizes the challenge he has created for himself, the challenge facing anyone who wishes to speak in a more connected way:

> Why is it that we cannot readily recover for our ordinary speech what is so tantalizingly offered by practice? Why is it that associations of humans

and nonhumans always become, once clarified, rectified, and straightened out, something so utterly different: two opposing sides in a war between subjects and objects? (266–67)

His solution has been elaborated more recently in *Politics of Nature*.[24] Here, Latour makes what would otherwise be a provocative argument: Political ecology (what we would call environmentalism) must let go of nature, indeed "has nothing at all to do with 'nature'—that blend of Greek politics, French Cartesianism and American parks" (4–5). Wait, isn't this exactly the option we need to guard against? Perhaps Michael Soulé and Gary Lease were right when they wrote of theorists like Latour for whom "Certain contemporary forms of intellectual and social relativism can be just as destructive to nature as bulldozers and chain saws"?[25] No, as suggested above, Latour is as weary of constructivism as he is of realism. Latour's way out is not to declare nature a cultural construct, look down his nose at scientists backing environmental causes, and retire to his ivory tower. He is attempting something different here, and his best way to describe it is, as foreshadowed in *We Have Never Been Modern*, by recourse to political metaphors (Constitution, Parliament) and hybrid/processual metaphors (collectives, circuits). Latour knows that there is no way to resuscitate a language already binarized, so he will not avail himself of the ready solutions proposed via the nature- (or culture-) based frameworks of meaning modernity offers. His preferred metaphors are messy, they emphasize a panoply of actors that are always more than two, they dwell on process as opposed to static substance, there are no trump cards of Nature or Culture (or Science or Religion) to offer some straight road ahead. All actors are provisionally included, all sources of expertise—the Scientist, the Economist, the Moralist, and others—given due consideration.

The details of Latour's solution in *Politics of Nature* are as elaborate as his Gallic style of argument, but what you perceive is an emphasis on connection, a new metaphysics, if you will, in which one could potentially recover the original, expanded-circle notion of environment, and revivify science and religion in the process. Once, in Latour's terms, the "mononaturalism" rendered by an objectivist caricature of science is released, once the inevitable "multiculturalism"—privatizing and subjectivizing religion and other sources of inspiration to keep them out of the way of science and politics—is understood for its fracturing implications, then what is left is the task of building a "common world": "If mononaturalism combined with multiculturalism strikes you as an imposture, if you really no longer dare to be modern, if the old form of the future really has no future, then must we not put back on the table the venerable terminology of democracy?" (227–28).

A New Vision: The Death and Rebirth of Environment

Latour's perspective provides both an ontology of connection, and a critique of the epistemology of disconnection. Greater connection is not, then, needed between people and the environment; this expression comes from binary seed, and will get us no further than we have already come, conceptually, politically, or otherwise. *Environment* is a way of recognizing the larger circle: It is not the natural stuff to which we must remember our connection, it is the connection itself, which includes, yet moves far beyond, this natural stuff. There are many connections we have forgotten, many interweavings we neglect; there is nothing special at all to environment in the narrow sense, but everything special to environment in the larger sense. There is, similarly, nothing special to science and religion, but everything special to the sense of engagement, understanding, and meaningful action offered through multiple scientific and religious pathways. If we can let go of binaries, we may get somewhere.

Recently, there has been something of a stir in contemporary American environmentalism, motivated in no small part by the shifting political winds ushered in by the Bush administration. One controversial argument that emerged is known as the "death of environmentalism": Its authors, Michael Shellenberger and Ted Nordhaus, are featured in the epigraph of this chapter.[26] The charges put out by Shellenberger and Nordhaus were serious: "Today environmentalism is just another special interest. . . . Most of the movement's leading thinkers, funders and advocates do not question their most basic assumptions about who we are, what we stand for, and what it is that we should be doing."[27] The rejoinders were equally strident. For instance, Carl Pope of the Sierra Club argued:

> Shellenberger and Nordhaus complain that "Most environmentalists don't think of 'the environment' as a mental category at all—they think of it as a real 'thing' to be protected and defended. They think of themselves, lit-

> erally, as representatives and defenders of this thing." So? Without being too precious, the environment is a real thing. There is a global carbon cycle, human interventions are a small if meaningful part of the evolutionary process, homo sapiens depend upon a complex web of both geochemical and biological processes.[28]

We hear clear overtones in these quotes of the very issue at hand in this essay. To Pope's defense, environment-as-thing has led environmentalism to some stunning discoveries and important, albeit generally limited, victories over the last thirty years. Though Shellenberger and Nordhaus use climate change as their prime example of the ineffectiveness of environmentalism, climate change is in some ways a perfect example of the kinds of issues that environment-as-thing is best at clarifying. Think of the amount of computer modeling, the careful studies of air-water interactions in climate cycles, the endless considerations of vegetative feedback loops, all the efforts of science that have all proven indispensable in gaining even an imperfect idea of where the global climate system seems to be heading—all very impressive in its explication of biophysical processes and human impacts. But, as Shellenberger and Nordhaus argue, global warming continues relatively unabated. And the hardest and possibly most important environmental problems demand a language and process that speak of both fact and value, much along the lines of Latour's preferred "factish" approach to blending fact and fetish—practices generally relegated to the binary of science and religion—in recognizing that culture and nature are interwoven to the point that these übercategories are useless, indeed harmful in the long run.

I am not sure whether Shellenberger and Nordhaus' argument over the death of environmentalism is primarily about means or ends: Certainly a great deal of critique has focused on their discussion of the effectiveness of various means, again reinforcing my concern that environmentalists have difficulty moving beyond strategy to think more deeply about just what sort of environment they struggle so hard to save. But the discussion is a good one. Will environmentalists craft a new vision of environment? Will they continue as a special interest, or will they realize that there is no trump card they or anyone else holds, and instead forge a vision of the larger circle, perhaps one founded in terms of democracy, as Latour prefers, or maybe a vision of environment molded in another fresh language? Or will

Shellenberger and Nordhaus' notion of postenvironmentalism take root, leaving environmentalists behind? It is up to those who care most about the environment to work for a rebirth of the vision of connectedness that aroused such a passion and dedication in the first place. Environment is about more than nature; it is more than some recent meeting ground between an agreed-upon division of intellectual labor between science and religion. New wine cannot be poured into old wineskins. It is time for a new—or at least renewed—vision of environment.

Acknowledgments

I acknowledge the helpful input of New Visions contributors at our May 2006 workshop, plus that of Jennifer Bernstein, Evan Berry, and Brendon Larson. It is always dangerous to attempt such a grand argument in such a small space, so I accept responsibility for all remaining lapses. Since this essay was prepared, several scholars have explored an "ecology without nature" in much more detail than I do here; see, for instance, Timothy Morton, *Ecology without Nature: Rethinking Environmental Aesthetics* (Cambridge, Mass.: Harvard University Press, 2007), and Slavoj Zizek, *In Defense of Lost Causes* (London: Verso, 2008).

Notes

1. Michael Shellenberger and Ted Nordhaus, "The Death of Environmentalism: Global Warming Politics in a Post-Environmental World," *Grist* magazine (2005), http://grist.org/news/maindish/2005/01/13/doe-reprint (accessed January 18, 2005).
2. This neologism has recently been promoted by American Honda Motor Co. to suggest the company's commitment to environmental leadership. See http://environmentology.honda.com (accessed September 25, 2006).
3. As the author of but one of a host of related arguments, Tim Ingold has noted how the now-ubiquitous notion of a global environment represents the culmination of the process of separation I allude to in this essay, where *our* environment has become *the* environment, writ large. See Tim Ingold, "Globes and Spheres: The Topology of Environmentalism," in Kay Milton, ed., *Environmentalism: The View from Anthropology*, 31–42 (London: Routledge, 1993). A very clear and straightforward discussion of the epistemological path I recommend here is provided by Katherine Hayles; see N. Katherine Hayles, "Searching for Common Ground," in Michael E. Soulé and Gary Lease, eds., *Reinventing Nature? Responses to Postmodern Deconstruction* (Washington, D.C.: Island Press, 1995), 47–63.
4. Neil Evernden, *The Social Creation of Nature* (Baltimore: Johns Hopkins University Press, 1993), xi.

5. There is an important story I am not fully recounting here as to how nature, itself a polyvalent concept, became generally understood as a unified biophysical reality; for some background, see e.g., Raymond Williams, "Ideas of Nature," in *Problems in Materialism and Culture* (London: Verso Press, 1980), 67–85. In one provocative argument, geographer Kenneth Olwig has maintained that our prevalent understanding of nature changed from generative process to biophysical reality in large part via the politicized European notion of landscape; see Kenneth Robert Olwig, *Landscape, Nature, and the Body Politic: From Britain's Renaissance to America's New World* (Madison: University of Wisconsin Press, 2002).
6. Entry from second edition, 1989, online at http://dictionary.oed.com (accessed May 31, 2005).
7. Entry from third edition, 2002, online at http://mwu.eb.com (accessed May 31, 2005).
8. Rachel Carson, *Silent Spring* (Boston: Houghton Mifflin, 1962).
9. For recent commentaries on Carson that extend this notion of environment-as-connection, see Philip Cafaro, "Rachel Carson's Environmental Ethics," *Worldviews: Environment Culture Religion* 6, no. 1 (2002): 58–80; Finis Dunaway, "The Ecological Sublime," *Raritan* 25, no. 2 (2005): 78–97.
10. Peter J. Bowler, *The Norton History of the Environmental Sciences* (New York: W. W. Norton and Co., Inc., 1993).
11. For one recent environmental issue featuring both of these groups, see James D. Proctor, "Whose Nature? The Contested Moral Terrain of Ancient Forests," in William Cronon, ed., *Uncommon Ground: Toward Reinventing Nature* (New York: W. W. Norton, 1995), 269–97.
12. For greater discussion of this one/two-domain theory, see James D. Proctor, "Introduction: Rethinking Science and Religion," in Proctor, ed., *Science, Religion, and the Human Experience* (New York: Oxford University Press, 2005), 3–23.
13. Stephen Jay Gould, *Rocks of Ages: Science and Religion in the Fullness of Life*, 1st ed., The Library of Contemporary Thought (New York: Ballantine Pub. Group, 1999).
14. In an address to the American Academy for the Advancement of Science, John Hedley Brooke, a contributor to this volume, has challenged the ready distinction between fact/value and science/religion as recommended by Gould; see John Hedley Brooke, "Shaping the Content of Science: Have Religious Beliefs Played a Role?" (AAAS Symposium: Non-Overlapping Magisteria? Washington, D.C., February 19, 2005).
15. Mission to Washington, "Declaration of the Mission to Washington," in Roger S. Gottlieb, ed., *This Sacred Earth: Religion, Nature, Environment* (New York: Routledge, 1996), 640–42.
16. National Religious Partnership for the Environment, *Earth's Climate Embraces Us All: A Plea from Religion and Science for Action on Global Climate Change*, www.nrpe.org. (accessed September 12, 2004).
17. As one example, see Paul R. Ehrlich and Anne H. Ehrlich, *Betrayal of Science and Reason: How Anti-Environmental Rhetoric Threatens Our Future* (Washington, D.C.: Island Press, 1996).
18. For a classic account, see James Lovelock, *Gaia: A New Look at Life on Earth* (Oxford: Oxford University Press, 1995).
19. Anthony Giddens, *The Consequences of Modernity* (Stanford, Calif.: Stanford University Press, 1990).

20. Bruno Latour, *We Have Never Been Modern* (Cambridge, Mass.: Harvard University Press, 1993), 2–3.
21. See Bruno Latour, "On Recalling ANT," in John Law and John Hassard, eds., *Actor Network Theory and After* (Oxford: Blackwell, 1999); *Reassembling the Social: An Introduction to Actor-Network-Theory*, Clarendon Lectures in Management Studies (Oxford: Oxford University Press, 2007).
22. The use of systems concepts is extremely broad, running minimally from the integrative work of Gregory Bateson (*Mind and Nature: A Necessary Unity* [Toronto: Bantam Books, 1979]) to the recent establishment of systems biology (see www.systemsbiology.org [accessed May 30, 2006]). What I present here emphasizes the general use of systems thought in environmental discourse and analysis.
23. Bruno Latour, *Pandora's Hope: Essays on the Reality of Science Studies* (Cambridge, Mass.: Harvard University Press, 1999).
24. Bruno Latour, *Politics of Nature: How to Bring the Sciences into Democracy* (Cambridge, Mass.: Harvard University Press, 2004).
25. Michael E. Soulé and Gary Lease, "Preface," in Soulé and Lease, eds., *Reinventing Nature? Responses to Postmodern Deconstruction* (Washington, D.C.: Island Press, 1995), xv–xvii.
26. An online summary and set of perspectives is provided by *Grist* magazine; see http://grist.org/news/maindish/2005/01/13/doe-intro (accessed January 18, 2005).
27. Shellenberger and Nordhaus, "The Death of Environmentalism: Global Warming Politics in a Post-Environmental World." Their argument has been elaborated in *Break Through: From the Death of Environmentalism to the Politics of Possibility* (Boston: Houghton Mifflin, 2007).
28. Carl Pope, "And Now for Something Completely Different," *Grist* magazine January 13, 2005, http://www.grist.org/news/maindish/200513/01//pope-reprint/index.html (accessed January 18, 2005).

Bibliography

Bateson, Gregory. *Mind and Nature: A Necessary Unity*. Toronto: Bantam Books, 1979.

Bowler, Peter J. *The Norton History of the Environmental Sciences*. New York: W. W. Norton and Co., Inc., 1993.

Brooke, John Hedley. "Shaping the Content of Science: Have Religious Beliefs Played a Role?" Paper presented at *AAAS Symposium: Non-Overlapping Magisteria*, Washington, D.C., February 19, 2005.

Cafaro, Philip. "Rachel Carson's Environmental Ethics." *Worldviews: Environment Culture Religion* 6, no. 1 (2002): 58–80.

Carson, Rachel. *Silent Spring*. Boston: Houghton Mifflin, 1962.

Dunaway, Finis. "The Ecological Sublime." *Raritan* 25, no. 2 (2005): 78–97.

Ehrlich, Paul R., and Anne H. Ehrlich. *Betrayal of Science and Reason: How Anti-Environmental Rhetoric Threatens Our Future*. Washington, D.C.: Island Press, 1996.

Evernden, Neil. *The Social Creation of Nature*. Baltimore: Johns Hopkins University Press, 1993.

Giddens, Anthony. *The Consequences of Modernity*. Stanford, Calif.: Stanford University Press, 1990.

Gould, Stephen Jay. *Rocks of Ages: Science and Religion in the Fullness of Life*. 1st ed, *The Library of Contemporary Thought*. New York: Ballantine Pub. Group, 1999.
Hayles, N. Katherine. "Searching for Common Ground." In Michael E. Soulé and Gary Lease, eds., *Reinventing Nature? Responses to Postmodern Deconstruction*, 47–63. Washington, D.C.: Island Press, 1995.
Ingold, Tim. "Globes and Spheres: The Topology of Environmentalism." In Kay Milton, ed., *Environmentalism: The View from Anthropology*, 31–42. London: Routledge, 1993.
Latour, Bruno. *We Have Never Been Modern*. Cambridge, Mass.: Harvard University Press, 1993.
———. "On Recalling ANT." In John Law and John Hassard, eds., *Actor Network Theory and After*, 15–25. Oxford: Blackwell, 1999.
———. *Pandora's Hope: Essays on the Reality of Science Studies*. Cambridge, Mass.: Harvard University Press, 1999.
———. *Politics of Nature: How to Bring the Sciences into Democracy*. Cambridge, Mass.: Harvard University Press, 2004.
———. *Reassembling the Social: An Introduction to Actor-Network-Theory*. Clarendon Lectures in Management Studies. Oxford: Oxford University Press, 2007.
Lovelock, James. *Gaia: A New Look at Life on Earth*. Oxford: Oxford University Press, 1995.
Mission to Washington. "Declaration of the Mission to Washington." In Roger S. Gottlieb, ed., *This Sacred Earth: Religion, Nature, Environment*, 640–42. New York: Routledge, 1996.
Morton, Timothy. *Ecology without Nature: Rethinking Environmental Aesthetics*. Cambridge, Mass.: Harvard University Press, 2007.
National Religious Partnership for the Environment. 2004. "Earth's Climate Embraces Us All: A Plea from Religion and Science for Action on Global Climate Change." www.nrpe.org (accessed September 12, 2004).
Nordhaus, Ted, and Michael Shellenberger. *Break Through: From the Death of Environmentalism to the Politics of Possibility*. Boston: Houghton Mifflin, 2007.
Olwig, Kenneth Robert. *Landscape, Nature, and the Body Politic: From Britain's Renaissance to America's New World*. Madison: University of Wisconsin Press, 2002.
Pope, Carl. "And Now for Something Completely Different." *Grist* magazine (January 13, 2005). http://www.grist.org/news/maindish/2005/01/13/pope-reprint/index.html (accessed January 18, 2005).
Proctor, James D. "Whose Nature? The Contested Moral Terrain of Ancient Forests." In William Cronon, ed., *Uncommon Ground: Toward Reinventing Nature*, 269–97. New York: W. W. Norton, 1995.
———. "Introduction: Rethinking Science and Religion." In Proctor, ed., *Science, Religion, and the Human Experience*, 3–23. New York: Oxford University Press, 2005.
Shellenberger, Michael, and Ted Nordhaus. "The Death of Environmentalism: Global Warming Politics in a Post-Environmental World." *Grist* magazine (January 13, 2005). http://grist.org/news/maindish/2005/01/13/doe-reprint (accessed January 18, 2005).
Soulé, Michael E., and Gary Lease. "Preface." In Soulé and Lease, eds., *Reinventing Nature? Responses to Postmodern Deconstruction*, xv–xvii. Washington, D.C.: Island Press, 1995.
Williams, Raymond. "Ideas of Nature." In *Problems in Materialism and Culture*, 67–85. London: Verso Press, 1980.
Žižek, Slavoj. *In Defense of Lost Causes*. London: Verso, 2008.

14

SHOULD THE WORD *NATURE* BE ELIMINATED?

John Hedley Brooke

> *I conclude that natural science as a form of thought exists and always has existed in a context of history, and depends on historical thought for its existence. From this I venture to infer that no one can understand natural science unless he understands history: and that no one can answer the question what nature is unless he knows what history is.*
>
> R. G. Collingwood, *The Idea of Nature*[1]

In one of the most potent images to accompany the new, experimental philosophy of the seventeenth century, Francis Bacon envisioned nature on the rack, forced to reveal her secrets. The degree to which Bacon personified *nature* and the many possible ramifications of his gendered language have generated a lively debate in which the exploitative and penetrative connotations of probing scientific methods have to be tempered with less rapacious readings. As Katharine Park has recently observed: "The new renaissance figuration of nature as naked and female certainly authorized a more exploitative attitude towards the natural world, although the dominant metaphor was consumption (of nature's inexhaustible 'milk' or bounty) rather than inquisition, dissection, or rape."[2] The role of metaphor in reenvisioning nature has been a prominent theme throughout this book, and Bacon's imagery provides a familiar example of how what is meant by *nature* cannot ultimately be torn from its cultural roots. According to Hans Jonas, Bacon's insistence that "the mind may exercise over the nature of things the authority which properly belongs to it" informed both a new vision of knowledge and a new vision of nature.[3] The nature of things was left with little or no dignity of its own. Having once been integrated within a cosmos, humankind stepped outside nature, seeking new powers of control. Much has been written on the consequences of this elevation of human presumption, which, ironically, Bacon grounded in the premise that humankind alone, and not nature, is made in the image of God.

The irony has not been lost on modern theologians, one commentator on Wolfhart Pannenberg's theology noting that we now find ourselves "on the rack with nature."[4] A duality between "humans" and "nature," sharpened in both Baconian and Cartesian philosophy, has inexorably contributed to its own dissolution. The collapse of dualities is to be one of the main themes of this essay because I want to suggest that some of the more persistent meanings of *nature* have been sustained by sets of antitheses in which the word *nature* is one of the terms. We are currently living through an era in which the traditional dualities are failing and our grasp of what it means to speak of nature at all is wavering.

That the concept of *nature* is deeply problematic is obvious from reflection on its simplest meanings.[5] One does not have to be a historian of science to recognize the multiplicity of its uses. It might describe a landscape and the components of what might be found in that landscape. And yet, as Martha Henderson shows in her contribution to this volume, there is the immediately complicating question: Do landscapes evolve as nature changes or as humans alter their beliefs? Conservation, she writes, is not a scientific project, but rather, a representation of coevolutionary processes between human nature and relationships with the natural world.[6] The word *nature* might also describe objects to be found in plant and animal kingdoms, as represented, for example, in still-life paintings (*nature morte*, as the French would say). Nature, as depicted in art and poetry, might have connotations of violence and terror, or of beauty and sublimity, as it did for Wordsworth.[7] We are then brought face-to-face with further ambiguity, because it becomes natural to ask how the nature of *nature* is understood! It might be understood as that which is untouched by human hands, when a contrast between *nature* and *art* might be presupposed. Or, again, the world of *nature* might refer to all that is, an entire system of nature. John Stuart Mill famously differentiated between these last two meanings when he wrote "On Nature" in his *Three Essays on Religion*.[8] The word can also carry connotations of a veiled or hidden world, as in *nature's secrets*, and a world to which access is sometimes given by scientific instruments, thereby raising the difficult question of whether the instrument merely reveals or actually produces an otherwise hidden entity. *Nature* might also mean that which is real and inviolable, that which puts constraints on human ambition, that which is itself bounded by *laws* of nature. In David Livingstone's chapter in this volume, and with particular reference to the extermination of the Maori, there is reference to the "iron laws of nature" as these were understood by the Irishman William Travers: When a "white race comes into contact with an indigenous dark race on ground suitable to the former, the latter must disappear in a few generations." As both Livingstone and Nicolaas Rupke show in their essays in this volume, there can be irony in the iron, locked in that one word *must*, which has so often gained its strength from local contexts, events, and perceptions.[9] Reflecting on the preconditions of the possibility of granting nature an authoritative moral voice, Lorraine Daston and Fernando Vidal have astutely observed that in the construction of a "must," the "necessities rarely prove ironclad; if they did, nature's moral authority would be superfluous." Their point is that "no one deliberates the rights and wrongs of obeying the law of gravity; vices necessitated by nature are not blameworthy."[10]

In what Robert Boyle described as a "vulgar" notion, nature was really Nature, capitalized because the implication was that nature was not merely a world "out there," but an agency, personified as in Mother Nature.[11] This journalistic usage continues unabated even if, in the interests of political correctness, the maternal reference has been deleted. We read in our newspapers that 2005 was the year in which Nature turned against humankind. Others have personified the word more generously, as when Freeman Dyson deemed Nature to have been "kinder to us than we had any right to expect": As we "identify the many accidents of physics and astronomy that have worked together to our benefit, it almost seems as if the universe must in some sense have known that we were coming."[12]

For a historian of science, the word is even more beguiling because questions about the constitution and interpretation of nature have been central to the scientific enterprise. Not only that, but changes in paradigm have meant that our understanding of the supposed realities beneath the surfaces of the nature we perceive has itself changed enormously. The title of Carolyn Merchant's well-known book, *The Death of Nature*, captured one such switch—from an animated to a mechanistic world picture in seventeenth-century Europe.[13] Underlying that change was another of the portentous "new visions" identified by Hans Jonas—one in which the new mechanics and astronomy would lead Blaise Pascal to articulate his unease, confined to a mere corner of an alienating universe.[14] No sooner was the real world redefined in terms of the primary qualities of matter, epitomized by the swirling vortices of Descartes' cosmology, than the mechanistic picture was itself displaced—certainly highly qualified—by Newton's introduction of invisible forces and "active principles."[15] However we interpret the later mathematical formalism of quantum mechanics, a scientific understanding of *nature* was to be transformed again in the twentieth century by the new physics of Max Planck, Albert Einstein, Niels Bohr, and Werner Heisenberg. For the science historian, there is the additional complication that *nature* is apprehended through idealization, through comparison of the "real" with the idealized mathematical model, a process that has been described as explaining the real by the impossible.[16] As the Oxford mathematician Roger Penrose has noted, the continued success of the mathematization of nature forces us to ask whether, in the last analysis, the concept of *nature* can be extricated from Platonist metaphysics.[17]

For a science historian also interested in the interpenetration of scientific and religious beliefs, the problems become deeper still. In explaining to undergraduates what is meant by *natural theology*, one usually has to say something like this: It is the attempt to achieve knowledge of the nature of God through reflection on the natural world, using only the natural faculty of reason. To all the ambiguities associated with the term *natural world*, we see here the meshing of two more senses of *nature*. The first is that of "character," or defining essence, here applied to God but applicable more generally to the objects of nature, as in Aristotelian philosophy, where it is in the nature of an earthy body that it will return to its "natural place" at the center of the cosmos. The second pertains to a human attribute (the faculty of reason) that, under the doctrine of *imago dei*, differentiated humans from all other creatures and that was also an aspect of an inherent "human nature."

Once theological considerations are brought into the picture, as they have to be for any sensitive understanding of how the "natural sciences" grew out of seventeenth-century "natural philosophy,"[18] even more questions arise about the signification of *nature*. For example, is nature autonomous or do the laws of nature simply summarize how God normally chooses to act in the world, the latter view asserted by no less a Newtonian than Samuel Clarke?[19] And as soon as one starts asking questions about God's relation to the natural world, one is confronted by a bewildering range of models devised to make sense of the concept of divine activity, ranging from the clockwork universe of the deists to panentheistic models in which *nature* is construed as if it were the body of God.[20] In theological discussion, it has been rare to grant full autonomy to nature, though one of the consequences of a successful scientific culture has been the retreat from models of an intervening deity toward statements in which God is primarily conceived to empathize with and suffer alongside human distress. It is worth recalling, however, that among the founders of modern science were those, such as Robert Boyle, who saw no contradiction between discovering the "rules" by which God's will was enacted in creation and allowing that the Creator might "intermeddle," where necessary.[21]

Boyle defended his position, as Newton would also do, by drawing an analogy between human and divine volition. If we can voluntarily move our limbs as we wish, this demonstrates that there can be physical consequences of mental agency. Why, then, should not the Divine will be capable of effecting changes in a physical world? My main reason, however, for reintroducing Boyle at this point is to record his plea for the abolition of the word *nature*. In *A Free Enquiry into the Vulgarly Received Notion of Nature* (1686), we find his radical claim that the word *nature* is so ambiguous that it has become a liability. His prescription is to replace it, for each of its denotations, with a precise description of what is intended by its use. Instead of *nature* in the sense of an active creative agency, *natura naturans,* the word *God* should be used in order to highlight the difference between the Creator and the created order. A proper and profound reverence requires this. Instead of *nature* in the sense of a defining characteristic, the word *essence* will do. Where the word denotes that which belongs to a living creature at its nativity or accrues to it by its birth, why not say that the creature is *born so* or has been *generated such*? Instead of *nature* in the Aristotelian sense of an internal principle of local motion, all we need say is that the body "seems to move spontaneously" upwards or downwards. Where *nature* is used to refer to an established order or "course of things corporeal," it is easy, says Boyle, to substitute what it denotes; namely, the established order or settled course of things. Some substitutions became more prolix. Where *nature* was used to mean an aggregate of the powers belonging to a body, especially a living one, Boyle recommended that "we may employ the *constitution, temperament,* or the *mechanism,* or the *complex of the essential properties or qualities,* and sometimes the *condition,* the *structure,* or the *texture* of that body."[22] We might begin to see why Boyle's radicalism in seeking to exterminate the word would run into difficulty; but he pressed on regardless. When *nature* was used to refer to the "world" or the entire "universe," it was no hardship to substitute those words instead. Accordingly, "the phenomena of nature" would simply become the phenomena of the world or of the universe. For his theological coup de grace, Boyle shot down an eighth usage altogether: As for *nature* taken to mean a "goddess," or a kind of semideity, "The best way is not to employ it in that sense at all."[23]

Boyle's determination to dispense with a troublesome word may not have fallen on deaf ears, but his project was scarcely successful. It does, however, raise questions concerning the forces that have kept the word alive, including the topical question whether now, at last, it might be rendered obsolete. My suggestion is that, along with the need to have a word for that "world out there other than ourselves," concepts of nature have been sustained by a series of polarities in which they have featured. The nature/supernature duality would be one. The contrast between two books—those of nature and of Scripture—would be another. The ancient contrast between nature and art would be a third. The common dichotomy between nature and nurture would have to be included, as would those features of human experience that underline the impotence of humankind when set against the power of natural events like hurricanes and tsunamis, earthquakes, and volcanic eruptions. If it is correct to say that the vitality of the word *nature* has depended on the sustenance of these dualities, and not least of the Baconian dichotomy between nature and humankind, it would follow that their collapse might create opportunities for new visions not so much of nature, as of its dispensability. In the remainder of this essay I shall propose that the dualities have indeed been corroded, if not destroyed, with the consequence that the concept of *nature* has become even more destabilized. Despite these transformations, there are reasons why the term remains remarkably resilient.

Some theologians, looking back to the Hebrew Scriptures, have said that a nature/supernature dichotomy was easily misconceived in the Christian tradition. This is because there is no concept of an independent "nature" in the Bible and because, on a sophisticated understanding of the doctrine of creation, what is described as *nature* is not an autonomous domain but continually dependent on and sustained by a transcendent power. According to this understanding, divine activity does not consist in "supernatural intervention," but in working through the physical powers built into the world. As Antje Jackelén reminds us in her essay in this volume, among the early church fathers, Basil saw no contradiction in conceptualizing divine action in this manner, as God worked "in, with and through" nature.[24] The problem, of course, is that the moment one refers to physical powers as "natural causes," one has presupposed the very duality that is judged to be misconceived. It is quite clear, however, that in early modern European societies the world was seen as open to marvels and miracles that were frequently interpreted as belonging to a supernatural, even (in the context of magical effects) a demonic realm.

The relationship between the new science and popular understandings of the supernatural was complex. One reason for this is that there was a third category of event, the "preternatural," in addition to the natural and the supernatural. The preternatural were unusual, marvelous phenomena, which did not, however, require any suspension of God's ordinary providence for their manifestation. As Lorraine Daston and Katharine Park, among others, have pointed out, the boundaries of the three realms, despite their metaphysical clarity, were extremely difficult to elucidate in practice.[25] Where the mechanization of nature was favored by Catholic scholars, it could be in the defense of the miraculous, not of its deletion. As in the reasoning of Marin Mersenne, a contemporary of René Descartes, one could only differentiate sharply between a marvel and a genuine miracle if one already knew the boundaries of the "natural." A mechanistic conception of nature could sharpen those boundaries. More rigorous criteria for the detection of miracles had been necessitated by discomfiting Protestant critiques of Catholic credulity.[26]

There is complexity again in that voluntarist theologies of nature—which stressed the freedom of God to make whatever world God wished and a freedom to act subsequently in ways that made a difference—could themselves blur distinctions between nature and supernature. When Samuel Clarke declared that the laws of nature simply express the way in which God normally chooses to act in the world, he was both eliding the nature/supernature distinction and reserving, as Newton did, too, the possibility of abnormal action by fiat. Clarke's contention, as paraphrased by an unconvinced Leibniz, was that "there is no difference, with respect to God, between natural and supernatural."[27] With respect to the world, however, the move toward modernity was surely one in which the laws of nature acquired a greater autonomy and inviolability until Charles Darwin could say, in his *Autobiography*, that "The more we know of the fixed laws of nature, the more incredible do miracles become."[28] When speaking of the secularization of knowledge, if not of the secularization of society, few would deny that advances in the sciences have shrunk the spaces once occupied by references to deity. It may only be a god-of-the-gaps that has been squeezed out of the world, but the fundamental duality between nature and supernature has collapsed in ways that have privileged the natural above the supernatural.

To express it like that, however, is to expose a difficulty with the claim that the "natural" might then be dispensable. *Nature*, in Boyle's sense of a "system," swallows up supernature and becomes all there is. But, precisely because of that, the word retains its resilience, since nature as an entirety as well as an entity has *traditionally* been one of its meanings. Ancient references to the "book" of nature remind us of that. This takes us to our second, vulnerable duality—the contrast between the book of nature and the book of Scripture. As long as one could speak of two books, the book of God's "word" and the book of God's "works," the use of the word *nature* might be relatively trouble-free. It was possible for Galileo to argue that God was no less excellently revealed in nature than in the books of the sacred scribes.[29] But that was already to tip the balance away from an earlier theology in which the scope of natural theology was more tightly circumscribed. Indeed, some have seen an intrinsic instability in the book metaphor, in that the image could be used to challenge the Augustinian precept that Scripture is the definitive source of revelation.[30] It was not until the nineteenth century that the "two books" analogy collapsed, a consequence, as James Moore has observed, of ever-more-stringent professional standards both in the sciences and in the academic study of theology.[31] Can we say that, with that collapse, the meaning of *nature* became less secure and even expendable? If, as Constant Mews has suggested, the image of the book of nature underscored an image of perfect order, then it may be no coincidence that it lost currency in the aftermath of Darwin's *On the Origin of Species* (1859), when natural selection could be described by John Herschel as the law of the higgledly-piggledy and when Darwin's own image of nature was that of a prolific but *tangled* bank.[32]

If the book of nature was eventually buried, another reason was the impropriety of comparing an evolving universe with a fixed and finished product. Yet nature was not buried with the book. A movement for nature study in schools, gathering momentum during the 1890s in the United States, would be one indication of its vitality.[33] The earlier rise of naturalistic theories of evolution, and Darwin's in particular, also gave *nature* a new lease on life. I say Darwin's "in particular" because of its personification of natural selection. Nature became a "selector," an agent fulfilling roles formerly ascribed to a divine artisan. This was Boyle's sense of *nature*, in which he complained it had usurped the correct word *God*. In Darwin's theory it did so with a vengeance, as the universe of the natural theologians was stolen from them.[34]

My third duality, the antithesis between nature and art, is an ancient one, finding expression in the belief that human artifacts could never reproduce exactly the workings of nature. Many were the debates in medieval alchemy concerning the status of artificial products. Did not true gold have an intrinsic "form" that the alchemist could never replicate? It was in the context of alchemy that the attempt to imitate "nature" gained a high profile. The hope was that the human agent, having mastered the right techniques, might speed up natural processes, such as those responsible for the production of metals beneath the Earth's surface. As William Newman has recently reminded us in *Promethean Ambitions*, a hard-and-fast distinction between nature and art was being eroded long before Francis Bacon dreamed of a science-based utopia.[35] The dream of surpassing nature through art was especially vivid, and recurrent, as recipes for creating life and achieving immortality were devised.

Although the pretensions of the magus could easily give rise to religious suspicion, not least because secretive practices could be identified with a quest for forbidden knowledge,[36] the dream of power over nature did become more respectable during the seventeenth century. A rigid distinction between nature and art had been compromised by early modern chambers of wonders, in which artisans would force nature into collaboration by shaping such suggestive natural forms as crystals, seashells, and certain stones into artifacts.[37] The distinction was then more deeply threatened by the mechanistic models characteristic of seventeenth-century mechanical philosophies, which had the effect of turning nature itself into a form of art. One of the mechanists, Henry Power, spoke of nature as full of "narrow engines" in which there was much "curious mathematics." The architecture of these "little fabrics," in his view, "more neatly set forth the wisdom of their maker." The existence of a natural theology here deserves special notice. In Power's account of the mechanical philosophy, causes were to be deduced by reconstructing nature. Ultimately, all things were artificial because "nature itself is nothing else but the art of God."[38] In his book *Toward a Theology of Nature*, Wolfhart Pannenberg has argued that one of the reasons for the vitality of the term *nature* has been its attraction for scientists wishing to dispense with the term *creation*.[39] Taking a broad view, there may well be truth in this. It would help to explain why the concept of *nature* has been seen as a problem for theology.[40] The French evolutionist Jean-Baptiste Lamarck spoke of living things not as creatures but as products of nature.[41] Nevertheless, as Power and other physico-theologians of the seventeenth and eighteenth centuries show, theology itself, insofar as it embraced arguments for design, actually made its own contribution to the blurring of the nature/art distinction. It was perhaps willing to do so because, through the study of the microscopic world, in particular, the perfection of nature's (i.e., God's?) art far transcended human artistry.[42]

It would be possible to write a detailed history of how, during the nineteenth century, in particular, great strides were made in simulating natural processes. The artificial synthesis of organic compounds makes an excellent case study because, when Friedrich Wöhler first synthesized urea in 1828, his reproduction of a natural product, exciting though it was, did not replicate the conditions under which "nature" did the work. The high temperature and concentrated acid employed by Wöhler would have been fatal to the living organism.[43] I mention this as a reminder that there was no smooth path to the surpassing of nature. As the Cambridge Platonist Ralph Cudworth had insisted almost two hundred years earlier, there was a "plastic nature" or "plastic principle" *within* nature that explains how natural effects are achieved so easily and discreetly compared with the noise and effort of human agency.[44] Such vitalistic ideas certainly helped to preserve a privileged status for the works of nature; but advances in both the chemical and life sciences gradually displaced them, until, in the twentieth century, a time was reached when, in certain respects at least, art had surpassed nature. We now have pharmaceutical products, different kinds of insulin, for example, that can function more quickly (and more slowly when necessary) than the naturally occurring compound. We have custom-designed genes that do not exist in nature. Computers have surpassed the human brain in memory power. Crucially, programs of genetic enhancement are raising the question whether the concept of a fixed "human nature" might not be obsolete. With the collapse of the nature/art distinction, are not the days of "nature" numbered?

The interfusion of nature and technology certainly gives rise to new visions, but not so much of "nature" as of an increasingly "manipulated nature." Would it be far-fetched to imagine a day when, down on the farm, cows choose when to be milked, make their own way to stalls where they are identified by a chip, receive an automatic rubdown, listen to classical music as a laser-guided machine attaches to their udders, and receive a head massage—all while the farmer takes a vacation, keeping in contact with his herd via a laptop? It is not far-fetched because it has already happened, with concomitant claims for the improvement of "nature" in that the cows, having learned how to relieve themselves through the system, are allegedly "a lot less stressed and more content."[45]

This vignette could stand for many in at least three respects. First, since it was through artifice that cows were bred for larger udders, the notion that a cow without a chip belongs to a pristine "nature" would be simplistic. Second, there could be (and has been) theological debate about the propriety of intervening in natural processes. I am not thinking here of religious gut reactions that complain of "playing god," but more sophisticated reflection on what it might mean to *collaborate with* the Deity in the improvement of the world. For the nineteenth-century Scottish evangelical Hugh Miller, the rendering of cows more useful to humans was a manifestation of an apposite spirituality in which God-given natural resources were malleable and legitimately molded.[46] Third, the innovative techno-nature might give cows more control over their lives, but it leaves ethical problems unresolved. Despite reassurances about the benefits of the robotic system, the prospect of expanding it into another branch of factory farming is an obvious cause for concern.

It is, however, with reference to human rather than bovine learning that claims for the annihilation of the nature/art distinction become most pressing. Analogies between human learning as a process of neurochemical self-organization and mathematical models of artificial intelligence in self-organizing computer systems have given rise to claims such as the following:

> Regardless of whether they occur within a software system or a human brain, material modification and learning become continuous processes. After this, it can either be said that "natural," human intelligence is "artificial" and constructed in the sense that its apparatus mutates as it learns, grows and explores its own potentiality; or that "artificial" intelligence is "natural," insofar as it pursues the processes at work in the brain and effectively learns as it grows. Either way the distinction between nature and artifice is collapsed.[47]

If this were truly the case, a new vocabulary to describe our prophetic visions would seem essential. The extent to which a composite "techno-nature" can be further improved is, however, a deeply unstraightforward question. To speak glibly of enhancing human capacities can be to miss the force of Francis Fukuyama's protest, in *Our Posthuman Future,* that natural selection has itself been a perfecting process and that to interfere unduly with the results of thousands upon thousands of years of human evolution may be to risk seriously adverse side effects.[48] Such a view restores something to nature, not least in the acknowledgment of continuing constraints. This sense of nature, not as a current system only but as a historical process, might be difficult to jettison, even if we presume to guide its future direction. On one reading, our capacity to manipulate, and our ability to have visions at all, are products of evolution, with the implication that what we are discussing is, in the last analysis, nature transforming itself. But in the more radical, less preemptive, readings, the vision of limitless self-modification spells the end not only of the words *human nature* but of the supposed reality to which they have routinely referred.[49]

The hypothesis that a fourth polarity—that between nature and nurture—has helped to keep *nature* in the frame could be explored in as great a depth as a history of the nature/art duality. Its social and political dimensions have been enormous, as reference to the single word *eugenics* would remind us. Many past investigations and debates appear to have centered on a simple either/or: Is this behavior a result of nature or of nurture? To weigh the respective contribution of each is at once a more challenging venture. And yet, as a recent study by Matt Ridley has shown, the antithesis between the two poles is misconceived.[50] It is the interplay between nature and nurture that counts and makes us who we are. This is a message endorsed by Barbara King in her contribution to this volume, when she discusses the thesis that "emotional signaling in the context of nurturing care drives the development of ever-higher levels of thinking."[51] It seems less clear, however, than in the case of the nature/art duality that a syncretic nature-nurture approach would necessarily compromise the category of *nature* itself. In fact, Ridley chose to end his book with the lines: "Nature versus nature is dead. Long live nature via nurture."[52] And one cannot miss the irony in his speculation that the very recourse to dichotomies is in our genes, is "in human nature."[53] Similarly, King concludes her essay with the assertion that "the human religious imagination developed from nature," though *nature* in her analysis refers to the complex patterns of ape socioemotionality.

The longevity of nature in the context of a nature/nurture antithesis may derive from the constraints we both ascribe to, and discover in, the materials of scientific inquiry. The dictum that we are what our genes make us would be an obvious example of how such constraints find expression in popular parlance. It is hard to overestimate the consequences of a paradigm dominated by a medical rhetoric in which specific genes are deemed responsible for specific maladies. The day on which I first wrote that last sentence saw the announcement of the discovery of a gene that correlates with a high risk of schizophrenia. It was already known that children missing a certain part of chromosome 22 were at risk, but it was now reported that having a particular version of the gene COMT worsens the symptoms.[54] For as long as the public is confronted only with a gene-based causality, it will be hard to dispense with that meaning of *nature* that refers to those material realities that set bounds to the possible. This particular sense of *nature* is evident in statements to the effect that to pursue a particular life, one's choice might be to "defy nature." An editorial in the *British Medical Journal* of September 16, 2005, asserted that women delaying pregnancy until the age of thirty-five or over are "defying nature and risking heartbreak." Above that age, fertility rates plummet. Nature might be malleable and open to remedial technology, but underlying material realities cannot be disregarded. The question then becomes whether we might prefer a word other than *nature* to describe the properties and disposition of material objects. In a departure from conventional philosophies of science in which primacy has been given to explanatory "laws of nature," Nancy Cartwright has returned to an almost Aristotelian focus on the *natures* of bodies, permitting the substitution of *capacities* for *natures*.[55] Interestingly, Cartwright's motivation stems in part from a belief that the concept of a *law* of nature is still theologically tainted, harking back to an older vision of divine legislation.[56]

I have not so far considered what some might see as the greatest challenge to an uncritical use of the word *nature*. This is the complicating fact that in our attempts to understand the material realities (spiritual, too, for that matter) underlying our concepts of the "natural," those very concepts are theory-laden and in important respects constructed rather than simply read off the face of "nature." The question raised by Nicolaas Rupke in his chapter in this volume becomes pressing here: Have the scientists brought us in touch with essential nature or have they been engaged, like the Humboldt biographers, in forms of appropriations, whereby nature has been made to serve as building blocks of cultural constructs?[57] Note another either/or in which an "essential nature" is permitted to survive, at least for the purposes of the argument! The challenge, to which many historians of science have ably responded, has been expressed thus: However natural categories are, we need to search whether and by what means they find their existence as natural categories.[58] For feminist writers, the most telling examples come from the manner in which a predominantly male medical profession has categorized *woman* gynecologically.[59] Both the naturalizing of women and the feminizing of nature have had their correlates in male domination—a point addressed in greater detail in Andrew Lustig's chapter.[60]

The constitution of such diverse categories as *race, species, organic,* and *creation* would provide other rich examples of naturalizing trends sometimes having overt political motivation. It must also be noted that the breaking down of dualities can itself be a political act, as when Joseph Priestley in the eighteenth century created the space for a science of the brain by rejecting a body/soul dichotomy in favor of a radical monism.[61] Accordingly, a greater sensitivity to the way in which the category of nature is deployed politically has been registered within the practice of theology as well as in the practice of the history of science. In the first volume of his trilogy, *A Scientific Theology: Nature,* Alister McGrath identified a new threat to the possibility of a natural theology.[62] In the absence of an unconstructed, neutral, objectively defined "nature," how could a theology drawn from nature be stabilized? As he puts it:

> To speak of "natural theology" is to make an implicit judgement concerning the "nature" which constitutes the foundation for any such theology. As we have stressed . . . , the notion of "nature" is intensely problematic, in that it is heavily conditioned by a series of prior assumptions, and mediated through a complex network of social constructions.

McGrath's solution is to substitute *creation* for *nature,* whereby *creation* also means a Trinitarian event. That might work for those sympathetic to his Christian theology, but, at the same time, it serves to underline the non-naturalness of the concept of *creation.*

I have been suggesting in this essay that to ask whether there can be new visions of nature is to ask an exciting but difficult question. Such are the ambiguities of the word *nature* that, for a science historian, the possibility of dispensing with the word altogether, as Boyle wished to do long ago, is attractive. I have suggested that the vitality of the word has been sustained by a set of dualities in which it has been embedded, and which, in their turn, have been qualified, even deconstructed, through our greater understanding. The analysis could be extended to incorporate the natural/unnatural duality, which has featured prominently in moral discourse; for are there not many examples where what might once have been deemed "unnatural," as with homosexual congress, has had to be recategorized? But in each case some sense of a constraining reality, a reified "nature," survives. As Martin Rudwick argued in the last chapter of his *Great Devonian Controversy*, even the most radical methodology favored by sociologists of science has to reckon finally with the fact that there is a "nature," which insists on having a say when it comes to the plausibility of our representational theories:

> It should be clear that the analogy of mapping yields a way of retaining the constructivists' insistence on the social processes that went into the making of a piece of scientific knowledge such as the Devonian, while also allowing the realists' insistence that the real natural world of rocks and fossils, and the real history of the earth that originally brought them into existence, had a more than marginal effect on that claimed knowledge.[63]

And yet the question might still be posed: May we not delete the word *natural* in the phrase "the real natural world of rocks and fossils"? Let us assume for a moment that we could. Our new visions could then be described as new visions of reality, however forbidding *that* word might be! To invoke it would not be to bypass the vexing questions addressed by Gregory Peterson in his essay in this volume—whether, for example, emergent properties of systems are as real as the components of the substrate from which they emerge.[64] Nor can it bypass the sophisticated moves necessary to defend a "scientific realism" from criticisms grounded in the historical realities of theory change.[65] But the proposal has clear resonances with James Proctor's thesis that one can analyze the real connectedness of phenomena without having to reify the word *environment*.[66]

New visions of "reality" are not in short supply. In a recent commentary on Richard Dawkins' use of language, Steven Pinker has envisioned a new scenario in which mental states are attributed to material objects. This contrasts with Dawkins' reserve when responding to accusations of impropriety in speaking of "selfish" genes. Whereas Dawkins will say that he was using the word *selfish* only as a mnemonic for technical concepts in evolutionary dynamics and never intended it as a literal attribution, Pinker questions whether such caveats are really necessary: "I sometimes wonder ... whether caveats about the use of mentalistic vocabulary in biology are stronger than they need to be—whether there is an abstract sense in which we can literally say that genes are selfish, that they try to replicate, that they know about their past environments."[67]

There is, of course, an immediate problem. We have no reason to believe that genes have conscious experience. But then comes the surprise: "A dirty secret of modern science is that we have no way of explaining the fact that humans have conscious experience either." And that being so, "If information-processing gives us a good explanation for the states of knowing and wanting that are embodied in the hunk of matter called a human brain, there is no principled reason to avoid attributing states of knowing and wanting to other hunks of matter." This is a vision of reality in which one may speak of the "intelligent agenda" of genes with the quotes removed.

I have introduced Pinker's mentalizing of matter not to evaluate it (it actually belongs to a long tradition of bestowing on material components properties that have been thought peculiar to higher-level systems), but to show that, in locating new visions of nature, the word *nature* can be dispensable—as long as nature's powers and constraints can be translated as the powers and constraints of objects that are ultimately real in the world. As Andrew Lustig notes in his essay in this volume, a certain "givenness" of nature, understood as an available resource, is assumed even in describing it as malleable.[68] Having earlier raised the question as to whether our new visions should be not of nature but of its dispensability, I would like to conclude with the very simple observation that in the most radical visions of a future society, the overarching duality is no longer that of man and nature, but of "metaman" and a reality that combines in intimate union what was once nature with technological wizardry. We are molded by our machines as we mold them, and the future that continues to emerge from this symbiosis is, in a nontrivial sense, not for us. With reference to a host of posthuman visionaries, Langdon Winner has cautioned against the current frenzy surrounding cyborgs, hybrids, transhumans, extropians, and the like. There is still a place for "rethinking what it means to be human in the first place,"[69] a place, one might add, to which theologians as well as scientists will feel they have a right of access. Even in Winner's critique, however, there is a loser; and it is the word *nature*. The relevant category, he suggests, is less that of *human nature* than of the *human condition*. In his magisterial history of the human sciences, Roger Smith observed that "the search for knowledge of human nature, which had its roots in the devout natural philosophy and Christian natural law arguments of the seventeenth century, led in the eighteenth century to La Mettrie's denial of immateriality, to Diderot's advocacy of sexual liberation and to Sade's spectre of natural violence." His point was that "the search for a science of man was not a neutral undertaking."[70] There is no reason to suppose that inquiries into the human condition might be any more successful in being so. But with an essential human nature in peril, there could still be the vision of a shared human condition, which, as Winner insists, is more likely to provide a sound anchor for ethical conduct and wise global policies than intoxicating divisions between humans and posthumans.[71]

The issues I have raised in this chapter—the protean meanings of *nature*, the status of the dualities by which it has been sustained, the desirability and possibility of its abolition, the sense of ineradicable constraint that lends plausibility to the substitution of the word *realities* for *nature*—have, of course, been discussed many times before and often with impressive sophistication. Daston and Vidal conclude their editorial introduction to *The Moral Authority of Nature* by considering the resistance of the word *nature* to critical annihilation. Their salutary conclusion is that "no amount of well-intentioned intellectual political hygiene" is likely to get rid of it. Their primary reasons complement those I have given by underlining the investment placed in *nature* by those who boost their own credentials by claiming to speak for it. Doctors, scientists, jurists, theologians, politicians, and activists all become the voice of nature in so many contexts.[72] It is a powerful observation because it reminds us that dialectical philosophical analyses can take us only so far in the attempt to categorize contending constructions of nature. It also provides a telling reason why substituting *reality* for *nature* will never work in practice because to invoke *reality* to bolster one's authority and sustain one's professional, social, or political agenda is a little too flagrant, making it too obvious what the game is. The advantage of *nature* is that it is a term of mediation, no less absolute in its decrees but more tolerant of the human voices that speak for it. To seek or dwell on new *visions* of nature may, in a crucial respect, be to privilege the wrong sense. New visions *of* nature, in the last analysis, will continue to be new voices *for* nature.

Notes

1. R. G. Collingwood, *The Idea of Nature* (Oxford: Oxford University Press, 1965 [orig. 1945]), 177.
2. Katharine Park, "Nature in Person: Medieval and Renaissance Allegories and Emblems," in Lorraine Daston and Fernando Vidal, eds., *The Moral Authority of Nature* (Chicago: University of Chicago Press, 2004), 50–73, on 71.
3. Hans Jonas, *The Phenomenon of Life: Toward a Philosophical Biology* (Evanston, Ill.: Northwestern University Press, 2001), 92.
4. Cornelius Buller, *The Unity of Nature and History in Pannenberg's Theology* (Lanham, Md.: Rowan and Littlefield, 1996), 9.
5. For an excellent overview and guide to an extensive literature, see Lorraine Daston and Fernando Vidal, "Doing What Comes Naturally," in Daston and Vidal, *Moral Authority of Nature*, 1–20, especially 3–7.
6. Martha Henderson, "Rereading a Landscape of Atonement on an Aegean Island," in this volume.
7. See the essay by Fred Ledley, "Visions of a Source of Wonder," in this volume.
8. For Mill see Mark Sagoff, "Nature and Human Nature," in Harold Baillie and Timothy Casey, eds., *Is Human Nature Obsolete? Genetics, Bioengineering, and the Future of the Human Condition* (Cambridge, Mass.: MIT Press, 2005), 67–98, esp. 75–77.
9. David Livingstone, "Locating New Visions," in this volume; Nicolaas Rupke, "Nature as Culture: The Example of Animal Behavior and Human Morality," in this volume.
10. Daston and Vidal, "Doing What Comes Naturally," 6.
11. Robert Boyle, *A Free Enquiry into the Vulgarly Received Notion of Nature* (1686), sections II and IV, in M. A. Stewart, ed., *Selected Philosophical Papers of Robert Boyle* (Manchester, UK: Manchester University Press, 1979), 176–91. For the full text of this essay, see Michael Hunter and Edward B. Davis, eds., *The Works of Robert Boyle* 14 vols. (London: Pickering & Chatto, 1999–2000), vol. 10. Hunter and Davis provide further contextualization in their "The Making of Robert Boyle's *Free Enquiry into the Vulgarly Receiv'd Notion of Nature* (1686)," *Early Science and Medicine* 1 (1996): 204–71.
12. Freeman Dyson, "Energy in the Universe," *Scientific American* 225, no. 3 (September 1971): 50–59, on 59.
13. Carolyn Merchant, *The Death of Nature: Women, Ecology and the Scientific Revolution* (San Francisco: Harper & Row, 1979).
14. Jonas, *Phenomenon of Life*, 214–16.
15. Stephen Snobelen, "'The True Frame of Nature': Isaac Newton, Heresy, and the Reformation of Natural Philosophy," in John Brooke and Ian Maclean, eds., *Heterodoxy in Early Modern Science and Religion* (Oxford: Oxford University Press, 2005), 223–62.
16. Alexandre Koyré, *Metaphysics and Measurement: Essays in the Scientific Revolution* (London: Chapman and Hall, 1968), 4.
17. Roger Penrose, *The Road to Reality: A Complete Guide to the Laws of the Universe* (London: Jonathan Cape, 2004), 1027–29.
18. Amos Funkenstein, *Theology and the Scientific Imagination from the Middle Ages to the Seventeenth Century* (Princeton, N.J.: Princeton University Press, 1986); Matthew Eddy and David Knight, eds., *Science and Beliefs: From Natural Philosophy to Natural Science, 1700–1900* (Aldershot, UK: Ashgate, 2005).

19. John Brooke, *Science and Religion: Some Historical Perspectives* (Cambridge: Cambridge University Press, 1991), 157.
20. Philip Clayton and Arthur Peacocke, eds., *In Whom We Live, Move and Have Our Being: Panentheistic Reflections on God's Presence in a Scientific World* (Grand Rapids, Mich.: Eerdmans, 2004), esp. 81–84.
21. Robert Boyle, *An Essay, Containing a Requisite Digression, Concerning Those That Would Exclude the Deity from Intermeddling with Matter* (1663), in Stewart, *Selected Philosophical Papers*, 155–75.
22. Boyle, *Vulgarly Received Notion of Nature*, in Stewart, *Selected Philosophical Papers*, 180.
23. Ibid.
24. Antje Jackelén, "Creativity through Emergence: A Vision of Nature and God," in this volume.
25. Lorraine Daston and Katharine Park, *Wonders and the Order of Nature 1150–1750* (New York: Zone Books, 1998), 121.
26. Robert Lenoble, *Mersenne ou la naissance du mécanisme*, 2nd ed. (Paris: Vrin, 1971); Peter Dear, *Mersenne and the Learning of the Schools* (Ithaca, N.Y.: Cornell University Press, 1988).
27. H. G. Alexander, ed., *The Leibniz-Clarke Correspondence* (Manchester, UK: Manchester University Press, 1956), 29–30.
28. Charles Darwin, *The Autobiography of Charles Darwin*, Nora Barlow, ed. (London: Collins, 1958), 86.
29. Galileo Galilei, "Letter to the Grand Duchess Christina (1615)," in Maurice Finocchiaro, ed., *The Galileo Affair: A Documentary History* (Berkeley and Los Angeles: University of California Press, 1989), 87–118, esp. 93.
30. I owe this point to Constant Mews, "The World as Text: The Bible and the Book of Nature in Twelfth-Century Theology," in Thomas Heffernan and Thomas Burman, eds., *Scripture and Pluralism: Reading the Bible in the Religiously Plural Worlds of the Middle Ages and Renaissance* (Leiden, The Netherlands: Brill, 2005).
31. James Moore, "Geologists and Interpreters of Genesis in the Nineteenth Century," in David Lindberg and Ronald Numbers, eds., *God and Nature: Historical Essays on the Encounter between Christianity and Science* (Berkeley and Los Angeles: University of California Press, 1986), 322–50.
32. Mews, "World as Text"; Adrian Desmond and James Moore, *Darwin* (London: Michael Joseph, 1991), 485; Charles Darwin, *On the Origin of Species by Means of Natural Selection* (1859), reprint of first edition (London: Watts and Co., 1950), 414.
33. Sally Kohlstedt, "Nature, Not Books: Scientists and the Origins of the Nature-Study Movement in the 1890s," *Isis* 96 (2005): 324–52.
34. Walter Cannon, "The Bases of Darwin's Achievement: A Revaluation," *Victorian Studies* 5 (1961): 109–34.
35. William Newman, *Promethean Ambitions: Alchemy and the Quest to Perfect Nature* (Chicago: University of Chicago Press, 2004), 256–71.
36. Peter Harrison, "Curiosity, Forbidden Knowledge, and the Reformation of Natural Philosophy in Early-Modern England," *Isis* 92 (2001): 265–90; and "Original Sin and the Problem of Knowledge in Early Modern Europe," *Journal of the History of Ideas* 63 (2002): 239–59.
37. Daston and Park, *Wonders and the Order of Nature*, 260–61; 279.
38. For the documentation here, see John Brooke and Geoffrey Cantor, *Reconstructing Nature:*

The Engagement of Science and Religion (Edinburgh: T&T Clark, 1998), 323 and 342.

39. Wolfhart Pannenberg, "God and Nature," in Ted Peters, ed., *Toward a Theology of Nature: Essays on Science and Faith* (Louisville: Westminster/John Knox, 1993), 50–71.
40. Gordon Kaufman, "A Problem for Theology: The Concept of Nature," *Harvard Theological Review* 65 (1972): 337–66.
41. Ludmilla Jordanova, "Nature's Powers: A Reading of Lamarck's Distinction between Creation and Production," in James Moore, ed., *History, Humanity and Evolution* (Cambridge: Cambridge University Press, 1989), 71–98.
42. Brooke and Cantor, *Reconstructing Nature*, 217–20.
43. John Brooke, "Wöhler's Urea and Its Vital Force?—A Verdict from the Chemists," *Ambix* 15 (1968): 84–114.
44. John Brooke, *Thinking About Matter: Studies in the History of Chemical Philosophy* (Aldershot, UK: Ashgate, 1995), chap. 4. The apposite discussion in Cudworth's *The True Intellectual System of the Universe* (1678) is available in Gerald Cragg, ed., *The Cambridge Platonists* (Oxford: Oxford University Press, 1968), 239–42.
45. *Sunday Times* (23 October 2005).
46. Hugh Miller, *The Testimony of the Rocks* (Edinburgh: Nimmo, 1869), 142, 198–201.
47. Sadie Plant, "The Virtual Complexity of Culture," in George Robertson et al., eds., *FutureNatural: Nature, Science, Culture* (London and New York: Routledge, 1996), 203–17, on 204–5.
48. Francis Fukuyama, *Our Posthuman Future: Consequences of the Biotechnological Revolution* (London: Profile Books, 2002), 97–98.
49. Sagoff, "Nature and Human Nature," 67.
50. Matt Ridley, *Nature Via Nurture: Genes, Experience and What Makes Us Human* (London: Harper, 2004).
51. Barbara King, "God from Nature: Evolution or Emergence?" in this volume.
52. Ridley, *Nature Via Nurture*, 280.
53. Ibid., 278.
54. *The Times* (24 October 2005).
55. Nancy Cartwright, *The Dappled World: A Study of the Boundaries of Science* (Cambridge: Cambridge University Press, 1999).
56. For the fullest and most recent study of the issues here, see Lydia Jaeger, "*Lois de la Nature et Raisons du Coeur: Les Convictions Religieuses dans le Débat Epistémologique Contemporain*," Doctoral thesis, University of Paris, 2005, esp. chap. 4. For a recent attempt to prove the existence of God from the kind of necessity enshrined in nature's "laws," see John Foster, *The Divine Lawmaker: Lectures on Induction, Laws of Nature, and the Existence of God* (Oxford: Oxford University Press, 2004).
57. Nicolaas Rupke, "Nature as Culture: The Example of Animal Behavior and Human Morality," in this volume.
58. N. Coupland, ed., *Styles of Discourse* (London: Croom Helm, 1988), 14, cited by Nelly Oudshoorn, "A Natural Order of Things? Reproductive Sciences and the Politics of Othering," in Robertson, *Future Natural*, 122–32, on 122.
59. Oudshoorn, "A Natural Order of Things?" 122–24.
60. Andrew Lustig, "The Vision of Malleable Nature: A Complex Conversation," in this volume.
61. Huw Price, "The Emergence of the Doctrine of the 'Sentient Brain' in Britain, 1650–

1850," D.Phil. dissertation, University of Oxford, 2006; Fernando Vidal, "Brains, Bodies, Selves, and Science: Anthropologies of Identity and the Resurrection of the Body," *Critical Inquiry* 28 (2002): 930–74.

62. Alister McGrath, *A Scientific Theology: Nature* (Edinburgh: T&T Clark, 2001), 249.
63. Martin Rudwick, *The Great Devonian Controversy: The Shaping of Scientific Knowledge Among Gentlemanly Specialists* (Chicago: University of Chicago Press, 1985), 454.
64. Gregory Peterson, "Who Needs Emergence?" in this volume.
65. John Worrall, "Structural Realism: The Best of Both Worlds?" *Dialectica* 43 (1989): 99–124, and reproduced in David Papineau, ed., *The Philosophy of Science* (Oxford: Oxford University Press, 1996), 139–65; cf. K. Stanford, "No Refuge for Realism: Selective Confirmation and the History of Science," *Philosophy of Science* 70 (2003): 913–25.
66. James Proctor, "Environment after Nature: Time for a New Vision," in this volume.
67. Steven Pinker, "Yes, Genes *Can* Be Selfish," *The Times Book Review* (4 March 2006): 12–13.
68. Andrew Lustig, "The Vision of Malleable Nature: A Complex Conversation," in this volume.
69. Langdon Winner, "Resistance Is Futile: The Posthuman Condition and Its Advocates," in Baillie and Casey, *Is Human Nature Obsolete?* 385–411, on 405.
70. Roger Smith, *The Fontana History of the Human Sciences* (London: Fontana, 1997), 232.
71. Winner, "Resistance is Futile," 406–7.
72. Daston and Vidal, "Doing What Comes Naturally," 11–12.

Bibliography

Alexander, H. G., ed. *The Leibniz-Clarke Correspondence*. Manchester, UK: Manchester University Press, 1956.

Baillie, Harold, and Timothy Casey, eds. *Is Human Nature Obsolete? Genetics, Bioengineering, and the Future of the Human Condition*. Cambridge, Mass.: MIT Press, 2005.

Boyle, Robert. "A Free Enquiry into the Vulgarly Received Notion of Nature (1686)," sections II and IV. In M. A. Stewart, ed., *Selected Philosophical Papers of Robert Boyle*, 176–91. Manchester, UK: Manchester University Press, 1979.

———. "An Essay, Containing a Requisite Digression, Concerning Those That Would Exclude the Deity from Intermeddling with Matter (1663)." In Stewart, ed., *Selected Philosophical Papers*, 155–75.

Brooke, John. "Wöhler's Urea and Its Vital Force?—A Verdict from the Chemists." *Ambix* 15 (1968): 84–114.

———. *Science and Religion: Some Historical Perspectives*. Cambridge: Cambridge University Press, 1991.

———. *Thinking About Matter: Studies in the History of Chemical Philosophy*. Aldershot, UK: Ashgate, 1995.

Brooke, John, and Geoffrey Cantor. *Reconstructing Nature: The Engagement of Science and Religion*. Edinburgh: T&T Clark, 1998.

Brooke, John, and Ian Maclean, eds. *Heterodoxy in Early Modern Science and Religion*. Oxford: Oxford University Press, 2005.

Buller, Cornelius. *The Unity of Nature and History in Pannenberg's Theology*. Lanham, Md.: Rowan and Littlefield, 1996.
Cannon, Walter. "The Bases of Darwin's Achievement: A Revaluation." *Victorian Studies* 5 (1961): 109–34.
Cartwright, Nancy. *The Dappled World: A Study of the Boundaries of Science*. Cambridge: Cambridge University Press, 1999.
Clayton, Philip, and Arthur Peacocke, eds. *In Whom We Live, Move and Have Our Being: Panentheistic Reflections on God's Presence in a Scientific World*. Grand Rapids, Mich.: Eerdmans, 2004.
Collingwood, R. G. *The Idea of Nature*. Oxford: Oxford University Press, 1965 (orig. 1945).
Coupland, N., ed. *Styles of Discourse*. London: Croom Helm, 1988.
Cragg, Gerald, ed. *The Cambridge Platonists*. Oxford: Oxford University Press, 1968.
Darwin, Charles. *On the Origin of Species by Means of Natural Selection* (1859), reprint of first edition. London: Watts and Co., 1950.
———. *The Autobiography of Charles Darwin*. Edited by Nora Barlow. London: Collins, 1958.
Daston, Lorraine, and Katharine Park. *Wonders and the Order of Nature 1150–1750*. New York: Zone Books, 1998.
Daston, Lorraine, and Fernando Vidal. "Doing What Comes Naturally." In Daston and Vidal, eds., *The Moral Authority of Nature*, 1–20. Chicago: University of Chicago Press, 2004.
Dear, Peter. *Mersenne and the Learning of the Schools*. Ithaca, N.Y.: Cornell University Press, 1988.
Desmond, Adrian, and James Moore. *Darwin*. London: Michael Joseph, 1991.
Dyson, Freeman. "Energy in the Universe." *Scientific American* 225, no. 3 (September 1971): 50–59.
Eddy, Matthew, and David Knight eds. *Science and Beliefs: From Natural Philosophy to Natural Science, 1700–1900*. Aldershot, UK: Ashgate, 2005.
Foster, John. *The Divine Lawmaker: Lectures on Induction, Laws of Nature, and the Existence of God*. Oxford: Oxford University Press, 2004.
Fukuyama, Francis. *Our Posthuman Future: Consequences of the Biotechnological Revolution*. London: Profile Books, 2002.
Funkenstein, Amos. *Theology and the Scientific Imagination from the Middle Ages to the Seventeenth Century*. Princeton, N.J.: Princeton University Press, 1986.
Galilei, Galileo. "Letter to the Grand Duchess Christina (1615)." In Maurice Finocchiaro, ed., *The Galileo Affair: A Documentary History*, 87–118. Berkeley and Los Angeles: University of California Press, 1989.
Harrison, Peter. "Curiosity, Forbidden Knowledge, and the Reformation of Natural Philosophy in Early-Modern England." *Isis* 92 (2001): 265–90.
———. "Original Sin and the Problem of Knowledge in Early Modern Europe." *Journal of the History of Ideas* 63 (2002): 239–59.
Hunter, Michael, and Edward Davis. "The Making of Robert Boyle's *Free Enquiry into the Vulgarly Receiv'd Notion of Nature* (1686)." *Early Science and Medicine* 1 (1996): 204–71.
———, eds. *The Works of Robert Boyle*, 14 vols. London: Pickering & Chatto, 1999–2000, vol. 10.
Jaeger, Lydia. "*Lois de la Nature et Raisons du Coeur: Les Convictions Religieuses dans le Débat Epistémologique Contemporain*." Doctoral thesis, University of Paris, 2005.

Jonas, Hans. *The Phenomenon of Life: Toward a Philosophical Biology*. Evanston, Ill.: Northwestern University Press, 2001.

Kaufman, Gordon. "A Problem for Theology: The Concept of Nature." *Harvard Theological Review* 65 (1972): 337–66.

Jordanova, Ludmilla. "Nature's Powers: A Reading of Lamarck's Distinction between Creation and Production." In James Moore, ed., *History, Humanity and Evolution*, 71–98. Cambridge: Cambridge University Press, 1989.

Kohlstedt, Sally. "Nature, Not Books: Scientists and the Origins of the Nature-Study Movement in the 1890s." *Isis* 96 (2005): 324–52.

Koyré, Alexandre. *Metaphysics and Measurement: Essays in the Scientific Revolution*. London: Chapman and Hall, 1968.

Lenoble, Robert. *Mersenne ou la naissance du mécanisme*, 2nd ed. Paris: Vrin, 1971.

McGrath, Alister. *A Scientific Theology: Nature*. Edinburgh: T&T Clark, 2001.

Merchant, Carolyn. *The Death of Nature: Women, Ecology and the Scientific Revolution*. San Francisco: Harper & Row, 1979.

Mews, Constant. "The World as Text: The Bible and the Book of Nature in Twelfth-Century Theology." In Thomas Heffernan and Thomas Burman, eds., *Scripture and Pluralism: Reading the Bible in the Religiously Plural Worlds of the Middle Ages and Renaissance*, chap. 5. Leiden, The Netherlands: Brill, 2005.

Miller, Hugh. *The Testimony of the Rocks*. Edinburgh: Nimmo, 1869.

Moore, James. "Geologists and Interpreters of Genesis in the Nineteenth Century." In David Lindberg and Ronald Numbers, eds., *God and Nature: Historical Essays on the Encounter between Christianity and Science*, 322–50. Berkeley and Los Angeles: University of California Press, 1986.

Newman, William. *Promethean Ambitions: Alchemy and the Quest to Perfect Nature*. Chicago: University of Chicago Press, 2004.

Oudshoorn, Nelly. "A Natural Order of Things? Reproductive Sciences and the Politics of Othering." In George Robertson et al., eds., *FutureNatural: Nature, Science, Culture*, 122–32. London and New York: Routledge, 1996.

Pannenberg, Wolfhart. "God and Nature." In Ted Peters, ed., *Toward a Theology of Nature: Essays on Science and Faith*, 50–71. Louisville: Westminster/John Knox, 1993.

Park, Katharine. "Nature in Person: Medieval and Renaissance Allegories and Emblems." In Lorraine Daston and Fernando Vidal, eds., *The Moral Authority of Nature*, 50–73. Chicago: University of Chicago Press, 2004.

Penrose, Roger. *The Road to Reality: A Complete Guide to the Laws of the Universe*. London: Jonathan Cape, 2004.

Pinker, Steven. "Yes, Genes *Can* Be Selfish." *The Times Book Review* (4 March 2006): 12–13.

Plant, Sadie. "The Virtual Complexity of Culture." In George Robertson et al., eds., *FutureNatural: Nature, Science, Culture*, 203–17. London and New York: Routledge, 1996.

Price, Huw. "The Emergence of the Doctrine of the 'Sentient Brain' in Britain, 1650–1850." D.Phil. dissertation, University of Oxford, 2006.

Ridley, Matt. *Nature Via Nurture: Genes, Experience and What Makes Us Human*. London: Harper, 2004.

Robertson, George, et al., eds. *Future Natural: Nature, Science, Culture*. London and New York: Routledge, 1996.

Rudwick, Martin. *The Great Devonian Controversy: The Shaping of Scientific Knowledge Among Gentlemanly Specialists.* Chicago: University of Chicago Press, 1985.
Sagoff, Mark. "Nature and Human Nature." In Baillie and Casey, eds., *Is Human Nature Obsolete?* 67–98.
Smith, Roger. *The Fontana History of the Human Sciences.* London: Fontana, 1997.
Snobelen, Stephen. "'The True Frame of Nature': Isaac Newton, Heresy, and the Reformation of Natural Philosophy." In Brooke and Maclean, *Heterodoxy,* 223–62.
Stanford, K. "No Refuge for Realism: Selective Confirmation and the History of Science." *Philosophy of Science* 70 (2003): 913–25.
Stewart, M. A., ed. *Selected Philosophical Papers of Robert Boyle.* Manchester, UK: Manchester University Press, 1979.
Vidal, Fernando. "Brains, Bodies, Selves, and Science: Anthropologies of Identity and the Resurrection of the Body." *Critical Inquiry* 28 (2002): 930–74.
Winner, Langdon. "Resistance Is Futile: The Posthuman Condition and Its Advocates." In Baillie and Casey, *Is Human Nature Obsolete?* 385–411.

Afterword

VISUALIZING VISIONS AND VISIONERS

James D. Proctor

New Visions?

In the foregoing essays, a diverse array of scholars has engaged with visions of nature, science, and religion. Has there been any commonality in their conclusions concerning this trilogy, or do the essays simply represent fourteen differing points of view, nothing more? And, perhaps more presumptuously, where is the space for new visions of nature, science, and/or religion amid these critical accounts?

Consider this brief essay to be not so much a conclusion as an invitation to reread the foregoing essays with an eye toward how they relate to each other, and ultimately what sorts of new visions possibly arise as a result. We want to create for the reader an opening rather than a closing. It proceeds from three assumptions. The first is that, for all their limitations, the five visions of nature presented as our point of departure—evolutionary nature, emergent nature, malleable nature, nature as sacred, nature as culture—enjoy significant scholarly and popular resonance, a strong degree of de facto legitimacy. Thus, any "new visions" we wish to discover or proclaim will necessarily entail some creative weaving of these, and possibly other, existing visions. Visions are not built *ex nihilo*, even by visionaries. The connections and differences between these visions, then, offer key insights toward situating ourselves in, and possibly beyond, them.

The second assumption is that, as there can be no vision without a visioner, comparison of visions are enhanced by comparing the visioners as well. Visions aren't free-floating ideas: They relate not only to each other, but to the preconditions of their own existence, which minimally entail a visioner—though whether this visioner is a visionary is up to others to judge. Our volume features five point-of-departure visions, but fourteen visioners—fourteen scholars who gave the best part of three years to engage with each other over nature, science, and religion. How they relate to each other matters as much as the ideas they embrace.

A third assumption is both methodological and metaphorical: New light may be shed on visions, and visioners, by means of visualization techniques, or methods of graphic representation. If a vision is a view, it seems difficult to capture or compare visions in the form of text alone—despite our textual efforts in this volume.

Visualization played an early role in this collaborative project: Our Website[1] and project graphics featured a spherical brain/earth image, representing biophysical and human nature, cast in the form of a dodecahedron, a twelve-sided geometric shape composed of pentagons, which approximates a sphere via these flat surfaces. A dodecahedron may be an appropriate symbol for the totality of human understandings of reality, as it has fascinated mathematicians and mystics alike since the time of the classic Greek philosophers. Mathematically, it is known as a regular polyhedron or Platonic solid; for ancient mystics, the dodecahedron represented the "fifth element" (after earth, air, fire, and water) of ether from which the universe was made.[2] In a three-dimensional world, one can only imagine three possible independent perspectives, as famously symbolized by the book cover of *Gödel, Escher, Bach*.[3] A cube, for instance, has six faces defining three planes of orientation (e.g., the top and bottom face of a cube define parallel planes of similar orientation); each plane corresponds to one of the three dimensions of ordinary space. But if orthogonality (right angles) is removed as the criterion of dimensionality, then the dodecahedron represents a doubling of perspectives relative to our ordinary perception of space, with twelve faces defining six planes of orientation.

Our means of visualization presented here relies in part on correspondence analysis, a method commonly used to represent in two-dimensional form how two variables relate to each other. Correspondence analysis is a complicated tug-of-war, in which each player holds a rope connecting to every player on the other side, and the final position of each player is based on the outcome of all tug-of-wars. Thus, the important result of correspondence analysis is the overall pattern more than the precise location of any individual element. We are working with a small amount of data, given only fourteen authors, as distinguished from one of the classics in applying this technique, Pierre Bourdieu's *Distinction*.[4] Nonetheless, the results suggest interesting patterns of alliance and difference between visions and visioners, and serve well our purpose of visualizing these patterns to provide readers with an incentive for further exploration.

Visions and Visioners

Points of Departure

As a point of departure, we know that the scholars participating in this volume represent a wide range of disciplinary specialties: theoretical ecology, biological anthropology, cultural geography, history of science, Christian theology, and so forth. But many of our contributors have significant advanced backgrounds in multiple fields: for example, Nicolaas Rupke's training includes both history of science and marine geology, and Willem Drees builds on advanced degrees in physics and theology. So it was not only the disciplinary connections between, but within, our contributors that offered important means of bridging fields.

The correspondence analysis result in Figure 1 is built on a survey in which we asked all authors to rate their level of disciplinary expertise in each of four general fields contributing to our collaboration. The benefit of correspondence analysis is that the result suggests not only which contributors broadly resembled each other in their suite of disciplinary backgrounds, but how the classes of disciplines related as a result. Like

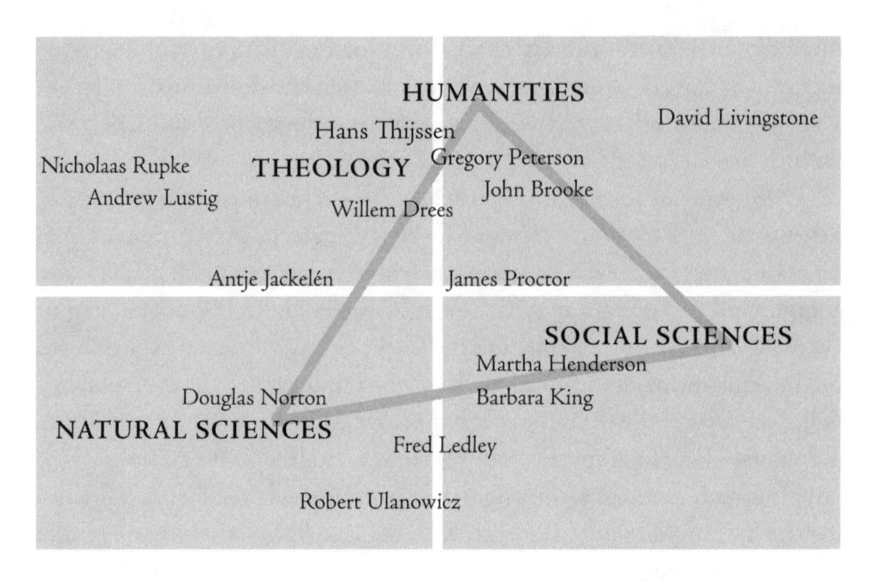

Figure 1. Correspondence analysis of contributors to this volume in terms of their disciplinary backgrounds.

other correspondence analysis results, the placement of each label should be understood with respect to all labels in the paired dataset: Those near the periphery (as defined by the intersecting horizontal and vertical lines) are marked as much by their distance from certain labels as their proximity to others. So, in Figure 1, David Livingstone resides in a common space occupying the humanities and social sciences and distant from the natural sciences, whereas Robert Ulanowicz is at once proximate to the natural sciences and (in terms of disciplinary expertise) relatively distant from the humanities.

Based on this visualization, we can reconstruct something of the raw intellectual material out of which our collaboration was built. The effort enjoyed expertise in all four of these overarching disciplinary classes, with a preponderance of backgrounds in the humanities and theology but with sufficient representation in all areas to afford engagement with related visions of nature. Yet no one of us possessed a complete background in all these areas: This would have placed such a participant in a position toward the center. We needed each other to arrive at a more complete picture.

The disciplinary areas themselves are arrayed in an interesting manner. First, the proximity of the humanities and theology is apparent, which may strike the American reader as odd, given that these two areas are no longer commonly associated with each other in higher education, but many of our European participants are found here, and some of our American contributors with expertise in theology also maintained expertise in related humanities fields, such as history or philosophy. Second, the general domains of the natural sciences, social sciences, and humanities are not arrayed in a linear fashion, as assumed in how our visions of nature were introduced, running from natural-science visions to social-science visions and ultimately to those diffuse in the humanities. In this result, the natural sciences are as close to the humanities as are the social sciences, yet each maintains its distinctive domain and, presumably, a distinctive contribution to envisioning nature, science, and religion. Additionally, none resides in the center: All are more or less equally "peripheral" to our joint intellectual backgrounds.

An invitation to reread the foregoing essays could proceed from this map of disciplinary expertise: Do those contributors with similar backgrounds have similar positions on nature, science, and religion? What of those with apparently differing backgrounds? Do those most distant from certain disciplinary areas maintain this distance from their affiliated visions—evolutionary nature, for instance, for the natural sciences, or nature as culture for the humanities?

Visions According to Visioners

A different picture emerges when considering the general level of agreement the contributors maintained with each of the five visions of nature (Figure 2). Bearing in mind that each author's placement in correspondence analysis reflects not so much an affinity with the proximate vision of nature as one piece of a larger pattern, we can nonetheless see some interesting differences among authors, and consequent differences among visions as viewed by these authors. Some near the periphery, like Willem Drees or me, are located as much by their concerns about certain visions (Willem's concerns about sacred nature and mine about evolutionary nature) as by their approval of others. Many of our authors are located quite close together in spite of disciplinary differences as revealed above; thus, for

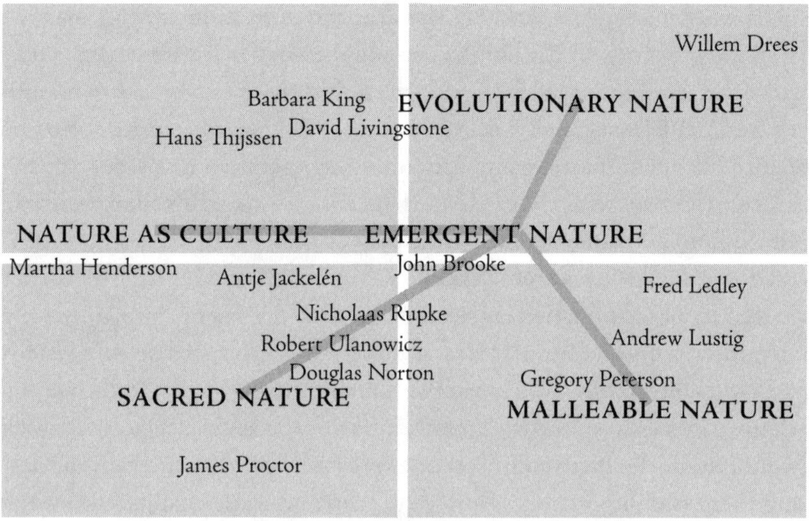

Figure 2. Correspondence analysis of the general level of agreement regarding our five visions of nature among contributors to this volume.

instance, Nicolaas Rupke and Robert Ulanowicz, or Barbara King and David Livingstone, are proximate. There is thus no tidy pattern between disciplinary backgrounds or specialties and the authors' take on these five visions, which is clearly underdetermined by disciplinary expertise. The reader may be interested in comparing the essays of near and distantly placed authors to consider how they relate to a particular vision.

A potentially provocative lesson is suggested by the placement of the five visions, remembering that this placement is a function only of what our contributors thought about them. There is something about emergent nature that led to a more shared sense of support among our authors, thus placing it near the center of the correspondence analysis diagram, while the other four radiate out from emergent nature in distinct patterns: evolutionary nature on the opposite end of nature as sacred, malleable nature in its own sector, and nature as culture likewise. Indeed, there was among a core group of contributors quite strong interest in emergent nature early on in our collaboration, with some seeing it as having the potential to bring together the other visions. Yet this view was not universally held; as a result, in related essays, such as those by Willem Drees or Antje Jackelén, we can see a healthy give-and-take with respect to emergent nature.

As suggested in the Introduction, some of these visions are more provocative than others, leading to both strong support and strong opposition among academics; two examples are evolutionary nature and nature as culture. In our case, many of the contributors were relatively comfortable with nature as culture, so the pattern that results between visions and authors is not necessarily indicative of the pattern among scholars or people in general.

Mixing Nature, Science, and Religion

Figures 1 and 2 compare the authors in terms of their disciplinary backgrounds and views on visions of nature. An entirely different way to compare them was adopted in the following graphic, which was based on the actual words they used in their essays. The process involved performing a count of the most frequent key terms from their essays, then asking them to examine each of these words in terms of their resonance with science, religion, and/or nature, all as broadly defined in this volume. Based on the resonance of their key terms with these three domains, each essay suggested a certain proportionality of emphasis, which is indicated as a point in Figure 3.

To understand each point of Figure 3, one must trace it back to the nature, science, and religion sides: An essay, for instance, whose key terms were exclusively engaging with science yet had little to do with nature or religion would be located at the lower-left corner, with a science score of 10 (100 percent), and nature and religion scores of 0. Thus, the more specialized the essay's key terms were along one or two of these domains, the more it could be expected to be located toward the periphery, while the more nature, science, and religion were all substantively engaged via essay key terms, the more likely it was to be located near the center of the triangle.

A major argument we are collectively making is that these domains are hard to separate; but are the actual words we use to make this argument similarly entangled? Figure 3 suggests that this is, to a large extent, true: Many of our essays utilized key terms that, individually or collectively, mixed nature, science, and religion. Some did not: Robert Ulanowicz placed many of his frequent key terms—*action, causal, complexity,* and

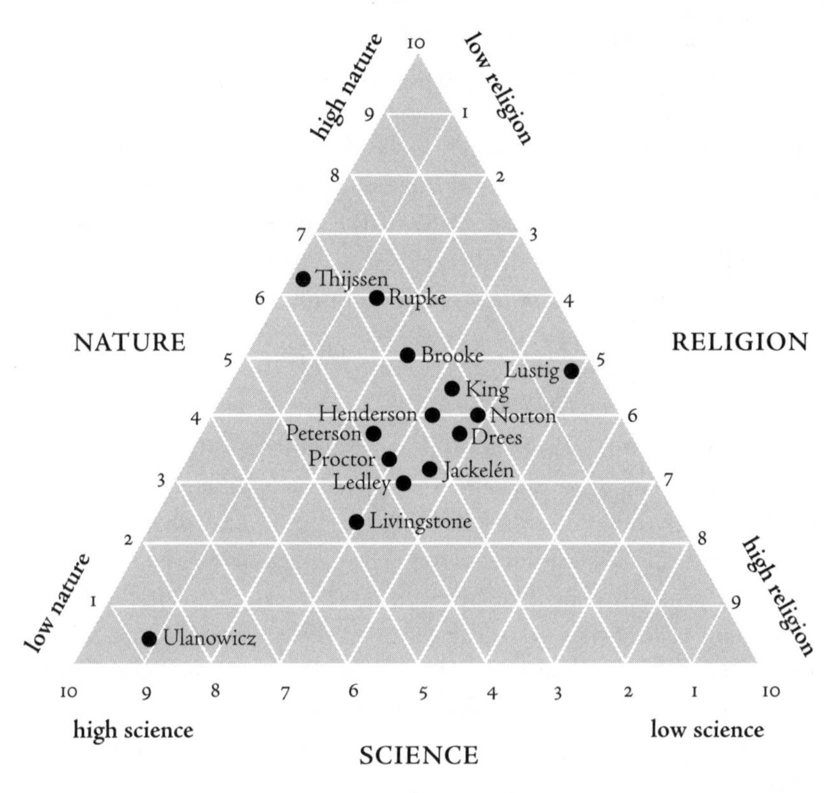

Figure 3. Diagram showing emphasis on nature, science, and/or religion of key terms in essays from this volume.

process are examples—in the domain of science alone, though his essay's argument engaged nature and religion from the vantage point of science. As another example, Johannes Thijssen's most common key terms—*animal, creature, human, reason,* and others—generally mixed nature and science, but did not immediately engage religion, though the implications of his study of human nature are of clear religious relevance. Yet these and other exceptions aside, not only did all essays have something of significance to say about nature, science, and religion, but the most important terms defining these essays were themselves entangled in many cases. We explored nature, science, and religion from differing disciplinary points of departure and with differing outlooks, but in a broader sense we utilized a similarly interwoven vocabulary.[5]

The Way Forward

Nature/Antinature Worldviews

What hope do we have for new visions of nature, science, and religion coming from this series of essays? The first assumption noted at the outset is worth repeating: Our conversations suggest that any "new" vision would by necessity and respect be crafted out of some features of existing visions. We do not propose some neologism for envisioning nature, from which some entirely new notion of the interrelationship of science and religion would emanate. There are plenty of popular volumes competing for this claim on the shelves of one's local bookstore, and we do not wish to enter the prophetic fray.

Consider, for a moment, that the substrate out of which any "new" vision would be created may already exist, as manifested in the tensions between existing visions. Evolutionary nature, emergent nature, nature as culture, and others are not merely different: They potentially differ in certain significant ways, and these specific axes of difference suggest paradoxes that should be explored, perhaps even embraced as a testament to the healthy tension between them. Paradox may be what one ultimately discovers when the search for truth arrives at last at its foundations, as suggested in the famous Niels Bohr quote: "The opposite of a correct statement is a false statement. The opposite of a profound truth may well be another profound truth."[6]

> Perhaps, then, what we see emanating from these essays in terms of a way forward is a playful yet tense interplay around certain key themes, and these key themes will remain no matter how new the vision. Building on the inspiration of John Hedley Brooke's essay, we decided to describe these themes in terms of four key binaries related to philosophical dimensions of worldviews founded on nature versus antinature, including ontology (what is real), epistemology (how we know), and ethics and aesthetics (what is good and beautiful). Each binary consists of a nature/antinature pair: a "naturalistic" ontology, epistemology, and ethics/aesthetics versus an "antinaturalistic" version. We gave the latter specific names to represent their antinature position. So, in the case of ontology, a naturalistic worldview understands reality as rooted in physical nature alone, whereas the embrace of most versions of *theism* rejects this strong nature pole. There are a myriad of ways scholars have written about whether or not a

naturalistic ontology and theism are compatible, yet the tension between the two polar viewpoints may convey something important about how views of nature constrain our understandings of reality.

Epistemology is slightly more complex, as two key approaches related to knowledge must be somewhat distinguished from each other. The first roots knowledge in objectively determined nature, which has processes and properties independent of the cultural and other filters through which we know it, and is the basis of epistemological realism. An antinaturalistic position in this first sense is broadly *constructivist* in arguing that knowledge is a cultural product with little discernable correspondence to objective nature. Whereas constructivism challenges the objectivity of naturalistic approaches to knowledge, the second antinature approach, *particularism,* challenges their claims to universality. The unity, hence universality, of nature is key to propounding scientific knowledge as itself generalizable, whereas epistemologies of a more situated, particularist bent find too much plurality in how reality unfolds in particular times, places, and other contexts to strive toward universal accounts of knowledge. Constructivist and particularist accounts overlap but are not entirely the same: A particularist claim can, for instance, be quite realist within its given context.

Finally, in the realms of ethics and aesthetics, nature has often been invoked as a way to root discourse on the good (or the right) and the beautiful. Here, what can generally be described as *antinaturalism* asserts that what is found in nature is not relevant to deciding what is good/right or what is beautiful. Antinaturalism suggests that we should not follow nature as a moral or aesthetic guide.

Author by Worldview

Given these worldview elements, contributors were asked to locate themselves along a continuum from nature to antinature for each of the four binaries—assuming that mixtures and creative combinations potentially existed in between each set of poles. Figure 4 summarizes the results of a correspondence analysis of their responses. As with other CA results, their positions should be understood relative to all other author positions, and to the location of the four antinature poles as represented on the diagram. Thus, for instance, Johannes Thijssen's position at the bottom is as much a function of his comparison to other authors' positions regarding

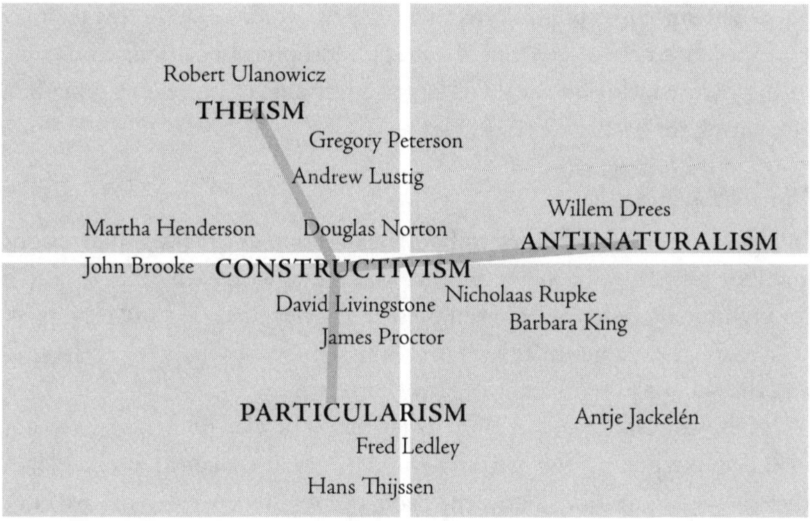

Robert Ulanowicz
THEISM
Gregory Peterson
Andrew Lustig
Willem Drees
Martha Henderson
Douglas Norton
ANTINATURALISM
John Brooke
CONSTRUCTIVISM
David Livingstone
Nicholaas Rupke
Barbara King
James Proctor
PARTICULARISM
Antje Jackelén
Fred Ledley
Hans Thijssen

Figure 4. Correspondence analysis of support for antinature worldview elements among contributors to this volume.

the theism pole at the top as a strong embrace of the particularism pole close to him.

One immediate observation comes from comparing Figures 1 and 4: There is some, but relatively little, connection between the disciplinary specializations of our contributors and their proclivities with respect to these worldview elements. Worldviews mix them up, offer distance and proximity, in quite different ways. Thus, for instance, Fred Ledley is close to Johannes Thijssen in terms of worldview, as are Robert Ulanowicz and Gregory Peterson, in spite of differing academic backgrounds.

Examining the pattern of worldviews resulting from this correspondence analysis, constructivism seems to have been the least divisive of the four elements, given its location near the center, whereas theistic ontologies, antinaturalistic moralities/aesthetics, and particularist epistemologies seem to provoke more difference among authors. To some extent, theism and particularism appear to have defined an axis of difference, suggesting interesting questions around whether each is tied to the nature pole of the other (i.e., theism corresponding to universalist epistemologies, and particularism corresponding to materialist ontologies). The reader may find specific traces of the four abstract philosophical positions, and come to her own conclusion on these larger questions, upon reexamining and comparing the essays based on Figure 4.

Vision by Worldview

In Figure 5, our contributors' embrace or rejection of the five initial visions and four worldview elements are then correlated with each other to search for preliminary patterns between these worldview tensions and visions. In this figure, the + lines refer to statistically significant positive correlations, and the – lines to significant negative correlations.[7]

These correlations refer only to our authors' viewpoints, and are not intended to represent any pattern beyond this group of contributors. Nonetheless, the connections between vision and worldview elements are both surprising and expected. The positive correlation between emergent nature and theism tells us that authors' interest in this vision rose or fell along with the strength of their theistic outlook. Similarly, the negative correlation between evolutionary nature and constructivism was expected, as objectivist, nonconstructivist epistemologies are much more prevalent in those areas of the life and behavioral sciences most closely associated with evolutionary processes. But the negative correlations between malleable nature and both constructivism and particularism are not immediately intuitive. And the positive correlation between evolutionary nature and antinaturalism is surprising as well: This suggests a worldview built on an evolutionary approach that stops short of endorsing any naturalistic foundation for moral or aesthetic judgments.

There are some interesting larger patterns between visions and worldviews worth further exploration. As an example, endorsement of evolutionary nature and nature as culture turned out to be (weakly) negatively correlated, as one would expect, given their differing epistemologies. But their corresponding worldview elements did not similarly contradict one another in the epistemological domain that would presumably separate these two visions. As Figure 5 suggests, embrace of evolutionary nature among our contributors involves a rejection of constructivism, but embrace of nature as culture involves support for particularism more than constructivism, of situating truth versus putting it in endless quotes. The assumed epistemo-

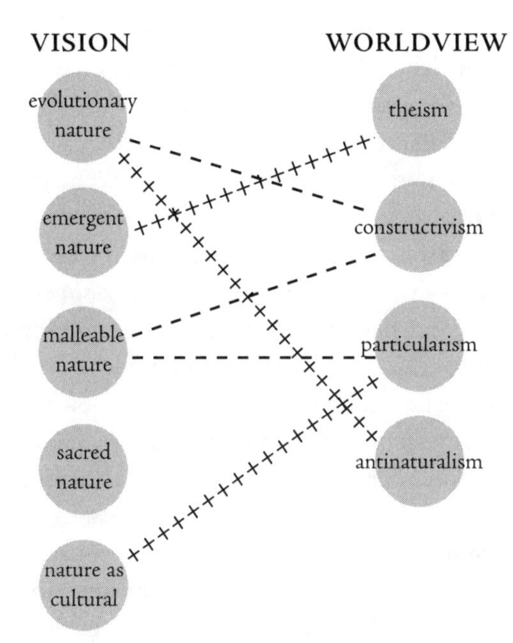

Figure 5. Statistically significant correlations (alpha = 0.05) between attitudes toward visions of nature and worldview elements among contributors to this volume.

logical divide between adherents of evolutionary nature and nature as culture is thus more subtle than generally believed, and indeed there may not be much of a divide after all, once greater clarification is obtained regarding the respective epistemological commitments and boundaries of these two positions.

What Figure 5 ultimately suggests is that the assumed worldview elements of each vision presented in the Introduction—say, evolutionary nature being antitheist or nature as culture being constructivist—are not monotonically reproduced among their scholarly adherents. There are indeed important nature/antinature binaries at work as nature, science, and religion are brought into conversation, but they do not parse out in entirely predictable ways, at least among our contributors. This suggests opportunities for dialogue and consensus along unexpected lines.

An Invitation

We urge you, then, to consider the findings above as less a conclusion than an invitation to reread the diverse essays constituting this volume, now with an eye toward envisioning nature in possibly new ways, new combinations of what had seemed to be irreconcilable differences. We have seen in this brief Afterword how the predictable becomes unpredictable, how familiar disciplinary boundaries fail to consistently define the distinctions between our contributors. We have seen how, in their own way, many of our contributors mixed significant doses of nature, science, and religion in their essays. We have seen important differences involving key nature/antinature binaries defining fundamental elements of a philosophical worldview, and yet these differences failed to parse out as they should have when considering implications for support or rejection of particular visions of nature.

In short, there is plenty of space here for you to creatively draw on the scholarly resources offered in this volume and recombine them in possibly new ways. It is not so much a matter of choosing whether to accept or reject a given vision, or where to place yourself along a given nature/antinature binary. The challenge instead may be much more to appreciate the richness and multivalence of our larger conversation, one in which we hope you will find a similarly rich and multivalent space for your own visions of nature, science, and religion.

Acknowledgments

I am grateful to Jennifer Bernstein for extensive data analysis and preparation of graphics for this essay.

Notes

1. See www.newvisions.ucsb.edu.
2. Peter R. Cromwell, *Polyhedra* (New York: Cambridge University Press, 1997).
3. Douglas R. Hofstadter, *Gödel, Escher, Bach: An Eternal Golden Braid* (New York: Basic Books, 1979).
4. Pierre Bourdieu, *Distinction: A Social Critique of the Judgement of Taste* (Cambridge, Mass.: Harvard University Press, 1984).
5. It is worth noting, in passing, that we spent considerable time exploring commonalities in these key terms across essays, yet discovered that there is such an abundance of key terms that, in general, only the most banal were shared, and other specific terms that were shared did not always seem to emanate from a common thematic use in the essays, so we discarded some hard work we had done on the resultant visualizations.
6. See James D. Proctor, "Geography, Paradox, and Environmental Ethics," *Progress in Human Geography* 22, no. 2 (1998); and James D. Proctor, "Solid Rock and Shifting Sands: The

Moral Paradox of Saving a Socially-Constructed Nature," in Noel Castree and Bruce Braun, eds., *Social Nature: Theory, Practice and Politics* (Oxford: Blackwell Publishers Ltd., 2001).

7. Alpha = 0.05. Given the small number of contributors, only very high correlations (those above approximately 0.5) were significant. Other interesting correlations are not included in this discussion due to the small size—even though our contributors do not, of course, represent a population sample.

Bibliography

Bourdieu, Pierre. *Distinction: A Social Critique of the Judgment of Taste*. Cambridge, Mass.: Harvard University Press, 1984.

Cromwell, Peter R. *Polyhedra*. New York: Cambridge University Press, 1997.

Hofstadter, Douglas R. *Gödel, Escher, Bach: An Eternal Golden Braid*. New York: Basic Books, 1979.

Proctor, James D. "Geography, Paradox, and Environmental Ethics." *Progress in Human Geography* 22, no. 2 (1998): 234–55.

———. "Solid Rock and Shifting Sands: The Moral Paradox of Saving a Socially-Constructed Nature." In Noel Castree and Bruce Braun, eds., *Social Nature: Theory, Practice and Politics*, 225–39. Oxford: Blackwell Publishers Ltd., 2001.

CONTRIBUTORS

JOHN HEDLEY BROOKE held the Andreas Idreos professorship of science and religion and directorship of the Ian Ramsey Centre at the University of Oxford from 1999 to 2006. His best-known books are *Science and Religion: Some Historical Perspectives* (Cambridge University Press, 1991); *Thinking About Matter: Studies in the History of Chemical Philosophy* (Ashgate, 1995); and (with Geoffrey Cantor) *Reconstructing Nature: The Engagement of Science & Religion* (T&T Clark, 1998; Oxford University Press, 2000).

WILLEM B. DREES is professor of philosophy of religion and ethics in the Leiden Institute for Religious Studies, Leiden University, The Netherlands, and editor-in-chief of *Zygon: Journal of Religion and Science*. After advanced degrees in theoretical physics (Utrecht University) and theology (Groningen University), he earned doctorates in theology (Groningen) and philosophy (Vrije Universiteit, Amsterdam). Among his publications are *Religion, Science and Naturalism* (Cambridge University Press, 1996); *Creation: From Nothing until Now* (Routledge, 2003); and *Religion and Science in Context: A Guide to the Debates* (Routledge, 2009).

MARTHA HENDERSON, a geographer interested in the cultural construction of landscape, is a member of the faculty at the Evergreen State College in Olympia, Washington. Her work in interdisciplinary education at Evergreen has encompassed environmental studies, exploration, ethnicity, and the Aegean. Her research and publications include a text on geography and ethnicity, and articles on the relationship between native American landscapes and U.S. federal policies.

ANTJE JACKELÉN is bishop of the Diocese of Lund in the Church of Sweden. Previously, she was associate professor of systematic theology and religion and science at the Lutheran School of Theology in Chicago as well as director of the Zygon Center for Religion and Science. She is the author of numerous articles and several books, including *The Dialogue between Religion and Science: Challenges and Future Directions* (Pandora, 2004) and *Time and Eternity: The Question of Time in Church, Science, and Theology* (Templeton Foundation Press, 2005).

BARBARA J. KING is chancellor professor of anthropology at the College of William & Mary in Williamsburg, Virginia. A monkey-and-ape watcher for twenty-five years, her recent books are *Evolving God: A Provocative View on the Origins of Religion* (Doubleday, 2007), and *The Dynamic Dance: Nonvocal Communication in African Apes* (Harvard University Press, 2004).

FRED D. LEDLEY is professor and chair of the Department of Natural and Applied Sciences at Bentley College in Waltham, Massachusetts. A physician by training, he has published extensively on molecular human biology and genomics, as well as issues that arise at the intersection of science, individuals, and society.

DAVID N. LIVINGSTONE is professor of geography and intellectual history at the Queen's University of Belfast, and a fellow of the British Academy. He is the author of several books, including *Nathaniel Southgate Shaler and the Culture of American Science* (University of Alabama Press, 1987), *Darwin's Forgotten Defenders* (Eerdmans, 1987), *The Geographical Tradition* (Blackwell, 1992), *Putting Science in its Place* (University of Chicago Press, 2003), and *Adam's Ancestors: Race, Religion and the Politics of Human Origins* (Johns Hopkins University Press, 2008).

ANDREW LUSTIG is the Holmes Rolston III Professor of Religion and Science at Davidson College in Davidson, North Carolina. His research interests include ethical issues posed by new biotechnologies, the relations between science and religion as discourses in the public square, and the justification of positive rights in international law. He has published ten books in bioethics, theological ethics, and health care policy, including, most recently, *Altering Nature: Religion, Biotechnology, and Public Policy* (2008), and is the author of nearly 170 other publication for both scholarly and general audiences. He is currently a member of the editorial board of *The Journal of Medicine and Philosophy* and of the advisory board of *Christian Bioethics*.

DOUGLAS NORTON is associate professor and chair of the Department of Mathematical Sciences at Villanova University in suburban Philadelphia. His mathematical research primarily concerns dynamical systems; his other interests include mathematical applications in the life sciences and interactions of mathematics with the arts and humanities. This last area has led full circle to consideration of implications of chaos theory in theology and views of the universe. He was a Fulbright lecture and research scholar at the University of Botswana in the 1996–1997 academic year.

GREGORY R. PETERSON is program coordinator of the Philosophy and Religion Department and an associate professor of philosophy and religion at South Dakota State University. He has published and lectured widely on issues of religion and science, especially in the area of theological anthropology and cognitive science. He is the author of *Minding God: Theology and the Cognitive Sciences* (Fortress Press, 2002).

JAMES D. PROCTOR is professor and director of the Environmental Studies Program at Lewis & Clark College in Portland, Oregon. His research primarily concerns contemporary environmental thought. He is the editor of *Science, Religion, and the Human Experience* (Oxford, 2005) and *Geography and Ethics: Journeys in a Moral Terrain* (Routledge, 1999).

NICOLAAS A. RUPKE is professor and director at the Institute for the History of Science, Göttingen University. He works on late-modern earth and life sciences and on science and religion. His most recent monograph is titled *Alexander von Humboldt: A Metabiography* (2008).

JOHANNES (HANS) THIJSSEN is professor of ancient and medieval philosophy at Radboud University Nijmegen (the Netherlands) and director of the Center for Medieval and Renaissance Natural Philosophy (www.ru.nl/phil/center/index). His research interests include premodern concepts of nature. He is the founder and a consulting editor of the journal *Early Science and Medicine* (Brill Academic Publishers) and of the book series *Medieval and Early Modern Science* (Brill Academic Publishers). He is chairman of the program called "From Natural Philosophy to Science," which is funded by the European Science Foundation (ESF).

ROBERT E. ULANOWICZ is professor emeritus of theoretical ecology at the Chesapeake Biological Laboratory, University of Maryland Center for Environmental Science in Solomons, Maryland. His research has centered on quantitative methods for the analysis of networks of ecological exchanges. He is the author of *Growth and Development: Ecosystems Phenomenology* (Springer, 1986); *Ecology, the Ascendent Perspective* (Columbia University Press, 1997); and *A Third Window: Natural Life beyond Newton and Darwin* (Templeton Foundation Press, 2009).

INDEX